More than Moore

G.Q. Zhang • A.J. van Roosmalen

More than Moore

Creating High Value Micro/Nanoelectronics Systems

Springer

Editors
Guo Qi Zhang
Delft University of Technology
NXP Semiconductors
Delft, Eindhoven
The Netherlands

Alfred J. van Roosmalen
Delft University of Technology
NXP Semiconductors
Delft, Eindhoven
The Netherlands

ISBN 978-0-387-75592-2 e-ISBN 978-0-387-75593-9
DOI 10.1007/978-0-387-75593-9
Springer Dordrecht Heidelberg London New York

Library of Congress Control Number: 2009921651

Printed on acid-free paper

Springer is part of Springer Science+Business Media (www.springer.com)

Preface

Since the first transistor was invented at Bell Laboratories in 1947, semiconductors rapidly evolved into a key enabler for providing solutions to societal, business, and consumer needs. Today, the semiconductor market has become the cornerstone of global high-tech economy, with more than 16% of the world economy built on semiconductors and with annual R&D budgets in the industry ranging up to 20% of revenue. At the same time, semiconductors have become the cornerstone of modern society, and pervaded and penetrated in numerous aspects of human lives.

Explosive technology development driven by Moore's law has been playing a vital role for the success of the semiconductor industry. In 2005, the strategic research agenda and vision for "More than Moore" (MtM) technology were formulated in a systematic manner by the European Technology Platform for Nanoelectronics ENIAC. Since then, we have witnessed the quick development of a new area of micro/nanoelectronics beyond the boundaries of Moore's law and into the area of "MtM," which creates and adds more nondigital functionality to conventional semiconductor technology, leading to virtually unlimited technology possibilities and application potentials for the semiconductor-based high-tech industry.

Based on the extensive results from the "MtM" technology domain team within ENIAC and the MEDEA+ working group ("Towards and beyond 2015"), this unique book presents the new semiconductor landscape by highlighting current and future semiconductor applications to meet various society needs, providing a systematic overview for the state-of-the-art of semiconductor technologies, especially the "MtM" technology, and summarizing the fast evolving business trends.

The semiconductor technology and business environment are very different today from what they were when G.E. Moore postulated the theorem now known as Moore's law, more than 40 years ago. We hope that this book can provide inspiration for the direction and future development of semiconductors in the decades to come.

It is our privilege, as the chair of ENIAC "MtM" technology domain team and the chair of ENIAC/AENEAS support group, to present this book. From this place, we like to thank all authors who have worked on the various chapters and also all technology domain team members in the ENIAC, EPoSS, and MEDEA+/CATRENE communities who have contributed to this book in one way or another.

Eindhoven, The Netherlands G.Q. Zhang and A.J. van Roosmalen

Contents

1 **The Changing Landscape of Micro/Nanoelectronics** 1
G.Q. Zhang and A.J. van Roosmalen

2 **Smart Integrated Systems: From Components to Products** 33
Thomas Gessner, Mario Baum, Wolfgang Gessner, and Günter Lugert

3 **RF Technologies and Systems** 63
W. De Raedt, G. Carchon, H.A.C. Tilmans, and E. Beyne

4 **High Voltage and Power** 85
G. Ricotti and G. Maggioni

5 **Sensors and Actuators on CMOS Platforms** 109
Mark Lantz, Christoph Hagleitner, Michel Despont, Peter Vettiger, Mario Cortese, and Benedetto Vigna

6 **Biochips** 179
C. Vauchier, and P. Puget

7 **Optoelectronics** 203
Jaroslav Kováč, František Uherek, Daniel Donoval, Jaroslav Kováč Jr, and Alexander Šatka

8 **Semiconductor Image Sensing** 239
N. Blanc, P. Giffard, P. Seitz, P. Buchschacher, V. Nguyen, and M. Hoheisel

9 **Heterogeneous Integration: Building the Foundation for Innovative Products** 279
Herbert Reichl, Rolf Aschenbrenner, Michael Töpper, and Harald Pötter

10 System-Level Design 305
Joel Björk, Peter Caldera, Paul Fugger, Karsten Einwich, Karl Hollaus, Wolfgang Pribyl, Jochen Reisinger, Willy Sansen, Andreas Wiesbauer, and Horst Zimmermann

Index 331

Contributors

Rolf Aschenbrenner Fraunhofer IZM, Berlin, Germany

Mario Baum Fraunhofer ENAS, Group Technologies/Waferbonding, Chemnitz, Germany

E. Beyne IMEC, Leuven, Belgium

Joel Björk Technikon Forschungs - und Planungsgesellschaft mbH, Villach, Austria

N. Blanc CSEM SA, Photonics Division, Zurich, Switzerland

P. Buchschacher Faculty of Electrical Engineering and Information Technology STU, Bratislava, Slovakia

Peter Caldera Infineon Technologies, Austria AG, Villach, Austria

G. Carchon IMEC, Leuven, Belgium

Mario Cortese STMicroelectronics S.r.l., Milan, Italy

W. De Raedt IMEC, Leuven, Belgium

Michel Despont IBM Zurich Research Laboratory, Rüschlikon, Switzerland

Daniel Donoval Faculty of Electrical Engineering and Information Technology STU, Bratislava, Slovakia

Karsten Einwich Fraunhofer Institute for Integrated Circuits, Dresden, Germany

Paul Fugger Infineon Technologies, Austria AG, Villach, Austria

Thomas Gessner Fraunhofer ENAS, Chemnitz, Germany

Wolfgang Gessner VDI/VDE Innovation + Technik, Berlin, Germany

P. Giffard CEA-LETI-Minatec, Grenoble, France

Christoph Hagleitner IBM Zurich Research Laboratory, Rüschlikon, Switzerland

M. Hoheisel Siemens AG, Forchheim, Germany

Karl Hollaus Institute for Fundamentals and Theory in Electrical Engineering, Graz University of Technology, Graz, Austria

Jaroslav Kováč Faculty of Electrical Engineering and Information Technology STU, Bratislava, Slovakia
International Laser Center, Ilkovičova 3, 81219, Bratislava, Slovakia

Jaroslav Kováč, Jr. Faculty of Electrical Engineering and Information Technology STU, Bratislava, Slovakia
International Laser Center, Bratislava, Slovakia

Mark Lantz IBM Zurich Research Laboratory, Rüschlikon, Switzerland

Günter Lugert Sensor & Actuator Systems, Siemens AG, Munich, Germany

G. Maggioni STMicroelectronics S.r.l., Milan, Italy

V. Nguyen CEA-LETI-Minatec, Grenoble, France

Harald Pötter Fraunhofer IZM, Berlin, Germany

Wolfgan Pribyl Institute for Electronics, Graz University of Technology, Graz, Austria

P. Puget CEA-LETI-Minatec, Grenoble, France

Herbert Reichl Technische Universität, Berlin, Germany

Jochen Reisinger Infineon Technologies, Austria AG, Villach, Austria

G. Ricotti STMicroelectronics S.r.l., Milan, Italy

Willy Sansen Department of Electrical Engineering, Katholieke Universiteit Leuven, Leuven, Belgium

P. Seitz CSEM SA, Research Center for Nanomedicine, Landquart, Switzerland

H.A.C. Tilmans IMEC, Leuven, Belgium

Michael Topper Fraunhofer IZM, Berlin, Germany

František Uherek Faculty of Electrical Engineering and Information Technology STU, Bratislava, Slovakia
International Laser Center, Bratislava, Slovakia

A.J. van Roosmalen Delft University of Technology, NXP Semiconductors, Delft, Eindhoven, The Netherlands

C. Vauchier CEA-LETI-Minatec, Grenoble, France

Peter Vettiger IBM Zurich Research Laboratory, Rüschlikon, Switzerland

Benedetto Vigna STMicroelectronics S.r.l., Milan, Italy

Andreas Wiesbauer Infineon Technologies, Austria AG, Villach, Austria

G.Q. Zhang Delft University of Technology, NXP Semiconductors, Delft, Eindhoven, The Netherlands

Horst Zimmerman Institute of Electrical Measurements and Circuit Design, Vienna University of Technology, Vienna, Austria

Chapter 1
The Changing Landscape of Micro/Nanoelectronics

G.Q. Zhang and A.J. van Roosmalen

Abstract In the past decades, the progresses of semiconductors are mainly powered by Moore's law focusing on IC miniaturization down to nanoscale, resulting in the transition from microelectronics into nanoelectronics. Recently, we have witnessed the quick development of a new area of Micro/nanoelectronics beyond the boundaries of Moore's law, called "More than Moore" (MtM). MtM creates and adds various nondigital functionalities to semiconductor products and focuses on creating high value micro/nanoelectronics systems, leading to virtually unlimited technology possibilities and application potentials. This chapter presents the changing global landscape of micro/nanoelectronics by summarizing the major ongoing technology and business development trends, especially the vision and strategy of MtM, by highlighting some potential applications, and by providing a systematic overview for the major MtM technologies.

Keywords Moore's s law • More than Moore • Beyond CMOS • Heterogeneous integration

1.1 Introduction

The first transistor was invented at Bell Laboratories on December 16, 1947, by William Shockley, John Bardeen, and Walter Brattain. This was perhaps the most important electronics event of the twentieth century, as it later made possible the integrated circuit and microprocessor that are the basis of modern electronics. Since then, semiconductors rapidly evolved into a key enabler for providing solutions to societal, business, and consumer needs for more than half century.

G.Q. Zhang (✉) and A.J. van Roosmalen
Delft University of Technology,
NXP Semiconductors, Delft, Eindhoven, The Netherlands

G.Q. Zhang and A.J. van Roosmalen (eds.), *More than Moore: Creating High Value Micro/Nanoelectronics Systems*,
DOI 10.1007/978-0-387-75593-9_1,

As of today, the semiconductor market has become the cornerstone of global high-tech economy. The worldwide market for electronic products in 2007 is estimated at $1,105 billion, and the related electronics services market at around $6,500 billion [1]. These product and service markets are enabled by a $280 billion market for semiconductor components and an associated $80 billion market for semiconductor equipment and materials. By comparison, the 2007 GWP (Gross World Product) is expected to reach $48,900 billion, implying that more than 16% of the world economy today is built on semiconductors (Fig. 1.1). In addition to its immediate economic value, the semiconductor industry is one of the biggest investors in R&D for the knowledge society, with typical annual R&D budgets in the industry ranging from 15 to 20% of revenue.

At the same time, semiconductors have become the cornerstone of modern society, and pervaded human lives in the past 50 years. Without them, the rich multimedia experience that we enjoy in today's world of CD, MP3, DVD, and the Internet would not have been possible. Without them, we would not be able to talk to people around the world, exchange messages, or share photographs and video clips via a personal portable device that fits into our pocket. Without them, our cars would do far fewer kilometers per liter of fuel, heavily pollute the environment, and cause more accidents. Gordon Moore estimated in 2003 that the number of transistors shipped in a year had reached about 10^{18}. That is about 100 times the number of ants estimated to be in the world. Semiconductors are with us everywhere and anytime.

The shift from the past era of microelectronics, where semiconductor devices were measured in microns (1 millionth of a meter) to the new era of nanoelectronics where they shrink to dimensions measured in nanometers (1 billionth of a meter) will make the semiconductor industry even more pervasive than it is today. It will allow much more intelligence and far greater interactivity to be built into many more everyday items around us, with the result that silicon chip technology will play a part in virtually every aspect of our lives, from personal health and traffic control to public security.

However, in the future, the semiconductor industry cannot be exclusively based on the same "business as usual" strategy. This is due to the facts that many aspects of the business, technology, design, and system level requirements are now

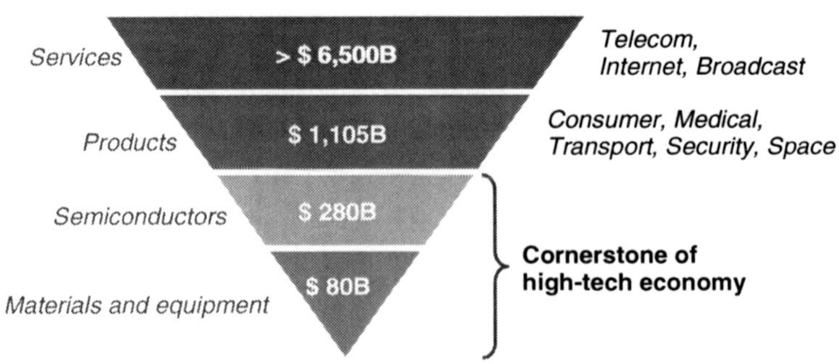

Fig. 1.1 Semiconductors underpin over 16% of the global economy

simultaneously changing when approaching fundamental limits at the nanoscale. The introduction of new materials and technology steps, increased process variability, tough reliability need, are all impacting system level design at the same time, confronted with extremely large and complex architectures and quasi-impossible-to-solve power density issues. On the other hand, the applications will also be different. Consumers and society at large demand new types of electronics products with more than digital function, short-time-to-market for new product creation, and continuous cost reduction. This chapter intends to draw an overview picture of the changing global landscape of semiconductors by highlighting some of the major development trends, covering both technology and business.

1.2 Technology Evolution

Since the invention of transistor in 1947, semiconductors have been undergoing many rapid changes empowered by numerous innovations and technology breakthroughs. In the following some of the major technology advances, both existing and emerging, will be discussed.

1.2.1 More Moore (MM)

On April 19, 1965, the Electronics Magazine published a paper by Gordon E. Moore in which he made a prediction that the number of transistors on a chip roughly doubles every 2 years. Known as Moore's law, his prediction has enabled widespread proliferation of technology worldwide, and today has become shorthand for rapid technological change.

Moore's law is about miniaturization, and about extreme miniaturization. As one example, the Intel® 45-nm high-*k* metal gate silicon technology packs more than 400 million transistors for dual-core processors and more than 800 million for quad-core. Intel demonstrated first 32-nm logic process with functional SRAM packing more than 1.9 billion transistors [2]. The total length of the interconnect lines connecting different transistors of a single IC can be as long as several kilometers.

Moore's law is about cost reduction, and about extreme cost reduction. The price per transistor on a chip has dropped dramatically since 1968 [3, 4]. Some people estimate that the price of a transistor is now about the same as that of one printed newspaper character. In 1978, a commercial flight between New York and Paris cost around $900 and took 7 hours. If the principles of Moore's law had been applied to the airline industry the way they have applied to the semiconductor industry since 1978, then that flight would now cost about a cent and take less than 1 s. It is this economic aspect of Moore's law that has made electronics so pervasive.

Moore's law characterized by extreme miniaturization and extreme cost reduction is not only valid for ICs; for backend technology, i.e., packaging and assembly, similar trends have also been observed. Taking some feature sizes of packaging and assembly as examples, one can see that wire diameters for bonding can be smaller than 10 μm; the interconnect pitch of wafer level packaging can be smaller than 20 μm; the thickness of copper film/trace in PCB can be smaller than 5 μm; the microvia diameters can be smaller than 20 μm; and the wafer thickness can be thinner than 20 μm. Clearly, Moore's law has not only driven the extreme miniaturization of the IC technology, but also pushed the packaging, assembly, and system level miniaturization, going beyond the visualization with our bare eyes.

Currently almost 70% of the total semiconductor components market is directly impacted by advanced CMOS miniaturization achieved by following Moore's law. This 70% comprises three component groups of similar size, namely microprocessors, mass memories, and digital logic. The analog/mixed-signal market largely relies on variants of CMOS technology that are less affected by the miniaturization race due to other constraints, such as the need to handle power and/or high voltage.

The "MM" development is defined as a relentless scaling of digital functions, and an attempt to further develop advanced CMOS technologies and reduce the associated cost per function along two axes. The first axis is a geometrical (constant field) scaling, which refers to the continued shrinking of horizontal and vertical physical feature sizes of on-chip logic and memory storage functions in order to improve integration density (reduced cost per function), performance (higher speed, lower power), and reliability values. The second axis of scaling relates to three-dimensional device structure improvements plus other nongeometrical process techniques and new materials that affect the electrical performance of the chip. This axis of "equivalent" scaling occurs in conjunction with, and also enables, continued geometrical scaling.

However, reaching this ultimate CMOS node at the deca-nanometer level around 2015 will not be business as usual [5]. It will require much more innovation in the coming decade than in the past ones. Indeed, today we have reached the end of classical Dennard scaling and we are confronted with a set of cumulative interrelated challenges at all levels of the value chain from system down to atomic level, requiring enormously innovative processing steps and new materials in contrast with the Dennard scaling of the past. The art will be to manage these processes such that 10 billion of these devices can be interconnected to create reliable architectures for applications. To summarize, the main challenges for "MM" technology are as follows [5]:

- In the process technology domain: massive introduction of new materials, introduction of new device architectures (FD mugfets), moving to EUV litho or nanoimprint litho, the increase of random device and interconnect variability especially in memories, reaching the limit of Cu interconnects (e-migration, cross-talk, etc.), and conflicting between dynamic and static power density.

- In the design domain: Ultimately, NRE cost may reach 1 billion € per platform if no drastic changes in design technology occur due to increased hardware-software interaction on multicore platforms. Furthermore, the design metrics have changed from maximizing raw performance (servers, microprocessors) to maximizing throughput (B/s)/W cm^2.
- The problem of ultimate scaling is shifting to both extremes of the value chain, i.e., coping with gigascale system-level design cost (hardware and software) on one hand and multiscale (from nano to macro) physical effects on the other hand. Most importantly, the challenges in the above-mentioned two domains are strongly inter-related, i.e., changing in process will affect the whole design process, from atomic to system level. To name a few they are as follows [5].
- Achieving the above power efficiency requires a joint optimization of concurrent hardware-dependent low-power software; heterogeneous multicore architectures; and sub-1 Volt digital, RF, and analog IP libraries using novel device architectures.
- Reaching lithographic limits will require the development of highly regular yet layout-efficient computing, storage, and communication structures amenable to automated design methodologies. Otherwise, IP development costs will become unacceptable.
- Random variability will affect parametric yield and will require, besides DFM techniques, novel ways to avoid corner-based design to cope with device uncertainty, and amenable to design automation. Designers need to be trained in their applications. This will require the development of self-healing, defect, and error-tolerant, yet testable design styles based on low-cost on-chip adaptive control systems.
- Reliable local and global on-chip communication in 22 nm and beyond technology will be a much more limiting factor than transistor scaling and will require, besides the investigation of optical-, wireless-, or CNT-based technologies, investigation of architectural solutions such as tile-based GALS architectures exploiting networks-on-chip. Three-dimensional integration and System-in-Package (SiP) should be studied as strong contenders to ultimate scaling for true system design, which is the ultimate goal of electronics.
- Analog and RF design will have to cope with ultimate digital scaling and further sub-1 Volt scaling. This will require extreme creativity in analog and RF system design by compensating analog deficiencies using digital techniques.

It is important to note that the "living apart together" relationship among system, platform, IP creators, and process development is no longer possible or appropriate. Currently only limited research activities are devoted to the link between the two extremes (gigascale system design and multiscale physical performance), without which the commercial exploitation of ultimate CMOS will fail. There is an urgent need for an interdisciplinary dialog and interdisciplinary teams of industry, research institutes, and academia to establish a common systemability view, roadmap, and joint effort, in order to cope with the hardware and software codesign gap (see Fig. 1.2).

ware and soft-
gap

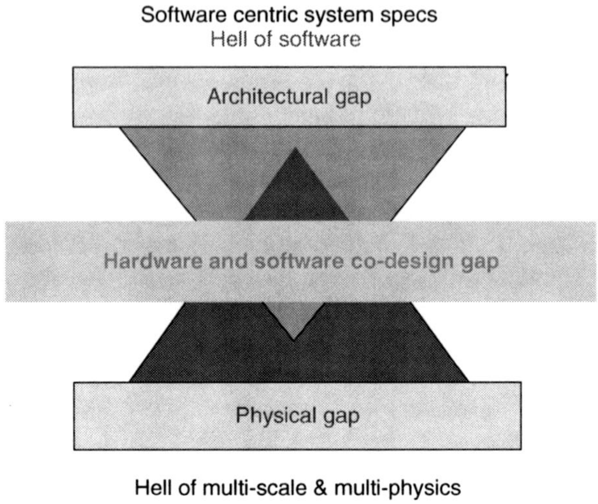

1.2.2 More than Moore (MtM)

In recent years, we have witnessed the emergence of an increasingly diverse area of micro/nanoelectronics that goes beyond the boundaries of Moore's law into the area of "More than Moore" (MtM) [1, 6–10]. From the technology perspective, "MtM" refers to all technologies enabling nondigital functions. They are based on or derived from silicon technology, but do not simply scale with Moore's law. From the application perspective, "MtM" enables functions equivalent with the eyes, ears, noses, arms, and legs of human beings, working together with the brain provided by microprocessor and memory subsystems. Figure 1.3 shows some typical "MtM" products, such as a mobile phone with nondigital functions (such as audio/video player, camera, Personal Health Monitor, GPS, Identification, Compass, etc.) and a great variety and quantity of sensors and actuators required in today's cars to monitor and control engine functions, safety, navigation, and comfort support. There are many other "MtM" products being able to realize a wide range of nonelectronic functions, such as mechanical, thermal, acoustical, chemical, optical, and biomedical. These nondigital functions provide means to sense, to interact with people, to interact with environment, and to power advanced semiconductor systems (see Fig. 1.4).

Generally speaking, "MtM" adds value to conventional semiconductors products via three routes [5]:

- *Interfacing (sensing and interacting) to the real world.* If the interaction is based on a nonelectrical phenomenon, then specific transducers are required. Sensors, actuators, displays, imagers, fluidic or biointerfaces (DNA, protein, lab-on-chip, neuron interfaces, etc.) are in this category.

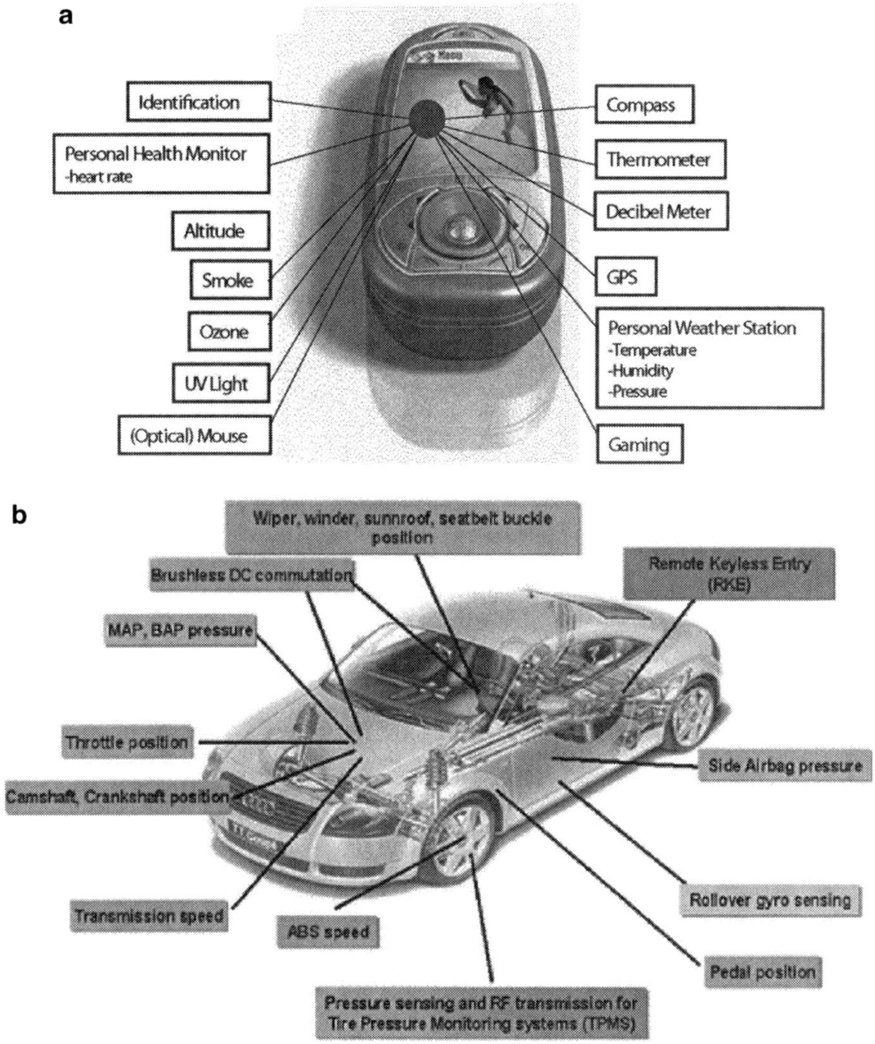

Fig. 1.3 Some "MtM" product examples

- *Enhancing* electronics *with nonpure electrical devices*. New devices can be used in RF or analog circuits and signal processing. Thanks to electrical characteristics or transfer functions that are unachievable by regular CMOS circuits, it is possible to reach better system performances. RF MEMS electroacoustic high Q resonators are a good example of this category.
- *Embedding power sources with the electronics*. Several new applications will require on-chip or in-package micro power sources (autonomous sensors or circuits with permanent active security monitoring for instance). Energy scavenging microsources and microbatteries are examples of this category.

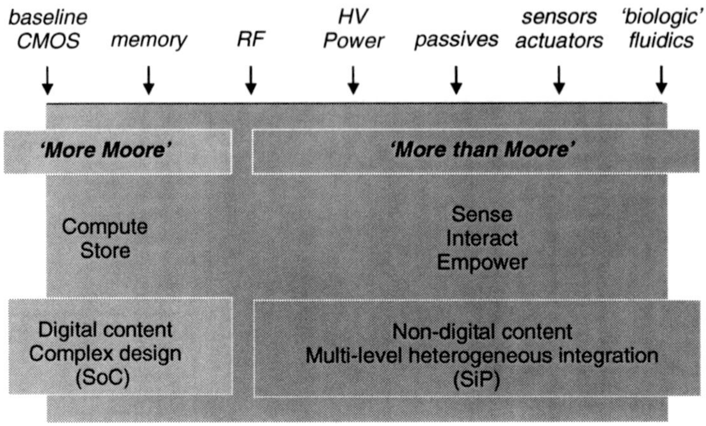

Fig. 1.4 Functions of "MtM" products

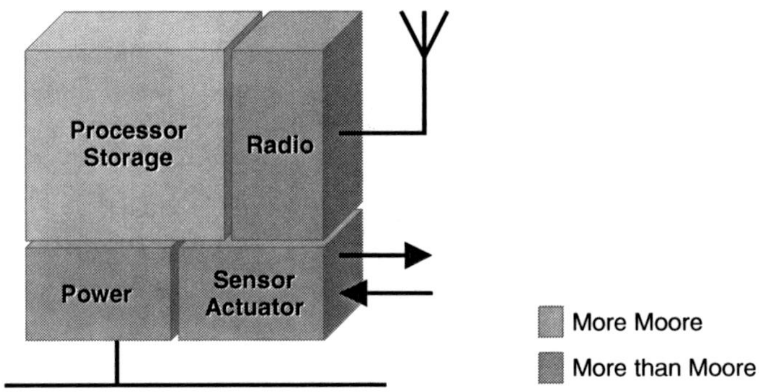

Fig. 1.5 Intelligent systems need "MM" and "MtM"

Clearly "MtM" technologies and products provide essential functional enrichment to the digital CMOS-based mainstream semiconductors. Along with IC technology, "MtM" becomes one of the key innovation drivers to meet current and future society needs with unlimited application potentials.

The emerging and rapid developments of "MtM" technologies and products are mainly driven by three factors given as follows [6].

First, the increasing social need for high-level system integration including non-digital functionalities, in order to interface to the real world in a wide range of societal relevant applications. The real world and the consumers are all analog, and digital functions alone are far from sufficient to meet the needs of human being. With industry entering into the nanoelectronics era, more and more consumers desire more functionalities beside the digital one. Figure 1.5 shows that intelligent system needs not only memory and processor, but also power, RF interfaces, and sensor and actuator functions.

Second, the need to create innovative products and broaden the product portfolio using less advanced wafer technology and production lines. Due to fierce competition and high investment costs of wafer fab, it is difficult to ensure business profitability by producing commodity ICs using less advanced wafer fab. However, for many applications, "MtM" products which are less geometric size dominant and allow a time-delay for miniaturization can still add value on top of the less advanced IC technology (see Fig. 1.6). It is worth to emphasize that although many novel "MtM" technologies and applications may not be size dependent as "MM," novel nanotechnologies will for sure provide important opportunities for the future development and success of "MtM" technologies. Besides, "MtM" technology will not only help to enlarge existing markets, but also drive the development of emerging ones – for example, Ambient Intelligence, Domotica, Lifestyle, Heath Care, Security, Food, Environment, and Energy. "MtM" is a unique opportunity for creativity, innovation, and new business creation for both small and large companies.

Finally, the need to overcome the cost and time-to-market limitations of System-on-Chip (SoC) development. Although many nondigital functions can theoretically be integrated onto these chips, doing so would involve prohibitive development time and cost. In addition, there is little prospect of a single practical IC technology that would allow integration with large number of diverse applications in the near future. It is therefore of paramount importance to balance the benefits of integrating some "MtM" functions on a chip, while integrating other functions in the same package to create SiP solutions.

It should be emphasized that "MM" and "MtM" are not competing with each others, but rather complimentary. To drive the future success of semiconductors, it is essential to integrate "MM" technology focusing mainly on digital functions with "MtM" technology focusing mainly on nondigital functions via heterogeneous system integration (see Fig. 1.7).

The technology challenges for "MtM" are application and product specific. Each type of technology, such as RF, sensors/actuators, biofluidics, HV and power,

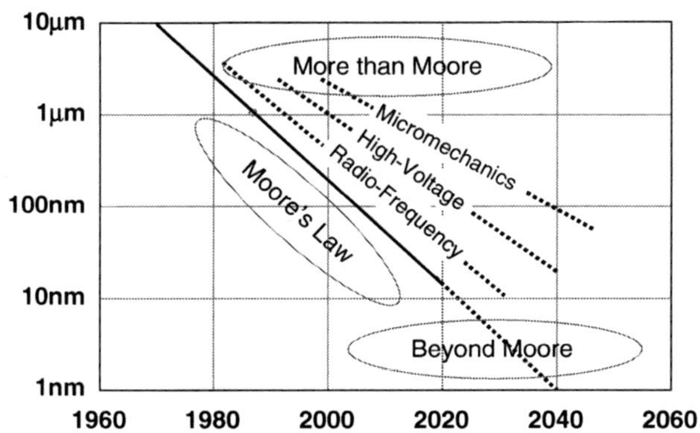

Fig. 1.6 Dimension delay of "MtM" products

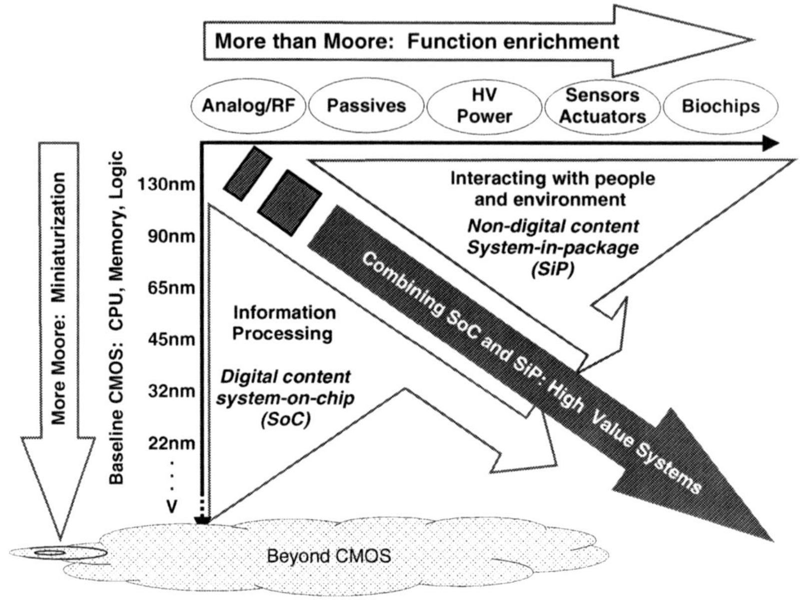

Fig. 1.7 ITRS nanoelectronics technology roadmap

Solid-State Lighting, MEMS/NEMS, has their own issues. Examples of some generic challenges are given below [11]:

- The introduction of new nanomaterials and nanotechnologies: To some extend, the success of nanoelectronics depends on the profound understanding of the properties and behavior of materials and their interfaces under manufacturing, qualification testing, and use conditions; and the capability to tailor the material design for the requirements of specific applications. This issue is already acute in the design of microelectronics. It is even more so for nanoelectronics and "MtM" technologies, wherein both multiscale size effect and multimaterial compatibility, stability, and reliability will be key to success. Among many challenges, characterization and modeling of material and their interface behavior need more attention, especially for multiscale, multiphysics, and time-dependent situations.
- Integrated process development via mastering the requirements and interaction among IC, package, and multifunctional systems.
- Establishing reusable design platform, process, and assembly environment for cost-effective mass implementation of a wide range CMOS compatible "MtM" devices, such as sensor, actuator, MEMS, and NEMS.
- Multiscale and multiphysical simulation, modeling, and characterization for "MtM" processes and products.
- Novel multifunctional system architecture and design.
- Designing for reliability, testability, compatibility, and manufacturability.

1.2.3 Beyond CMOS

"Beyond CMOS" covers the most advanced research activities to allow scaling of logic and memory functions to continue beyond the physical limits of Silicon-based CMOS technology [1, 5].

After more than 40 years of scaling according to Moore's law, we are rapidly approaching the CMOS scaling limit because we are reaching a point where an increase in power consumption coincides with an insufficient increase in operating speed. These highly undesirable effects are caused by a decrease in channel mobility and an increase in the interconnection resistance for smaller process geometries. The power consumption is largely due to increased leakage currents, short channel effects, source-drain tunneling, and p/n junction tunneling. Moreover, interconnections are increasingly becoming a limiting factor: the decrease in the pitch of interconnections and in the size of contacts and vias is causing an increase in overall resistance, while the reduced spacing is increasing propagation capacitances. The consequence is an increase in propagation delays and in the power consumption related to charging and discharging of interconnects to a point that already in today's 90-nm logic devices, a significant portion of transistors is dedicated only to driving interconnection lines, without playing any computational role. Physical limits of existing materials have been reached, and no significant progress can be expected in this area.

On top of that, the increasing impact of defects and the high level of complexity in both lithography and design have resulted in manufacturing costs rising dramatically. At the same time, the variability induced by the process variation at this nanoscale impacts also the yield and thus the cost. Even without taking into account the physical limits, all these combined effects push us closer to a point of reaching the limit of CMOS scaling.

There are a large numbers of different "beyond CMOS" options – but most are in an embryonic preindustrial phase. They can be realized via two different approaches. One is a gradual and evolutionary approach, by introducing new concepts and structures into the conventional CMOS technology. Another is the disruptive technology for replacing CMOS for some specific applications.

The evolutionary approach is based on the introduction of a new device, architecture, material, or process step inside the conventional CMOS technology, to solve one specific issue. It imposes the important constraint of CMOS compatibility, but has the advantage of reusing the huge amount of know-how developed till now. Examples are carbon nanotube transistor, nanowires, single electron devices, Resistance Change Memory, Ferroelectric FET Memory, defect and variation tolerant architectures, optical interconnect, RF interconnect, III-V compounds, magnetic materials, etc.

For disruptive approach, one can take the future computation scaling as one example, which requires capability using new state variables, efficient information transfer, and managing the heat transfer more efficiently.

New State Variables Many different information carriers need to be explored in addition to charge and none of them currently stands out as a clear winner. Examples include spin, molecular state, photons, phonons, nanostructures, mechanical state, resistance, quantum state (including phase), and magnetic flux.

Spintronics (spin-based electronics) has many potential advantages, including low power operation, nonvolatility, and co-localization of data processing and storage. Metal-based spintronics is likely to be first introduced for data storage applications using either spin torque switching or domain wall effects. Semiconductor-based spintronics could find application in data processing, though major breakthroughs are needed in materials (e.g., semiconductors with a higher critical temperature), devices (e.g., injection/detection trade-off), cointegration with CMOS, or in exploring promising physical phenomena (e.g., "dissipationless" spin current). Spintronics using half-metals and molecules also need to be explored. It should be stressed that no clear information processing device has so far emerged as a promising candidate to replace or supplement CMOS logic.

Molecular electronics is targeted at creating functional blocks at the molecular or supramolecular level that could be assembled in more complex functions. Fully molecular-based complex systems including interconnected molecular logic and molecular memory devices have still to be demonstrated. Limited molecular logic, memory, and interconnect functions have been shown, based on different types of molecules, but their integration into a single chip is still an issue. The first potential application is using the bistable behavior of certain molecules to produce memories with an extremely high density. We are still at a very early stage, where the reproducibility of reported results is not always evident. Specific issues, such as contacting the molecule, carrying enough current to provide noise immunity and a reasonable fan-out, and the addressing and read out of specific blocks remain to be solved.

Information Transfer Although significant research is carried out worldwide on "alternative" devices, no significant technological breakthrough has been achieved so far on information transfer in an integrated circuit. One of the more likely contenders to replace electromagnetic communication (i.e., information transfer through charge current in a metallic wire) is photon communication (i.e., light in the visible or IR range). More specifically, the very dynamic fields of nanophotonics, including plasmonics, allow the confinement and interaction of photons and electrons in a small volume, opening up the possibility of processing data at high frequency without compromising integration density.

It should be stressed that significant progress will come not only from breakthroughs in materials and device research, but also more significantly from the creative interaction of technology progress with progress in layout, design, software, and system research. For example, an optimized architecture through a better coding and localization of data is likely to bring significant improvement in information transfer techniques. Finally, more disruptive approaches such as stochastic resonance need to be explored.

Heat Transfer Management The emerging field of phononics aims to control phonon movement by using engineered nanostructures. It brings new opportunities in the interaction between quasiparticles (electrons, photons, spins, etc.) and phonons, potentially allowing better heat removal, isolation from thermal noise, and better carrier mobility.

Beside the computation scaling, the following issues are also important for the successful implementation of disruptive "beyond CMOS" technologies.

System Architecture At the device level, it is important to pay attention to the "systemability" of emerging devices, i.e., the capacity of a device to be integrated into a complex system. Moving up to functional block level, some emerging devices may offer new information processing paradigm by performing "dissipationless" computation in limited domains where information carriers will not encounter scattering (in a "ballistic" regime) or where phase information is maintained (as in quantum computing before decoherence occurs).

Emerging devices are expected to be more defective, less reliable, and less controlled in both their position and physical properties. It is therefore important to go beyond simply developing fault-tolerant systems that monitor the device at runtime and react to error detection. It will be necessary to consider error as a specific design constraint and to develop methodologies for error resiliency, accepting that error is inevitable, and trading off error rate against performance (speed, power consumption, etc.) in an application-dependent manner.

Using a similar approach, analog blocks of low complexity built with emerging devices may eventually find more extended use in balancing power consumption, in analog-digital partitioning and in signal restoration.

Von Neumann architectures – or more generally, programmable digital systems – will have to be reconsidered, especially with respect to optimizing the localization of data processing and storage and in coengineering the software and the architecture (e.g., parallel processing), without underestimating the legacy of more than 40 years of continuous development in classical electronic systems.

Open issues such as giving up deterministic computation (e.g., in neural networks or DNA computing) or addressing emergent behavior in complex systems are new research fields where multidisciplinarity is the key.

Physicists, designers, and system researchers cannot afford to work in isolated mode any more, focusing on their own field and having well-defined interfaces and handover mechanisms to other areas. The main challenge is to close the triangle between applications, emerging devices, and design resource constraints in order to manage complex interaction between the different levels of system development. It is therefore essential to develop real multidisciplinary cooperation between all those stakeholders who play a part in optimizing the overall performance of a system.

Manufacturing Opportunities As we cnter into the nm scale, the ability to manufacture billions of devices on a chip while maintaining full control over their properties is an overwhelming challenge that will probably lead to unbearable development and production costs. While it is difficult to predict which new processes will make their way into future manufacturing lines, there may be a comeback for chemistry, especially as development of the so-called "supramolecular toolbox" progresses and selective processes (e.g., surface functionalization) become more commonly used. Directed self-assembly (a "bottom-up approach") and possibly bioinspired and templated assembly are attractive concepts for low-cost manufacturing that need further investigation, although the fabrication of complex nonregular integrated systems has still to be demonstrated. Bioinspired manufacturing processes may be useful to address defect-resiliency and the self-repair of defective systems.

Future successful technologies may have to combine novel bottom-up and more traditional top-down manufacturing to achieve increased performance and cost effectiveness. Finally, as discussed in the previous paragraphs, research into new architectures may also help to relax the need for a deterministic approach to controlling the properties of the elementary devices.

Enabling "MtM" While a longer time frame is expected to implement disruptive "beyond CMOS" technology for IC, a significant achievement can be realized in a short or middle term, by developing "beyond CMOS" technology for novel "MtM" applications. At the moment, one can actually buy a handful of electronic products made with carbon nanotubes (CNT). Examples are CNT sensors (see Fig. 1.8), probe tips, and transparent conductive films. As one of the novel solid materials, nanowires have also received much attention from the R&D community as components for electrical circuits, sensors, or light emitting sources based on CMOS compatible processes. Although the R&D activities for CNT and nanowires were initiated to address the future need of IC technologies beyond the physical limits of CMOS, more and more R&D activity nowadays is devoted to using CNT and

Fig. 1.8 Cell interacting with nanotube structure

nanowires to create "MtM" products. Other examples are using spin-torque for RF detection, plasmonics for more sensitive optical sensors, nanodevices for molecular recognition, or nanostructured materials for enhanced energy efficiency. It is expected that increased effort in developing disruptive "beyond CMOS' technology for "MtM" applications will eventually speed up the development and implementation to replace CMOS.

Today there is no acceptable candidate to replace CMOS devices in terms of the four essential metrics needed for successful applications: dimension (scalability), switching speed, energy consumption, and throughput. Moreover, when adding reliability, design-ability, and mixed-signal capability as other key metrics, CMOS dominance is even more obvious. ITRS Emerging Research Devices [12] proposes criteria to evaluate the potential of emerging devices and circuits, wherein continuity is largely dominant, and these criteria are to evaluate the potential of emerging technologies in terms of their added value compared to scaled CMOS, and clearly oriented toward dense computing and memory applications. The analysis presented in the ITRS-ERD document is based on defining a set of criteria for logic and another for memories, and applying them to potential technologies such as given below.

- *Logic.* Scalability, performance, energy dissipation, gain, operational reliability, operating temperature, CMOS technological and architectural compatibility
- *Memories.* Scalability, performance, energy dissipation, OFF/ON ratio, operational reliability, operating temperature, CMOS technological and architectural compatibility

The metrics above primarily address the CMOS and thus concern the pursuit of the "MM" approach and target a "beyond CMOS" technology corresponding to "an information processing technology that enhances the scaling of functional density and performances while simultaneously reducing the energy dissipated per functional operation and that could be realized using a highly manufacturable fabrication process."

However, it is also essential to consider complementary criteria that take into account the system-related issues, such as, system-ability; ability to co-host memory, logic, and communication; handling of device variability; existence of an appropriate architecture; handling of analog and mixed signals; and potentials for combining "MM" and "MtM."

Finally, attention is needed to the timeline of "beyond CMOS" development. Emerging devices are likely to be introduced initially in the form of low complexity blocks integrated into complex systems, before moving to more complex regular structures and, at a later stage, to complex random logic computing blocks. After the basic functionality of an emerging device ("proof of concept") has been demonstrated, it is expected that a few optimized components integrated into a system would be the next logical step. At this point, major issues will still need to be resolved regarding manufacturing techniques and the reproducibility of performance, as well as in terms of design methodologies and system architectures.

1.2.4 Systems Architecture and Design

System architecture and design are increasingly important for semiconductor solutions [5]. The increasing complexity of solutions driven by the creation in many areas of new industrial and consumer applications requires an increasingly modular design approach, developing more innovative building blocks and flexible system architectures with higher integration capabilities. The semiconductor value chain is becoming more differentiated and traditional players are moving forward along the entire value chain to provide systems solutions as required by the market. The trend includes moving from components to solutions, e.g., in radio frequency (RF), from modules to fully integrated solutions and from components business to systems business (e.g., camera systems with sensors, mechanics, lenses; telecom systems; and architectural systems).

The increasing gap between nanoelectronics hardware technology and complex system design is an urgent problem to be solved yet. As a consequence, we are now facing the reality of being able to develop new technology nodes every 2–3 years while the time frame for new design tools is in the order of 10 years. This discrepancy is unacceptable in the future and a profound change in people's mindset and substantially increased effort in development are needed.

Moving to nanoelectronics and emerging of "MtM" technologies enable a tremendous increase in the functionality of electronics systems and in the applications of SoC and SiP products. The demanding solutions must be capable of capturing formal design specifications provided by system houses, allowing high-level system and architecture exploration within the underlying constraints of available implementation technologies. All aspects of product development must be embraced, including digital, analog/mixed signal, power electronics, and embedded software in conjunction with nonelectrical components like MEMS and NEMS. This will require expertise drawn from the many different disciplines involved in product design (system level design, HW and SW co-design, IC-packaging-system co-design, product/process and equipment co-design, verification, and physical implementation with constraints for test, reliability, and manufacturability). Several challenges should be solved with respect to design [1, 11]:

Design for Heterogeneous System The complex and heterogeneous system assemblies will increasingly be interconnected three-dimensionally (wafer level packaging, TSV), and controlled by an ever increasing amount of software. The design of these smart and heterogeneous systems will require new methods and approaches to compose these compact systems. Interfaces linking digital/analog, hardware/software, electrical/mechanical, etc. need to be handled at different abstraction levels. Another important issue is the design platform and reuse technology. The current trend toward compact and heterogeneous systems such as SiP requires extension of the classical SoC-centric design flow to efficiently support the "MtM" needs, which is, by definition, extremely diverse. To design compact systems in the "MtM" domain, an "MtM" integration platform will be required, which can serve as a "virtual prototyping" environment. It is economically not sustainable without developing a design platform and reuse technology to account for the needs of very diverse "MtM" product characteristics.

Design for Manufacturability Design for manufacturability (D4M) is aimed to accelerate process ramp-up and to enhance process yield, robustness, and reliability. To be able to do so, effort is needed to develop and enable random and systematic yield loss estimations from design through yield models and process-aware design flows, enabling yield optimization early in the design flow and reducing costly iterations. In future, estimation of the effect of systematic yield hazards on the final yield of a complex SoC or SiP will become extremely difficult. Given the very large expected increase in NRE and development costs, this will have to be done at an early stage of the design cycle and will be an indispensable step.

Design for Reliability Design for reliability (D4R) is aimed to predict, optimize, and design upfront the reliability of products and processes. It is also called virtual prototyping method. D4R requires a range of research activities including gaining a basic understanding of material behavior, degradation and failure mechanisms under multiloading conditions, by accelerated reliability qualification tests and advanced failure analysis, in combination with various accurate and efficient multiphysical and multiscale simulation models in order to predict the failure evolution. Other issues are increasing occurrence of soft errors, variability, and soft failure mechanisms beyond 90-nm technology, which demand research into self-repairing circuits and self-modifying architectures. The issues that need special attention are as follows:

- Integrated multiscale (from atomistic to macro, considering the strong size and surface effect), multiphysics (electrical, mechanical, thermal, physics, chemical, etc.), multidamage (cracks, delamination, fatigues, electromigration, voids, creep, degradations, etc.), and multiprocess (wafer, micromachining, packaging, assembly, qualification and application profile) modeling incorporating the important loading history in order to understand and predict the performance and reliability. Herein, new algorithms and simulation tools have to be developed.
- Advanced failure analysis techniques and correlation methods to localize the multiphysics-based failure modes and the associated process for multifunctional SiP, and to understand the failure mechanisms and their interaction.
- Novel experimental methods and techniques to extract material/interface and total system behavior, in order to provide inputs for modeling and simulation on one hand, and to verify the modeling results and design rules on the other hand, covering both nano- and macroscales.
- Reliable and efficient reliability qualification methods and physics of failure-based correlation models to accelerate reliability qualification tests.

Design for Testability Design for testability (D4T) is to secure functionality, quality, and reliability before release. It is a challenging issue especially for "MtM" applications with multifunction products, wherein test strategy, methodology, and equipment need to be further developed. Other reasons of D4T concern are the growing costs for Automatic Testing Equipment systems and test programming that are needed for testing high-volume, high-speed, and high-quality products. According to the ITRS roadmap, testing's share of the total manufacturing cost tends to approach 50% if no effective counteractions are taken. Besides the cost issue, the increasingly tightening time-to-market windows, the multifunction design targets, and the growing customers' expectations for further improvement of product quality cause equally serious problems. Key test development activities include test specifications/test bench conception and diagnostic; test program development (test vector conversion/test program synthesis); test program debug and optimization (test characterization/optimization); and test pattern generation. Several challenges associated with the testability of compact "MtM" systems are identified as follows:

- Testers are typically specialized to different functions. With compact "MtM" system, all technologies can be present in the package which makes conventional testing with just one tester type a big challenge. Multiple insertion testing is an option.
- A compact "MtM" system is truly a system and it is easy to put components into such a system in a way that is inaccessible from the outside pins, which means that testing of a chip may have to be done through another device, most likely in a "functional" way instead of a "structural" way. Thus test time and test completeness are big challenges.
- For the multifunctional SiP, the test problems potentially get more difficult, depending on the test requirements of the MEMS. Some types of MEMS may require a physical stimulus as part of the test process (pressure, acceleration, etc.), which can lead to test fixturing difficulties.
- For software testability means that configurable functions exist that can perform self-tests. In addition, functions should have clear semantics and results that do not depend too much on the order of calling. Separation of concerns in the software is crucial. Advanced component- and aspect-oriented development methods will aid to this.
- Testing strategies become increasingly platform based. Driven by the specific application market, they are becoming more local and more supported by dedicated application engineers. As a consequence, the development and implementation of such test platform systems require increased flexibility and ease of access globally. The flexibility of platforms will allow their being more widely used in probe, final test, or engineering environments.

1.2.5 Software

Software is playing more and more important role in semiconductors. The know-how of dedicated semiconductor applications is increasingly implemented in software. Despite the portable nature of software, the close interaction between software and hardware in embedded control is still essential for effective implementation. The functionality and market appeal of "MM," "MtM," and their integrations strongly depend on the contribution of the software that is embedded in it. Superior hardware technology alone cannot guarantee business success of future semiconductors. As a well-known example, huge effort is made to lower dielectric constant k in the IC backend stacks from 2.3 to 2.0, which is a 15% improvement, but this effort so far leads to serious yield issues while better software optimization can improve power efficiency and performance by factors between 2 and 10 [5]. It becomes obvious that simultaneous changes are needed in process technology, materials, device architectures, and design technologies coping with the need to create software-dominated platforms for future applications.

There is a clear trend for more and more software with greater functional diversity and architectural complexity. Taking the automotive industry as an example, it is estimated that 70% of future innovations will be software related and most other sectors are moving in a similar direction. The growing need for embedded software is fuelled by a need for additional and heterogeneous functionality, real-time performance, distribution across subsystems, reuse for multiple systems or in a platform, and long life time reliability needs. As systemability is becoming one of the key success criteria for future technology development, there is an urgent need to integrate nanoelectronics technology with embedded systems and system level design.

Another trigger for hardware and software codesign is due to the fact that the dramatically increased number of design tasks and their complexity are already leading to a phenomenon known as the "design gaps" - the difference between what should theoretically be integrated into systems and what can practically be designed into them, and what should be manufactured vs. what has been designed.

The main characteristics of this software trend are as follows [13]:

- Any complex semiconductor system embeds electronic parts and related software. There is a clear correlation between the attractiveness of the product and the total amount of electronics it embeds (hardware and software). The software controls the user-visible part and determines the price point and margin of the product. The software part is often used to create differentiation among products with similar dedicated hardware.
- The increasing complexity and focus of semiconductor applications on dedicated solutions and systems integration, implying sophisticated software and human interface functionalities, requires increasingly specialized software capabilities that only providers with a critical mass of skills and service resources are able to offer.
- Semiconductor vendors are required to provide the software ecosystem. Therefore, software providers increasingly offer full integration and services packages. Besides, the semiconductor industry is also required to provide more and more reference design and platforms. This means that the semiconductor industry provides not only ICs but also the systems software on top of which the OEMs can customize their products, including hardware and system software. The implementation of a complex platform already requires more software designers than hardware designers.

- As the variety of applications is broadening, the development of more advanced SW tools and packages is becoming increasingly sophisticated.
- In information processing, the trend is toward multiple cores. This will likely push design challenges even more toward SW. As a consequence of the increasing role of SW development by the semiconductor industry, design and embedded software R&D costs are rising faster than any other costs.

1.2.6 Heterogeneous Integration and Packaging

In the past several decades, semiconductors have been spending tremendous effort in developing and commercializing the Moore's law, leading to not only many breakthroughs and revolution in ICT, but also noticeable changes in the way of living of human being. While this trend will still be valid, reflected in the "MM" and "beyond CMOS," there are ever-increasing awareness, R&D effort, and business drivers to push the development and application of "MtM" enabling various nondigital functionalities. The future business opportunities and technology challenges will be the integration of Moore's law focusing mainly on digital functions with "MtM" focusing on mainly nondigital functions via heterogeneous integration and packaging.

Heterogeneous integration and packaging, at the center of any micro/nanoelectronics system creation, and the bridge between nano/micro semiconductor technology with macroscale applications, is the final manufacturing process transforming semiconductor devices into functional products for the end users. It provides electrical and multiphysical connections (e.g., bio connection for bio-SiP) for multisignal transmission, power input, and voltage control. It also provides for thermal dissipation, construction carrier, and the physical protection required for reliability. In the process of "MM" and "MtM" integration, heterogeneous integration and packaging play an essential role as the key enabler governing the multifunctional performance, size, weight, cost, and reliability of the final products. Heterogeneous integration will not only bring various multifunctional components together into one package but also provide an interface to the application environment. It therefore represents the glue between the world of micro/nanoelectronics devices and systems that humans can interact with. Heterogeneous integration has to ensure not only the integration of components based on different technologies and materials, but also the targets for miniaturization.

As one of the important packaging technologies, SiP refers to (multi-) functional systems built up using semiconductors and/or in combination with other technologies in an electronic package dimension. SiP focuses on achieving the highest value for a single system package, by extreme miniaturization, heterogeneous function (such as electrical, optical, mechanical, bio-, etc.) integration, short-time-to-market, and competitive function/cost ratio. Its concept applies to quite diverse technologies, such as semiconductors, sensors, actuators, power, RF modules, solid-state lighting, and various healthcare devices. To distinguish between

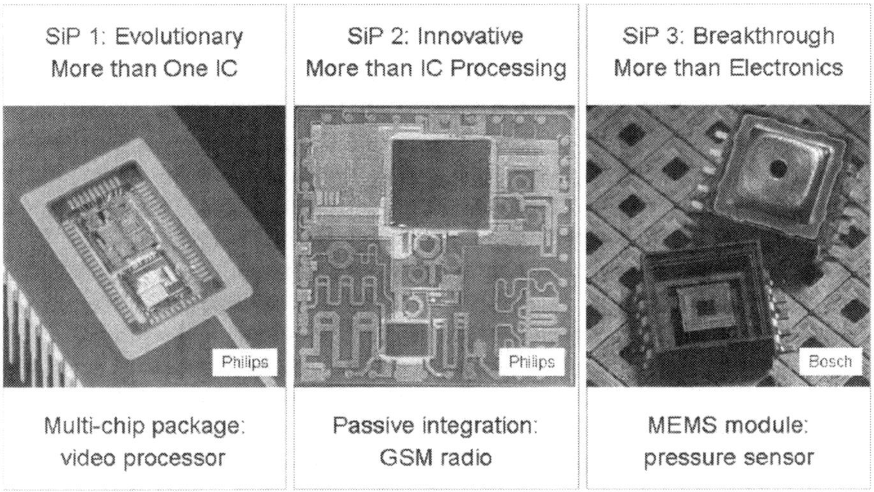

Fig. 1.9 Examples of three types of SiP

various SiPs, one can characterize SiPs into three categories (see Fig. 1.9). The first category refers to packages with multi-dies, such as McM, PiP, and PoP. The second category refers to subsystems built up using more than just IC process, such as passive integration. The last, the most challenging one, refers to compact system with more than electric functions, built up using multitechnologies and heterogeneous integration.

In the current semiconductor business, unfortunately heterogeneous integration and packaging are undervalued and underdeveloped. Many people consider packaging as low-tech and easy manageable process. There is a large gap between strong CMOS-based semiconductor device innovation capability and limited heterogeneous integration and packaging knowledge and know-how. It is also reflected by the fact that there are no industry-implementable roadmaps for nano-interconnect, nanopackaging, nanoassembly, and multifunction integration platform. For the coming decade, heterogeneous integration and packaging will be the bottleneck for the success of semiconductor industry, wherein packaging design concept, packaging architecture (often equivalent to system architecture), materials, manufacturing process, equipment, and system integration technology all need innovations and breakthroughs in an unprecedented speed.

On top of the above-mentioned roadmap issue, industries are confronted with ever-increased design complexity, dramatically decreased design margins, increased chances and consequences of failures, decreased product development and qualification times, and increased difficulties to meet quality, robustness, and reliability requirements. Other challenges for heterogeneous integration and packaging are given below:

Heterogeneity. From material perspective, future micro/nanoelectronics will be made of very different materials, which have to coexist despite their differences in behaviors (e.g., thermal expansion and biochemical interactions). From functional perspective, different functionality will be needed which requires large heterogeneity of processing, memory, communication resources, and input and output devices. From process technology perspective, various process platforms and building blocks are needed. These building blocks are prepared using various processes nodes, and will be exposed to a broad range of environmental constraints.

Complexity. Most of these systems are designed and built to embed intelligence and to enable products with the ability to react to their environment and to provide relevant and ergonomic information to their users. The amount of multimodal data to be processed by the system is very large. The user interfaces have to cope with complex and variable environments while taking into account the context of use and rapidly changing user behaviors.

Autonomous solutions. Most of the basic functions embedded in the systems have to be designed according to strong power requirements exploiting the most relevant energy sources. In many applications, energy resources are often the main bottleneck preventing larger market penetration of autonomous devices in both industrial and consumer products.

Multiscale. Multiscale nature (in both geometric and time domains) will have a very strong impact on the whole value chain of product creation process, from technology development to industrialization. Various constituting elements of SiP cover a very large-scale difference of geometric features, ranging from nanometers to millimeters. Moving from "Micro" to "Nano," and from nano- to multiscale will come with a paradigm shift.

Multidisciplinary. Micro/nanoelectronics innovation requires a large body of know-how. Taking a biosensor SiP as an example, it needs knowledge of not only electronic engineering, but also chemistry and bioengineering. Chemistry, thermal, metallurgy, physics, electrical, mechanics, optics, electromagnetics, and biology may all be involved in future product creation, which will require both hardware and software engineering.

Stochastic in nature. For micro/nanoelectronics, it is virtually impossible to design and manufacture products, and to process them with deterministic performance. For design parameters, such as material/interface properties, geometric dimensions, process window, and loading intensities, the deviations represented by different statistic characteristics and magnitudes are inevitable. With the IC technology moving toward "beyond CMOS" domain, control of multivariability at different scales become vital, especially if the performance at an atomistic level has to be linked with micro- or macrolevel requirements.

Product/process and equipment codesign. Yet another important challenge for the success of heterogeneous integration is product/process and backend equipment codesign.

The first challenge for backend equipment is driven by Moore's law. Nano IC needs backend equipment to be able to work on nanopackaging, nanointerconnecting, and nanoassembly, which are not available yet. As an example, wire bonding on this nanochip is a critical issue in the packaging and assembly processes, mainly because of the following reasons:

- IC (especially the Cu/low-*k* chip beyond 90 nm) and packages (especially SiP) become weaker
- Bonding pitch and wire diameter become smaller, and material will move from Au to Cu
- Product and process design windows are dramatically shrunken
- New and "green" materials are introduced with less structural robustness

Another challenge for backend equipment is driven by "MtM," which requires that backend equipment enable heterogeneous integration and assembly. Packaging can be difficult enough when the package contains the entire system integrated onto a single silicon die – the so-called SoC approach. It becomes even more difficult when multiple die and multifunctional devices are integrated into a single package to incorporate functions that are technically difficult or commercially inconvenient to incorporate into a SoC. These multiple device implementations normally contain a heterogeneous mix of silicon and non-silicon technologies, which further complicates package equipment design.

The ultimate challenge for backend equipment is to enable "MM" and "MtM" integration. It will be needed in the near future to integrate front-end equipment with backend equipment to meet the needs of high value heterogeneous system integration. Bringing "MM" and "MtM" together in semiconductor products is the battlefield of "Heterogeneous Integration," where nanoscale and microscale technologies meet. One cannot win this battle, without breakthrough in the associated equipment development.

1.3 Business Development Trends

The semiconductor business history in the last two decades shows unique characteristics with high average growth rate and cyclical pattern. Although it is not easy to highlight the key business development trends under the fast evolving macroeconomic situation, the following observations can be summarized [1, 13].

Economic Scale It has been shown that the top ten companies control about 40% of global markets share, and top 50 companies control more than 75% of the global market share. Number 1 or 2 in any market segment can make money, below number 5 it is very difficult, due to the clear correlation between market share and margins.

Cost for Operation First of all, the wafer manufacturing becomes extremely expensive. The set-up costs for a fabrication line increase substantially over time.

The steadily growing cost of investments in new, advanced CMOS manufacturing plants based on the most advanced process technologies is likely to continue. The costs for a leading-edge manufacturing line double between two CMOS technology generations. Today a wafer fab for 65 nm and 800k wafers/year costs around 4–5 billion US dollars. Even after a fabrication line has been built, rapid technological advance makes it likely that it will need to be upgraded several times during its productive life. Hence the most important cost factor in wafer production is the depreciation of equipment, buildings, and facilities. Semiconductors have been and remain highly R&D intensive. Semiconductor companies on average spend on the order of 18% of their sales revenues on R&D. The costs for process nodes, libraries, IP blocks, and system architectures also increased dramatically. Currently each new system platform technology costs about 300–500 million US dollars, process technology about 500 million US dollars per node, and libraries and IP blocks are needed for each new node. Costs for packaging and testing have also increased in recent years, and will continue to increase with a much higher rate to cope with the application needs of "MtM" and heterogeneously integrated multifunctional compact systems.

Devirtualization Before, many electronic system providers had own IC manufacturing, in order to ensure supply security and to keep competitive advantage for special and proprietary processes. In the last decade, this model became impractical due to rising capital spending, growing complexity of technologies, and subscale manufacturing capability. Almost all system houses have spun-out their IC division (e.g., Siemens – Infineon, Motorola – Freescale, Lucent – Agere, Philips – NXP Semiconductors), or sold/JV-ed their IC group to a semiconductor company (e.g., Alcatel – STMicroelectronics, Sony – Toshiba).

Specialization In the early days of semiconductors, Independent Device Makers (IDMs) could handle the entire value chain, sometimes even extending their business into manufacturing equipment and materials at one end and electronic products and services at the other. To address the exponential expanding in costs, critical know-how, and "MtM"-related application needs, semiconductors companies are changing from extremely broad portfolio to break-up and spin-out of product divisions. Today, IDMs typically outsource shareable tasks to more recently established businesses such as Original Design Manufacturers (ODMs), Electronics Manufacturing Services (EMS), and Design Houses. Many successful fablite and fabless companies (semiconductor companies relying totally on third-party foundries for manufacturing) have emerged. As a consequence, this trend is lowering the entry threshold for new companies into the semiconductor market.

In response to the consolidation toward common design and process platforms and their relative commoditization, competitive pressure is increasing for semiconductor companies to focus their R&D on accelerating the differentiation of their devices. This trend impacts the strategic priorities given to types of products and application segments. Choices concerning research investments made in-house or outsourced under collaborative funding schemes in order to optimize the use of dedicated process technologies and design activities are thus also affected. It is a clear and key trend of increasing specialization and pace of differentiation of semiconductors by product and/or market type. This trend applies primarily to sizeable IDMs keen to differentiate their product portfolio and to offer specific sets of products in order to maintain differentiating competitive advantages. The examples of spinning out of memory specialists are Micron, Spansion, Elpida, Hynix, Qimonda, and Numonyx. The examples of microprocessor specialists are Intel and AMD. The standard and multimarket IC specialists are ON, Fairchild, and NXP. One of the main challenges, however, is achieving this target cost effectively by using common platforms, standards, and reusable process while retaining key differentiators.

Marketwise, semiconductors are more than ever becoming the key enablers for forward-looking innovations in areas such as environmental controls, ambient intelligence, energy management, and biomedical applications. These new opportunities will accelerate and amplify dedicated R&D and engineering efforts to achieve more differentiated knowledge on a very large scale.

Advanced CMOS Manufacturing Has Become a Commodity With the cost of developing a next-generation CMOS technology platform increasing faster than the revenues of the semiconductor industry, many IDMs have also entered into industrial alliances in order to jointly develop common design, process platforms and their relative commoditization, so that they can get access to advanced CMOS technology platforms (at 32 nm and below) at affordable costs. The standard IP blocks provide little differentiation and advanced CMOS manufacturing has become a commodity.

Since it becomes increasingly difficult for any single company to bear the cost of advanced production facilities and to balance both increasing investment and ROI requirements, companies pool such production resources to achieve economies of scale. This also gives rise to increased specialization along three distinct manufacturing models based on emerging technology trends, market characteristics, and companies' strategic choices [13]:

- Fabrication lines for memories and standard products characterized by very high volume products such as microprocessors: Here one or two similar processes run with very few mask sets. These fabrication lines are maintained fully loaded, and the product cost is just a function of manufacturing efficiency, including investment cost and cost of capital, manpower, overhead and taxes, and size.
- Fabrication lines for logic products made using standard CMOS processes: Here a few processes run with a high number of products (mask sets). These fabrication lines are maintained fully loaded. In this foundry model, key cost parameters are: investment cost and cost of capital, manpower, overhead and taxes, size, flexibility on product mix within few processes, and cycle time.

- Fabrication lines for dedicated products with differentiated processes: These fabrication lines run different processes in parallel with a high number of products. The processes are largely tuned to each product and this is where the design and device interaction is maximal. Key cost parameters are cost of capital, manpower, overheads and taxes, flexibility on product mix within a large number of processes, cycle time, and design and control of a large number of different process routes. The investment cost is less than an issue in niche markets like MEMS where the standard is still 6-in. wafers.

System Knowledge System knowledge will continue to provide differentiation. The opportunities for semiconductor product providing dedicated functions and technologies focused on specific application areas are closely linked to end-user industries. Semiconductor companies may benefit strongly from access to, and proximity of, end-user industries.

Except the direct business alliance or partnership with application companies, system knowledge can also be developed by engagement with cluster or center of excellence with different critical innovation areas (due to less synergies between different application areas). In order to gain access to knowledge continuously with nonconventional semiconductor nature and new markets, it is essential for semiconductor industry to actively engage in clusters or center of excellence that include suppliers and end-users and focus on new applications and solutions. Forming such centers of excellence based on common interests and complementary capabilities enables IP generation while ensuring competitive differentiation and capturing new market opportunities through standardization.

Function Integration Customer expectations from semiconductor suppliers shift toward platforms, system integration, and services. This is mainly driven by many emerging "MtM" applications and the blurring boundaries between semiconductor players and OEMs. As specialized semiconductor applications increasingly provide full-systems solutions to end-user customers, there is a tendency for OEM systems design to migrate to semiconductor suppliers, opening up new collaborative opportunities for providing full R&D development and engineering for such solutions. This means that more systems knowledge is being integrated into semiconductor products.

Continuing disparity between life cycles for technology innovation (as much as 3 years) and application innovation (as low as 6 months), increasing market demand for first-time-right and zero-defect products, and the need for semiconductor companies to provide complete hardware/software reference designs have drastically changed the position of IDMs. No longer "arms-length" suppliers to their customers, semiconductor companies are now at the very heart of the innovation process in System Houses and OEMs. As a result, the formerly linear high-tech supply chain has expanded into a series of multiple interconnected ecosystems, all of which have the semiconductor industry as an essential common element.

Fablite and Fabless Design complexity caused by "MM" and "MtM," cost constraint, and time-to-market requirements lead to increased differentiation of providers offering IP blocks, specialized designs, and market functional blocks that are incorporated in SoC and SiP. The traditional business model of a semiconductor company as an IDM is shifting toward a model that increasingly seeks to optimize the combination of in-house manufacturing and service activities with outsourced ones. This dynamic is contributing to reshaping the semiconductor landscape, moving progressively away from the traditional IDM model and increasingly contracting related IC design and manufacturing services and/or acquire IP blocks from external suppliers who are not considered critical from a competitive differentiation point of view on the open market. Fablite and fabless are the business models emerging from this trend.

Ecosystems Ecosystems for nanoelectronics innovation are evolving in various parts of the world, yet it is clear that the only ones to survive will be those in which all players (industry, academia, and public authorities) cooperate with one another. For companies operating in the nanoelectronics value chain, optimized alliances and R&D involvement are critical to staying in the race. Long-term commitments, both in terms of money and people, are required by private and public stakeholders. Successful ecosystems must also recognize that SMEs and academia (universities and institutes) are fertile breeding grounds for new ideas.

1.4 Making it Happen

Future micro/nanoelectronics will be driven by two engines, namely "Moore's law" and "MtM," and they will be much more than miniaturization. They are becoming multifunctionality, multidesign requirements, multidiscipline, multiscale, multi-technology, multimaterial/interface, multiprocess, multidamage and failure mode, and multivariability. Future micro/nanoelectronics will be much more than technology. They are becoming multiapplication and market, multi-infrastructure, multi-innovation complexity, multibillon investment, multisupply chain, and multibusiness model.

The success of future micro/nanoelectronics depends on not only the availability of the required technologies, design, and the associated competencies, but also the existence of industrial visions, strategies, and business models that allow optimization of entire value chains to fit the characteristics and needs of specific applications. It is worth to mention that for all technology building blocks of "MtM," beside the functional requirements, cost, reliability, and time-to-market are dominating factors that will determine the ultimate success of the future semiconductors. Other aspects that deserve more attention are as given below.

Several elements need to be considered in defining new "MtM" business models. First, with more functions being integrated into a single compact system, partitioning in the value chain changes both in terms of the involvement of the different parties and in terms of their profitability. This may lead to changes in the supply chain management. Another challenge associated with supply chain management is the immaturity of many "MtM" markets and little consolidation in the number of industrial players involved. Second, new and quantitatively reliable cost models have to be developed. Currently some of the nonmature "MtM" technologies have unclarified cost consequences. Even for the nearly mature "MtM" technologies, the cost of heterogeneous integration is not always well defined. This is the case, for example, with the "known-good-die" requirement. The winning business models will be those that integrate and optimize all these important, interlinked, sometimes controversial elements.

It is important to know that due to the tremendously broad application scopes, it is not cost effective to develop individual technology for each specific application. The business success of future micro/nanoelectronics depends on the capability to combine the application-specific functional needs with cost, time-to-market, and reliability requirements. Therefore, it is vital to develop system architecture, codesign methods and tools, generic design platform and design flow, standardization, reuse technology, and modular processes.

The success of Moore's law has been enabled by an excellent ecosystem consisting of public awareness; the availability of resources (man-power, materials, finance, etc.); the existence of R&D, manufacturing, and supply chain infrastructures; and market maturity. To create effective and efficient ecosystems in the "MtM" business, partnerships will be essential from different aspects.

First, partnerships among the semiconductor sector should be further strengthened. In the past decade, many of the semiconductor companies have abandoned their vertical operation model and focused on their core business. More than ever before, they are now interlinked with each other in the industrial value chain. Collaboration on precompetitive content is a common practice, despite the fact that the business competition between them has been intensified. For the "MtM" business, there is a clear need to standardize and commoditize some of the required technologies and designs in order to enable product manufacturing to be quickly ramped up to an economically feasible scale. This can only be achieved by establishing structure cooperation within the semiconductor sector.

Second, partnership between semiconductor companies and system houses (including nonelectronic sectors, such as automotive, energy, health care, etc.) will be vital. For "MtM"-related technology and new product development, it is important to know the market requirements and trends in order to master the required application knowledge, share the R&D costs, and gain access to markets beyond the traditional operating scope of semiconductor companies. By joining forces with system houses, it will be possible not only to enlarge existing markets (for example, automotive and lighting), but also to drive the emergence and growth of new markets (for example, health care, personal wellness, energy).

Finally, partnership between industries and academia should be enhanced. The associated benefits are twofold. First, it will help to speed up the creation of fundamental knowledge in order to solve the gap between application requirements of future semiconductors (such as "beyond CMOS" and "MtM") and the availability of the underlying fundamental knowledge. Without the desperately needed fundamental knowledge, industry has to rely on the trial-and-error methodologies in technology development. Due to the ever-increasing competition pressure, companies worldwide are spending less effort on long-term research and predevelopment. More than ever, industry depends on academic institutions to deliver the needed fundamental knowledge. Collaboration between industry and academia will improve the efficiency and industrial relevance of fundamental research. Second, it will greatly enlarge the success rate of research and innovation conducted in academic institutions. In the past two decades, the academic community has invested a lot of effort in developing certain "MtM"-related technologies and even in product creation (for example, sensors and MEMS). However, these achieved technological feasibility and IP do not lead to business success, because academic institutions usually do not have suitable industrialization infrastructure, marketing and supply chain expertise, and know-how for business and cost modeling. This has hampered the beneficial commercialization and economic impact of some excellent R&D results.

Cooperation between multinational corporations and SMEs is also important for future micro/nanoelectronics. Due to the application diversity and many nonsolved challenges of "MtM" technologies and large number of alternatives of "beyond CMOS" technologies, it is a perfect playground for highly innovative SMEs to play an essential role in developing IPs by exploring both evolutionary and disruptive technologies. SMEs can also contribute to market scooting and market development by quickly bringing a small volume of prototypes and products to the users, and to serve some low-volume application needs. However, SMEs alone cannot push "MtM" technologies and products into mass consumer markets without the leading electronic industries committing their resource and capabilities. Teaming up between multinational corporations and SMEs will speed up the development of new technology and new markets and broaden the application scopes.

1.5 Conclusions

Gordon E. Moore ever stated, "The most exciting thing about new applications going forward will be the surprises I cannot predict. The most important things are usually the ones that people within the industry do not see. They tend to develop outside the industry. I do not know. I just wait to be surprised with the next one that comes along." However, without attempting to predict exactly what will happen in the future, by reviewing the history of semiconductors, one could have some inspiration for possible direction of future development.

The history of semiconductors can be roughly divided into three phases. The first phase is from 1970s to 1985. In this period of time, applications of semiconductors are mainly mainframe computers, TV, VCR, etc., with very limited penetration in consumer products. The second phase is from 1985 to 2005. In this period of time, semiconductors have gained wide applications in consumer products, such as PC, CD, mobile phone, digital camera, Internet, video and audio devices, etc. However, semiconductors have only touched the tip of the iceberg of potential consumer applications. The third phase, starting from 2005 and driven by the two engines (Moore's law and "MtM"), will focus on the creation of heterogeneous and intelligent systems, with unlimited potentials of applications, covering Ambient Intelligence, Domotica, Lifestyle, Heath care, Food, etc., which have never been experienced and imaged before. The third development phase will be characterized by a paradigm shift of semiconductors' technology and business from conventional IC centric to application system driven "MM" and "MtM" integration, wherein cost will remain the key for all consumer applications.

The future of semiconductors is bright, if and only if one can keep following Charles Darwin's statement: "It is not the strongest of the species that survives, nor the most intelligent, but the one most responsive to changes."

Acknowledgments The authors would like to thank for the valuable contribution from our colleagues, Mart Graef, Merten Koolen, and Peter Zegers.

We are grateful to all members of the ENIAC technology domain teams, especially Livio Baldi, Eddy Blokken, Michel Brillouët, Guy Dubois, Roger de Keersmaecker, Norbert Lehner, Harald Pötter, Herbert Reichl, Lothar Schrader, Dominique Thomas, Gilles Thomas, Kees Beenakker, Peter Caldera, Alain Cappy, Walter De Raedt; and to all members of Scientific Community Council of the ENIAC, especially Patrick Dewilde, Adrian Ionescu and Hannu Tenhunen.

We would like to express our gratitude to all members of the MEDEA+ working group ("Towards and beyond 2015"), especially Hugo De Man, Thomas Gessner, Frans Widdershoven, Marc Belleville, Thomas Skotnicki, Roland Thewes, and Harry Veendrick.

Appreciations also go to colleagues active in the IEEE-CPMT society, especially Willem Chen, Bill Bottom, Rao Tummala, C.P. Wong, and Rolf Aschenbrenner.

References

1. Strategic Research Agenda of ENIAC, 2007 Edition, Budapest, November 28, 2007.
2. http://www.intel.com/technology/architecture-silicon/45nm-core2/
3. http://www.intel.com/pressroom/kits/events/moores_law_40th/?iid=SEARCH
4. http://download.intel.com/museum/Moores_Law/Printed_Materials/Moores_Law_Perspective.pdf
5. Towards and Beyond 2015: Technology, Devices, Circuits and Systems, MEDEA+, 2006.
6. G.Q. Zhang, F. van Roosmalen, M. Graef, The paradigm of "More than Moore". Proceedings of ICEPT2005, ISBN 0-7803-9449-06, 2005, IEEE Cat number 05EX1194, pp. 17–24.
7. G.Q. Zhang, M. Graef, F. van Roosmalen, The rationale and paradigm of More than Moore. Proceedings of the 55th Electronic Components and Technology Conference, San Diego, CA, USA, June 1–3, 2006, IEEE Cat number 06CH37766, pp. 151–157.
8. G.Q. Zhang, M. Graef, F. van Roosmalen, Strategic research agenda of "More than Moore". Proceedings of the Seventh International Conference on Thermal, Mechanical and Multi-Physics Simulation and Experiments in Micro-Electronics and Micro-Systems, ISBN 1-4244-0275-1, Como, Italy, April 23–26, 2006, pp. 4–10.
9. G.Q. Zhang, W.D. van Driel, X.J. Fan, "Mechanics of Microelectronics" Series: Solid Mechanics and Its Applications, Vol. 141, XIII, hardcover ISBN: 978-1-4020-4934-7, 2006, 563 pages.
10. G.Q. Zhang, Nanoelectronics landscape - Applications, technology and economy, in James Morris and Debendra Mallik (Eds.) Nanopackaging, ISBN 978-0-387, Springer, New York, NY, 2009, pp. 1–19.

11. G.Q. Zhang, R. Begeer, R.A. Hartman, Strategic Research Agenda of Point-One, Eindhoven, 2007.
12. International Technology Roadmap of Semiconductors, http://www.itrs.net
13. Mastering Innovation, Shaping the Future, ESIA 2008 competitiveness report, 2008.
14. Strategic Research Agenda of ENIAC, 2006 Edition, Monte Carlo, November 30, 2006.
15. Strategic Research Agenda of EPoSS, European Technology Platform of Smart System Integration, 2007.

Chapter 2
Smart Integrated Systems: From Components to Products

Thomas Gessner, Mario Baum, Wolfgang Gessner, and Günter Lugert

Abstract Customer demands for devices with new increasingly complex functionalities, higher quality, lower cost, long-term reliability and usability both for investment and consumer goods constitute new technological challenges [EPoSS, the European Technology Platform on Smart Systems Integration (http://www.smart-systems-integration.org)]. The key factor for successful products will be system integration, miniaturisation and last but not least transdisciplinary technology approaches. Bridging the gap between the nanometre scale and the customer's macro-world will affect a multitude of technologies, materials and processes and their combination.

Smart systems technologies and their integration will therefore have a significant impact on the competitiveness of entire sectors such as automotive, aeronautics, medical technology, logistics, security, and process engineering. It will consequently contribute to solving major socio-economic problems in the health, environment, mobility and other domains. Smart systems will provide solutions for assistive technologies for the handicapped and elderly, will be the enabling technologies for new solutions in medical care and will help to reduce CO_2 emissions. Smart systems development will evolve from current first generation smart systems as, for example, object recognition devices through Smart systems of the second generation as artificial human organs up to third generation Smart systems combining technical and cognitive functions.

The following considerations will address the various aspects related to smart systems integration taking into account the major socio-economic megatrends.

Keywords Smart systems • Compact systems • Nanoelectronics applications • More than Moore • System integration • EPoSS • Heterogeneous integration • Ambient intelligence • Automotive electronics

---

T. Gessner (✉), M. Baum, W. Gessner, and G. Lugert
Fraunhofer ENAS, Chemnitz, Germany
e-mail: thomas.gessner@enas.fraunhofer.de

G.Q. Zhang and A.J. van Roosmalen (eds.), *More than Moore: Creating High Value Micro/Nanoelectronics Systems*,
DOI 10.1007/978-0-387-75593-9_2,

2.1 Introduction

Already today intelligent functions based on microsystems technologies can be found in many applications of everybody's everyday life. Mobile phones are called smart phones and technologically evolve together with digital cameras, music players, game consoles and PDAs. Cars are getting more and more "intelligent" due to self-controlled operation and adaptive safety systems. Minimal invasive therapy is inconceivable without microsystems including sensing functions, signal processing and actuators. The smart house concept and the so-called ambient-assisted living solutions bring smart sensors, actuators, and controlling as well as emergency systems to everyone. In most cases, measurement and actuation systems have to be able to interface, interact and communicate with other sub-systems and/or the macrosystem they are incorporated in.

Smart systems, however, go beyond microsystems for single physical, biological or chemical parameter measurements combined with signal processing and actuating functions. Smart systems integration addresses the demand for miniaturised multi-functional devices and specialised connected and interacting solutions. Multi-disciplinary approaches featuring devices for complex solutions and making use of shared and, increasingly, self-organising resources are among the most ambitious challenges.

The EPoSS Strategic Research Agenda is proposing a multi-level approach incorporating various technologies, functionalities and methodologies in order to respond to the social demand of new and visionary products. Rather than solving problems piece-meal at component level, systems approaches integrative concepts following a problem-solving approach are suggested. EPoSS will build upon new multi-disciplinary technologies and is neither dedicated to any specific research discipline nor does it restrict its activities to a certain scale or size of devices.

The vision is that smart systems will be able to take over complex human perceptive and cognitive functions. Smart devices will frequently act unnoticeably in the background and intervene visibly only when human capabilities to act and to react are reduced or cease to exist. Examples for such smart systems and related integration challenges include object recognition devices for automated production systems, devices for monitoring the physical and mental condition of vehicle drivers or integrated polymeric RFID systems for logistics packaging.

According to EPoSS members, the foremost general R&D priorities related to smart systems integration are:

- Robust systems, compatible and adaptive to environment and lifetime requirements
- Autonomous energy efficient and networked smart systems
- Smart systems with advanced functionality and performance

In 2006, the world market for microsystems technology amounted to estimated 40 billion USD. A market analysis by SEMI and Yole Developpement projects an increase of this market to 72 billion USD by the year 2011 [1]. Europe needs to raise its game in order to exploit its existing potential and to keep and increase its share in the world market for smart systems. In the years to come, the share of suppliers in the value chain will be rising because OEMs more and more search for partners with the ability to integrate knowledge across classic segments. Specialised suppliers of smart systems technologies will be able to provide this expertise for application fields of major importance. Thus, specialised SME suppliers will be important partners in fostering the promotion of smart systems and in advancing technological developments in Europe.

2.2 State of Technology

2.2.1 Automotive

The automotive industry is amongst the most important industries in Europe. Its global competitiveness relies strongly on continuous innovations to cope with increasing requirements in terms of mobility, safety and resource efficiency. These emerging challenges need comprehensive and interlinked action; at the sub-system level of the vehicle this is the domain for smart systems.

The automotive industry represents 3% of Europe's gross domestic product and 8% of the member states' total revenues. It makes a central contribution to providing the European public and economy with increased mobility: In 1970, the average European travelled 17 km daily, while today the corresponding figure is 35 km. In addition to mobility and flexibility in general, the automotive industry underpins the lifestyle that Europeans enjoy by facilitating safe social interaction and access as well as the reliable distribution of goods across the continent. With about 20% of the European industrial R&D the automotive sector is the largest R&D investor in Europe. It constitutes a major driver for the development and diffusion of new technologies and innovations throughout the economy.

At the same time, the automotive industry faces serious challenges in terms of road safety, CO_2 emissions and energy efficiency. Still, more than 40,000 people are killed in road accidents in Europe every year. And, exhaust gases from vehicle engines account for about 20% of greenhouse gas emissions. Innovative technologies will be required to tackle these issues, and smart systems can be expected to play the role of an enabling technology.

The general challenges that EPoSS has identified are in agreement with the strategic objectives defined by EUCAR, the European Council for Automotive Research[1]; ERTRAC, the European Road Transport Research Advisory Council[2] and CLEPA, the European Association of Automotive Suppliers[3]:

- Increasing number of automotive sub-systems
- Demanding development challenge of multi-domain systems

[1]EUCAR – European Council for Automotive R & D (http://www.eucar.be).

[2]ERTRAC – European Road Transport Research Advisory Council (http://www.ertrac.org).

[3]CLEPA – European Association of Automotive Suppliers (http://www.clepa.com).

- Complex interaction between components and systems
- Permanently increasing functionality
- Very high quality demands compared to other consumer goods
- Many safety-critical issues
- Need for gains in energy efficiency
- Harsh environment
- Decreasing time to market
- Permanent pressure on cost
- Growing global competition

In future smart systems integration will address various challenges from both the technological and the process side. The cross-border integration of technologies and competences as well as the linkage of different scale solutions are key strategies. Therefore, current R&D in smart systems for automotive applications is focused on intelligent materials, sensors, actuators, and interconnect technologies as well as on management methodologies for the multilevel integration of technological and human resources, on production methodologies for fast and efficient manufacturing and on standards and rules for global knowledge sharing and systems interoperability. This requires careful attention to the relevant systems scale: Over 2 km of wiring, 80–100 sensors, 50–80 processors and 80–100 actuators are common in cars (Fig. 2.1). Overall engine efficiency has improved by an average of 1% annually over the last 20 years, mostly due to the development of microtechnologies which allow precise control of fuel injection.

Simultaneously, adding of new functionalities has lead to an average 15% weight increase of all car segments as well as growing electrical power demand of an average of +100 W/year over the last 15 years. Rather than solving a problem at the component level, a system's approach is envisioned and a comprehensive solution is targeted for this issue. Improved engine efficiency, convenience and safety might then, for example, be linked to a substantial decrease in the use of primary energy and materials or to completely new power train concepts.

Fig. 2.1 Miniaturised and integrated sensors for automotive applications (by courtesy of Robert Bosch GmbH, Germany)

2.2.2 Aeronautics

Over the past century aeronautics, the art of "navigating" the "air" has evolved tremendously. Starting with the invention of simple machines capable only of flying, this field has created highly complex systems and devices making the distribution and exchange of goods and people worldwide possible. A modern world without aeronautics would simply not be possible. However, this enormous growth also poses problems that have to be dealt with. Aeronautics will experience a dramatic change in the coming years. In order to cope with the growing demand for air transport and air transport systems, completely new pathways have to be established.

As demand grows, air transport will have to become more individual, allowing for the fast, efficient and affordable transport of individuals and goods. Air transport will develop into a mass transport system, with increasing demands for speed, safety, security, environmental cleanliness and efficiency. Rising oil prices will add particular challenges in terms of energy efficiency.

Smart systems will have a decisive impact on the development of aircraft and spacecraft structures, performance monitoring, engine operations, navigation and safety systems. They will enable more precise control of aircraft while reducing the size, mass and power requirements for operational and safety functions as well as energy efficiency. They will be decisive for diagnostics and predictive-maintenance management of engine, aircraft structure and electronic sub-systems (Fig. 2.2).

Advisory Council for Aeronautics Research in Europe (ACARE)[4] – the European Technology Platform of the aeronautics industry – has defined a series of technological challenges in order to obtain progress in the following areas: quality and affordability, the environment, safety, air transport systems efficiency and security. Related to these objectives, EPoSS has identified the following future scenarios:

- The electrical aircraft
- The connected aircraft
- The intelligent aircraft
- The efficient aircraft

Smart systems technologies are expected to play a decisive role for these concepts.

2.2.3 Information and Telecommunication

Advanced smart systems integration technologies will ensure global competitiveness for European telecommunication players. The telecommunication sector embraces many areas, including the distribution of data, audio, video and other information

[4]ACARE – Advisory Council for Aeronautics Research in Europe (http://www.acare4europe.com).

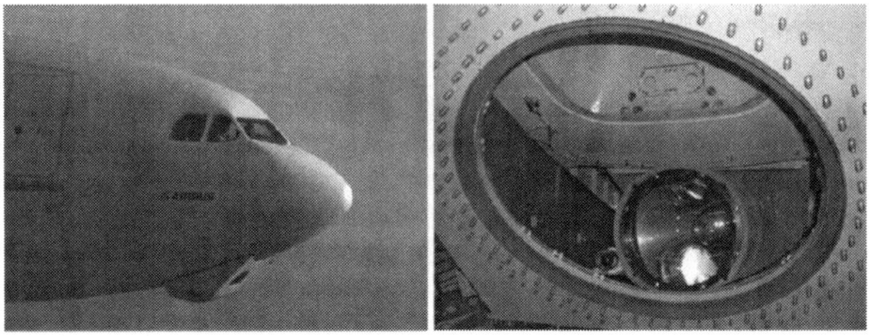

Fig. 2.2 An example for a smart system for improved safety and comfort in flight is the optical radar system (LIDAR) for turbulence detection, which was developed within the European AWIATOR Research project. For the tests, the LIDAR was contained inside a radome beneath the nose of an Airbus A340 test aircraft (by courtesy of Airbus)

via cable or wireless. The management and logistics of networks as well as service provision over networks has become a significant element of national economies. IT and communication infrastructure is of strategic importance. Telecommunication strongly influences the prosperity of other industrial sectors, e.g. aerospace, automotive and medical, wellness or logistics/retail, which are the most important ones for Europe's competitiveness at present and in future. Continuous product innovation cycles are important and essential for technological and economic progress. Developing sophisticated new products makes high demands from the application side, including:

- Signal conversion, transformation, adaptation, processing, computation and protection
- Communication
- Position-finding and localisation
- Energy scavenging
- Adaptive self-control
- Modelling, design and test
- Security and environmental safety

A range of technologies must be combined to optimise solutions and allow innovative sophisticated new product generations incorporating new materials, components, sensor and actuator technologies and packaging solutions.

2.2.4 Medical Technologies

There is broad agreement that the European Union faces major healthcare challenges in the years ahead due to an ageing population, negative environmental effects on personal health and demand for improved personal health care. These challenges will have significant effects on the healthcare system, as the need for medical assistance increases with age, the incidence of chronic diseases rises and demand for individualised health care grows. It will be a major challenge for our societies to cope with the expected cost increases. Healthcare expenditures presently account for about 10% of gross domestic product in industrialised countries and are growing at an average rate of about 6% per annum in most countries.[5] This trend is likely to accelerate drastically if nothing is done. There is also a broad expectation that the quality of the health care supplied to citizens rises at least with the rate of growth of overall wealth in a society. Increasingly, health is viewed as a matter of personal responsibility. More personalised health care is therefore likely to lead to a growing market for health products and services. The ageing society especially will have an enormous influence on this health market (Fig. 2.3).

To cope with these challenges and to ensure that the European medical technology sector remains competitive in the global economy, Europe must mobilise its scientific and technical knowledge to develop a range of new solutions that meet the demands of the healthcare system while containing the overall cost explosion. Various major approaches here are based on smart systems integration to provide technological solutions at reasonable cost and in developing technological solutions to provide more effective health care meaning shorter hospital stays, increase in outpatient procedures and preventive care to avoid severe pathologies (Fig. 2.4).

Fig. 2.3 ACURIS™ – new generation of an binaural hearing system with Siemens wireless e2e (ear-to-ear) technology (by courtesy of Siemens AG, Germany)

[5]OECD Health Data 2005.

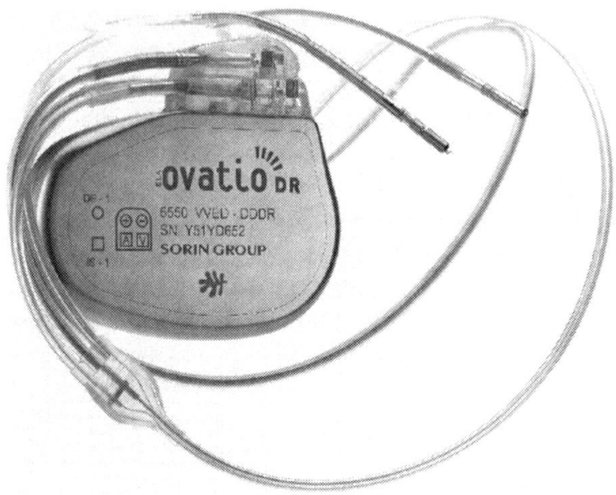

Fig. 2.4 Ovatio – new generation of an implantable heart defibrillator (by courtesy of Sorin Group, France)

2.2.5 RFID

Logistics and communication are considered to be the backbone of globalisation. They play an important role in the development of modern economic systems and sustainable economic growth. Radio Frequency Identification (RFID) has become a key technology in the ICT sector, because RFID tags can identify individual goods and commodities in real time and close the gap between the physical flow of things and the related information flow in IT systems. Smart system integration is crucial for increased functionality of RFID – and yet RFID technology can provide added value to smart systems integration in many other industrial sectors.

RFID is one of the evolving high-tech markets with an expected volume of 24.5 billion USD in 2015.[6] There is a strong RFID industry in Europe as well as a strong potential of industrial and private RFID users. Both technology providers and users may benefit from growing RFID markets and technologies. The international competition is strong, however, and Europe suffers from several drawbacks. Most importantly, there is a lack of standardisation, frequency harmonisation and research support.

Companies in logistics and retail have already launched the first RFID roll-outs. Other industrial sectors like automotive, aeronautics and pharmaceuticals are working on the adoption of RFID for their processes. Main applications will be the tracking and tracing of goods as well as anti-counterfeiting (Fig. 2.5). The public sector, too, has already set up several RFID-based projects like the E-passport

[6]EPoSS Strategic Research Agenda (Version 1.3, 2007).

Fig. 2.5 VarioSens – smart active RFID label (by courtesy of KSW Microtec AG, Germany)

Fig. 2.6 SIMATIC RFID Baggage System – The new way of baggage identification (by courtesy of Siemens AG, Germany)

or the tickets for the 2006 FIFA World Cup in Germany. The coordination of research and development at the European level is urgently required to improve the cross-sectoral application of RFID technology (Fig. 2.6).

In the long run, RFID is a core technology for ambient intelligence. Using RFID technology, everyday commodities and goods will become "smart objects". Elderly and disabled people will be supported by intelligent devices. The close tracking and monitoring of goods in the food chain will improve food safety. Smart industrial goods will store information about their components and their use. Waste disposal management will be switched from today's inefficient mass-oriented approach to an individual recycling process. In sum, RFID will enable the "Internet of things" proving more efficiency, more security and convenience.

Europe's competitiveness depends on the development and successful use of important key technologies like RFID. By initiating the i2010 strategy,[7] the European Commission has underlined the importance of ICT to growth and employment as set forth in the Lisbon Goals.

2.3 Future Trends and Application Examples

2.3.1 Automotive

In order to highlight the future role of smart systems in the automobile, the focus is put on the sub-systems of a car. Here, five areas of particular interest can be distinguished: safety, driver assistance, convenience, smart power train and cross-over topics.[8]

Safety includes active and passive vehicle systems to protect the driver, the passengers as well as road users. Here the major challenges within the next 15 years will be driver information on vehicle dynamics limitations, adaptive human machine interface systems and a personalised safety system adapted to characteristics of the individual. Additionally, driver drowsiness monitoring and pedestrian protection systems belong to the safety applications of the future. Furthermore collision mitigation systems, emergency braking systems for unavoidable accidents, vision enhancement systems including night vision and blind spot monitoring and vehicle interaction systems for cooperative driving will develop.

Driver assistance systems support the driver in steering the vehicle. Consumer demands, technical limits and legal issues all require the driver to retain full control of the vehicle. Taking account of human ability to deal with complex situations, a synergetic solution aimed at extending driver abilities is the mid-term perspective for vehicle control. The major fields here are lateral and longitudinal vehicle guidance systems including lane-keeping and lane-change support, automatic cruise control (ACC) stop & go and ACC for urban areas. Also semi-autonomous driving for defined situations like automated parking, automatic following and guided driving as well as personalised driving based on individual driving patterns, constitution and appropriate vehicle adjustments counts to driving assistance. The smart aspect of the mentioned systems is the adaptive approach, as compared to the kind of fixed solution which is common today. In near future, technical systems will provide optimal

---

[7]i2010 is the EU policy framework for the information society and media. It promotes the positive contribution that information and communication technologies (ICT) can make to the economy, society, and personal quality of life.

[8]EPoSS – Strategic Research Agenda (Version 1.3, 2007).

driver support taking account of vehicle and driver capabilities and characteristics. The system integration approach will build upon networked functionalities and shared computational power to analyse and interpret situations and decide on appropriate measures.

Convenience addresses secondary driver and passenger requirements beyond vehicle guidance. Convenience is one of the major decision factors for vehicle purchase. To feel good in a car, first of all the vehicle has to be safe. Therefore appropriate warnings and automatic interference for safety critical situations is one of the major objectives for manufacturer. Second, a car should fulfil the transport requirements – maybe a challenging demand for an increasingly crowded traffic scenario. Individual dynamic routing and supporting overall traffic measures address this issue. True convenience functions follow only in third place, e.g. automated secondary functions like non-fogging windscreens, anti-dazzle systems and automated light and wipers; user-identification systems; adaptive control elements and human-machine interfaces including situation-specific interaction, scalable and auto-adjusting vehicle control elements. Here, too, the system integration approach allows for changing the automotive technology approach from vehicle static to user centric. It is not the user who will adapt to the car or learn how to operate the systems, instead the vehicle will be expected to adapt to the user's needs and capabilities.

Smart power train addresses the overall objective of a clean and powerful transport system. Future concepts will make use of smart systems like high pressure direct injection and exhaust after treatment, smart energy strategies, alternative fuel concepts, hybrid and adaptive power train solutions, comprehensive energy management, as well as electric motors and active wheels. Scavenging–harvesting solutions like heat recovery can provide primary power to sensors and actuators, reduce complexity and may even yield primary power to the power train. Current and future fuel savings roadmaps on engine development will require more sensing and processing, electrical motors and actuators and sensors to improve engine performance including efficiency. Today's 12–15 sensors are expected to increase to over 30 in few years. An integrated approach including wireless and autonomous sensors/actuators will be necessary to handle this complexity.

The fifth cluster addresses the *crossover functionalities and methodologies*. They include data fusion and management, advanced human-machine interface concepts, manufacturing and design methodologies, integration, security, privacy and robustness.

2.3.1.1 Application Example: Smart Tire Pressure Sensor with Integrated Power Generators

by P. Ramm, Fraunhofer IZM (Munich) and H. Rödig, Infineon Technologies AG, Germany

To advance microsystem technologies by allowing the cost-effective realisation of highly miniaturised, truly autonomous systems for ambient intelligence, a special European project was set up called e-CUBES. The focus of e-CUBES is put on

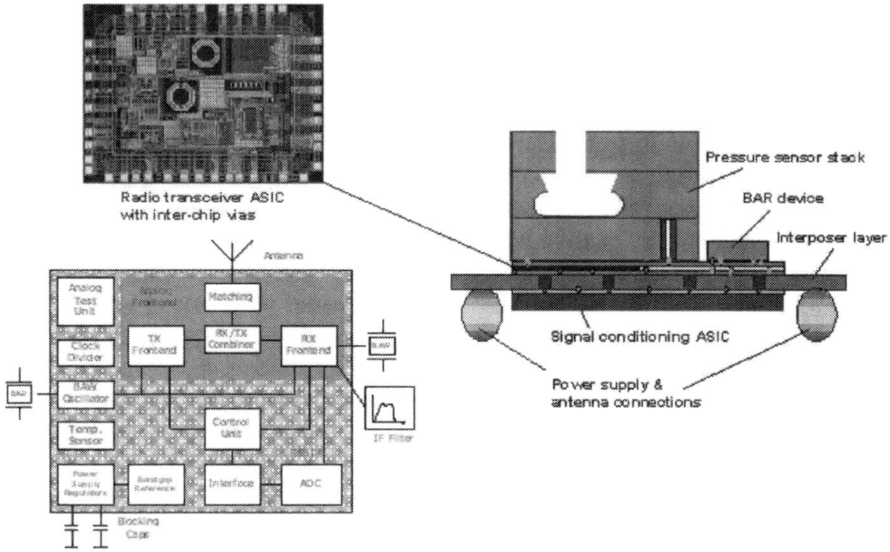

Fig. 2.7 Schematic of e-CUBES automotive system demonstrator for tire pressure monitoring with energy harvesting power supply (by courtesy of Infineon Technologies AG, SINTEF and Fraunhofer IZM)

using the third dimension for integration. The e-CUBES automotive system demonstrator depicted in Fig. 2.7 is a tire pressure monitoring system with an energy harvesting device. It exploits various 3D integration technologies and is expected to fulfil all the challenging automotive requirements at low cost (estimated at 2–5 € per unit) (Fig. 2.7).

2.3.1.2 Future Trends

The short- and mid-term perspective of smart systems for automotive applications will be driven by customer, legislative and competitive requirements. The four application clusters of safety, driver assistance, convenience and smart power train reflect these needs. To improve safety, three major approaches will be followed: a better understanding of the situation based on sensors to collect information about the environment; the merging of passive and active safety towards personalised and situation-specific accident mitigation measures and the active interacting of road users towards a collaborative traffic approach. Driver assistance calls for comparable measures or standards. Based on a comprehensive understanding of the traffic, vehicle and driver situations, appropriate supporting measures are chosen. Subsequently, the human–machine interface becomes of prime importance aiming for an adaptive and individual system. In parallel, convenience aspects will gain further in importance. The objective is to develop personalised vehicles – adapted to the driver and changing their attributes to personal and situation-specific needs.

Finally, yet importantly, smart power train applications are a major R&D focus in order to respond to the challenges of less fuel consumption and fewer emissions. It is a long road to a hydrogen infrastructure and new propulsion systems. Intermediate and competing solutions require further high attention. High-pressure injection systems and exhaust after-treatment are short-term perspectives; advanced hybrid concepts towards the full electric vehicle are a mid-term perspective.

To meet short-term requirements the development of auxiliary systems is an appropriate methodology. This does not require any changes to established systems, while still promising potential benefits of high relevance. The second, more mid-term approach, is optimisation of the original system, perhaps also including additional auxiliary systems. Particularly, smart systems are expected to play an important role as enabling technologies for the full electric vehicle. Examples include smart systems for the management of accumulators, active control units for electric motors and wheels and intelligent power electronics devices.

2.3.2 Aeronautics

The aeronautics industry has defined five major application fields of aeronautic smart systems. Principal trends are seen in relation to the customer, to cost and time efficiency and to environmental and security issues.

The passenger will expect the same services and the same comfort that he is used to at home or in the office. In-flight entertainment will have to provide high quality video and audio on-demand. Passenger comfort will have to be improved by, for example, reducing the perceived internal noise and through personalised passenger climate control. In addition, the cabin will have to become a flying office offering high-speed data connections and mobile (video)-telephony communication options. This development will be the *highly customer-oriented* air transport system.

The time needed for boarding and de-boarding, cargo and catering operations, cleaning and fuelling has to be reduced in order to decrease the overall passenger flight time and aircraft ground-time. This will reduce costs and increase customer satisfaction, and is called the *highly time efficient* air transport system.

As in other industries too, the costs are one of the main factor for successful products. The *highly cost efficient* air transport system should concentrate on maintenance expenditures that are among others the main costs for air systems operation. Maintenance strategy will have to evolve from the present time-scheduled approach to a more cost-efficient on-demand maintenance or predictive maintenance, increasing aircraft availability and safety.

The main challenges to reduce the environmental impact of operating, maintaining and manufacturing aircraft and associated systems relate to fuel saving, noise reduction inside and outside the aircraft and fewer emissions of harmful substances. These objectives require a better understanding of flight physics, new techniques for flow controlling, new lightweight materials, improved propulsion, optimised aircraft handling, etc. The main targets may be formulated as follows:

- To reduce fuel consumption and CO_2 emissions by 50%
- To reduce perceived external noise by 50%
- To reduce NO*x* emissions by 80%
- To make substantial progress in reducing the environmental impact of the manufacture, maintenance and disposal of aircraft and related products.

Then we will have reached the *ultra green* air transport system. Last but not least we all know the risk and the fear of passengers in planes especially regarding terrorism and attacks on airplanes. The main points that would be addressed for increasing security in air transport are the detection of dangerous goods or risky events in cargo operations; the detection of dangerous persons in passenger operations, e.g. through spectrometer analysis and automatic surveillance; reduce to zero the possibility of hijacking and real-time and remote aircraft telemetry. This development will be described as the *ultra secure* air transport system.

2.3.2.1 Application Example: An Integrated Inertial Navigation System for Aircrafts

by M. Hafen, Northrop Grumman LITEF GmbH and M. Baum, Fraunhofer ENAS, Germany

The accelerometer B-290 Triad has been developed for applications in attitude heading reference systems and inertial measurement units. It consists of three all-silicon micro-mechanical accelerometers with analogue and digital electronics and interfaces in a small housing. The sensitive element of the accelerometer is an elastically suspended pendulum, manufactured in silicon using micro-mechanic techniques. A batch process is used on wafer level. To achieve a high accuracy, the pendulum is force rebalanced by the electrostatic forces of the capacitive bridge. The digital restoring loops for the force rebalance are realised with a digital signal processor. The loop characteristics are defined in software that also performs the temperature compensation of bias and scale factor. Sensor data are scaled, fully temperature compensated and provided over a digital serial interface.

The MEMS acceleration sensors were developed in cooperation with Chemnitz University of Technology and transferred to LITEF GmbH, where the system development was performed. At least the reference system leads to more safety in airplanes and is able to define the airplane's position for a certain time even without GPS signal[9] (Fig. 2.8).

---

[9]Datasheet B-290 Triad.

Fig. 2.8 Three-axis inertial measurement unit for applications in attitude and heading reference and inertial navigation systems in aeronautics (by courtesy of Northrop Grumman LITEF GmbH, Germany)

2.3.2.2 Future Trends

In the world of smart systems technologies for aeronautics applications, four major future scenarios can be defined:

Within less than one decade the *electrical aircraft scenario* should have been achieved. This scenario envisages the replacement of, for example, pneumatic and hydraulic components with electrical actuators controlled by networked and flexibly managed devices. Smart sensors and actuators will permit by-wire functions. Within less than 10 years, fuel cell auxiliary power units are expected, using sensors, process monitoring and control. A new generation of optical components and optical sub-systems will then permit new by-light functionalities.

The *connected aircraft scenario* is focused primarily on the integration of the aircraft into an overall and global communication system. Communications technologies for aeronautical applications and for space will be interoperable with terrestrial systems and span the communications arena from the physical, data-link and network layers to the transport communications layer.

The *intelligence of the aircraft* will develop gradually and will peak with the all-freight, fully automated aircraft. Over one decade a series of intelligent features will be realised of which the first will be cabin control functions achieved by applying new sensing devices. Combined sensor and actuator devices will help to realise defence assistance sub-systems. Advanced ground operations will be possible by guidance, and a reduction of turn-around time will be achieved. Within one decade, it is expected to realise a fully situational awareness of the aeroplane, which includes all relevant information gathering and processing of the aircraft's environment and dircct-coupled feed forward control.

The *efficient aircraft* describes the future of the aircraft in terms of cost and performance improvements. Significant breakthroughs will be obtained in the near future already in passenger monitoring and guidance and tracking of luggage and goods, based on wireless technologies. By the beginning of the next decade, significant progress in aircraft all-weather capability is expected. New actuators and forward control will allow effective power control and hence power optimisation. Significant progress is furthermore expected in adaptive aerodynamics, resulting ultimately in morphing aircraft structures, particularly morphing wings. Flow control by integrated micro-sensors and actuators will contribute to these developments. The development of active-material-based embedded actuators in the aerodynamic lifting composite structure will enable effective load control devices for a series of new air vehicles (Fig. 2.9).

These four identified fields of application demonstrate a high potential for synergetic effects. They also show that smart systems will play a vital role in future aeronautics and will contribute to increased flight security, reduced environmental impact and raised performance of transport in general.

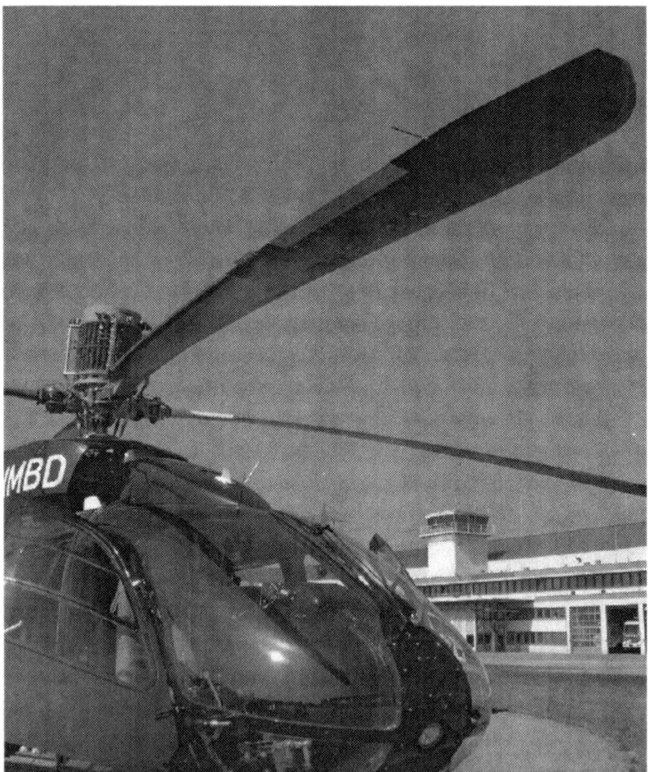

Fig. 2.9 Eurocopter with adaptive rotor blades for efficient operation, low noise and low vibration (by courtesy of EADS Deutschland GmbH)

2.3.3 Information and Telecommunication

The future of telecommunications will offer services or products that will enable the ultimate wireless-user experience: portable appliances universally connected to the best available broadband service, under all mobility conditions, indoors or outdoors. The achievement of this goal, however, crucially depends on the availability of the enabling wireless technology. There is a broad necessity for research in the field of advanced wireless components and systems to make the dreams of an always-connected, multi-standard, multi-service communication environment come true. Particularly, higher-order modulation schemes will require highly linear architectures and will increase the power consumption of the power amplifiers: efficiency and linearisation are therefore key areas of continued research and development.

The application fields of information and telecommunication (ITC) can be divided into the main areas of mobility and networking, against demands for increased performance and security. Some key issues and challenges for mobile communication technology are outlined below.

The functionality and features of mobile phones, base stations and multimedia devices are dramatically increasing. The pace of innovation is immense, forcing integration technologies, new materials, power management, new system concepts, miniaturisation and cost reduction. Integration and embedding technologies must be able to handle a lot of heterogeneous technologies, e.g., different substrates, materials and components.

The trends in the development of mobile phones include more and more multi-band, multi-standard, e.g. GSM, UMTS, CDMA, 3G, 4G, WiMax, etc. operation for global communication and for global and local positioning applications. To attack this challenge with respect to performance and costs, re-configurable, tunable and frequency-agile front-ends are required. This will require too the development of tunable filters and duplexers, tunable matching of antenna and power amplifier, frequency-agile filter banks and re-configurable transceiver ICs as well as heterogeneous 3D packaging technologies of high performance.

2.3.3.1 Application Example: Next-Generation Base Station Radio Frequency Architecture Applying RF-MEMS Components

by S. Leidich, Chemnitz University of Technology and S. Kurth, Fraunhofer ENAS, Germany

A significant potential to improve the performance and to lower the costs of cellular base stations is related to the radio frequency (RF) units which are the duplexers, the power amplifiers and the radio cards. However, Moore's law is not applicable to RF units, as the cost-related form factor is often a function of wavelength, which prevents a higher level of integration. Hence, significant form factor reduction in RF devices requires architectural innovation. Recent trends in spectrum regulation mandate a high level of flexibility from infrastructure equipment in terms of frequency

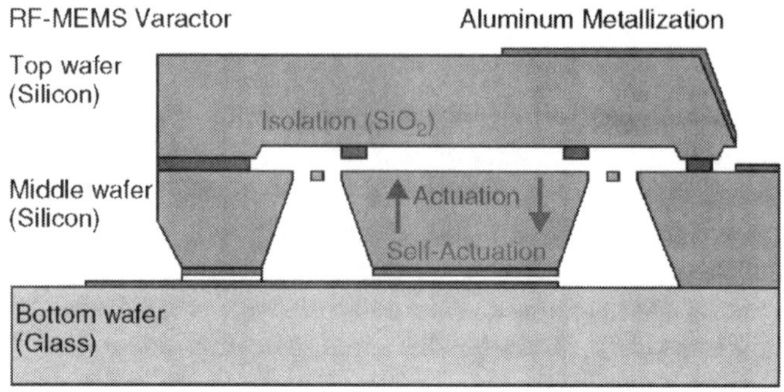

Fig. 2.10 Continuously tunable high power RF-MEMS varactor [3]

agility. Currently, a change in frequency requires an exchange of all RF-related units: the radio cards, the power amplifiers and the duplex filters. Radio frequency micro-electro-mechanical systems (RF-MEMS) are considered the key components for the alteration of analogue properties and can provide superior RF performance, in terms of low loss and high linearity compared to solid-state devices [2].

Tuneable capacitors (varactors) represent a vital component of frequency agile RF circuits. An important requirement of base station suitability is high power handling capability, since more than 20 W of RF power is transmitted. Self-actuation originating from electrostatic forces generated by high RF signal amplitudes is more critical to RF-MEMS than thermal budget considerations due to inherently low losses. A continuously tunable high power RF-MEMS varactor for frequencies from 0 Hz –4.0 GHz was presented, by Leidich et al. [3] Using silicon bulk technology and wafer bonding techniques, RF and electrostatic actuation electrodes are arranged vertically, such a way that oppositely oriented forces counteract attractive electrostatic forces generated by high RF signal amplitudes (see Fig. 2.10). The device is specified for 0.8–1.6 pF analogue tuning range and provides a Q-factor of >100 for frequencies $f < 2.0$ GHz. Intrinsic stability against high power RF bursts is achieved by using large capacitor plates and a narrow electrode separation, leading to a highly damped mechanical response. Consequently, mechanical oscillation at narrow tone spacing and accompanying intermodulation generation is suppressed (Fig. 2.10).

2.3.3.2 Future Trends

The mobile communication market has evolved rapidly. Standards have been extended or replaced continuously by new generations of standards. More and more new products and features have been introduced, ranging from base station to mobile phone or WLAN. Concurrently, communication systems migrate to support more and more multi-mode, multi-in/multi-out radio functions. Technical challenges increase almost every day.

The required R&D faces both technical and financial barriers. There is demand for new smart system technology in order to push innovation forward and create growth. Reductions in size, weight, cost and power combined with shorter product life cycles and increasing complexity cannot be achieved without technical innovation. An example of an important technology enabling significant improvements in these parameters is System-in-a-Package (SiP) technology. SiP technology allows flexible integration of different elements such as MEMS, optoelectronics, active and passive components and bio-electronics into the packaging, improving performance and reducing system cost.

2.3.4 Medical Technologies

Smart integrated systems will enable a shift of health care from today's reactive, specialised and provider-centric approach towards proactive, integrated and patient-centric solutions.

Smart integrated systems will help medical personnel to get deeper and quicker insights into the patient's status by opening additional windows to view the internal parts of the body on a microscopic up to a molecular imaging scale and on a continuous time scale. There will be personalised health monitoring of patients outside of the hospital or surgery right through to integrated delivery of care at home. This area is called *assisted health checking*.

Within the area *assisted therapy and therapy control* integrated bioinformatics medical data management will help medical staff to analyse data aggregated from various sources to extract clinical significance and ensure seamless connections and data exchanges between patient, healthcare provider, healthcare insurance companies and pharmaceutical suppliers. Long-distance teletransmission of patient data will enable continuous therapy control and monitoring. Therapy decisions like surgery, minimal-invasive and biocompatible methods based on smart robots will minimise discomfort for the patient.

Smart integrated sensor/actuator systems and neuronal coupling will replace body functionalities lost due to disease, accidents or violence, enabling the patient to continue normal life within the area *full functional substitution and rehabilitation*. Smart robotic tools will assist the elderly to maintain an autonomous life much longer time than now.

A network of self-organising sensory tools will continuously *monitor* the environment we live in to diagnose detrimental *external influences* to our health. This will include precautionary measures to ensure that we consume food that is beneficial to us.

The four identified application areas point to several research priorities. The most important areas for smart systems applications in the medical technology field are BioMEMS (Bio-micro-electro-mechanical systems including optical sensors), functional and cell biological imaging, aids for therapy and therapy monitoring, E-health, telemedicine and networking, data management, medical technology for

Fig. 2.11 Integrated electronics for heart assistance devices (by courtesy of SORIN Group, France)

regenerative medicine, modelling cardiac hemodynamics and electrophysiology and environmental influences (Fig. 2.11).

2.3.4.1 Application Example: Integration Aspects of a Polymer-Based SPR Biosensor with Active Micro-optical and Micro-fluidic Elements

by J. Nestler and K. Hiller, Chemnitz University of Technology, Germany

For point-of-care (PoC) diagnostics, fast and competitive diagnostic results are desired at an affordable price. The integration of sensing functionality as well as assay functionality into a compact device would allow transferring complexity from the macrosystem (readout-system) to the microsystem, leading to a "Lab-on-a-Chip", i.e., a smart system. As for a high level of integration device costs may become an issue, polymer-based solutions are preferred for most of the components of such a device.

However, the integration of a micro-optical biosensor together with surface functionalisation and fully integrated micro-fluidic functionality for the controlled transport of sample and chemicals is a complex task with respect to system design, interfaces, micro-fabrication and packaging.

As one example, a new platform for polymer-based *s* urface *e* nhanced *m* icro-*o* ptical *f*luidic *s*ystems is developed within the European project SEMOFS.[10] Critical integration aspects of active micro-optics like polymer LEDs (pLEDs), active sensing

[10] SEMOFS – European project for the development of surface enhanced micro optical fluidic systems (http://www.semofs.com).

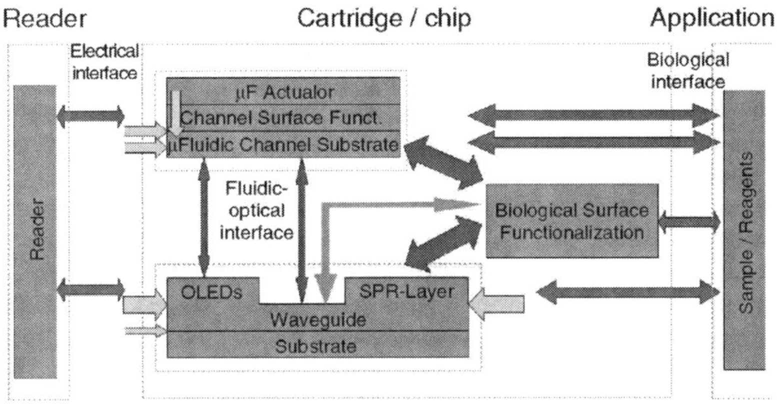

Fig. 2.12 Schematic overview of integration aspects and interfaces of SEMOFS project

surfaces, active polymer-based micro-fluidics (pumps, valves), surface modification and bonding are investigated, as well as the interfaces between the different elements inside a fully integrated biosensor system and to the "macro-world" [4] (Fig. 2.12).

2.3.4.2 Future Trends

The future trends for smart medical systems and applications will be shown in the same way like above, according to the four identified application areas. In the field of *assisted health checking*, medical imaging technologies are of primary importance as a diagnostic tool and as a key tool for delivering therapy and tracking the progress of treatment. Molecular diagnostics are important for point of care applications. These tools will be supported by bio- and nano-sensor technology including MEMS and microelectronics into one simple but reliable device for multiple application scenarios. The combination with wireless data transfer, including high-level medical opinion, will shift medicine from providing reactive treatment after symptoms appear to a proactive discipline that treats and prevents disease before it occurs. Non-invasive sensors using advanced MEMS technology can simplify disease diagnosis and monitoring particularly for elderly people or children and reduce expenditure compared with traditional health care. The combination of intelligent IT technology, e.g. data management, telemedicine, etc. with individual diagnostic systems will lead in 2015 and beyond to a totally integrated home care system.[11]

A major technology for *assisted therapy or therapy control* will be the development of medical data analysis software which will allow direct data exchanges

[11] EPoSS Strategic Research Agenda (Version 1.3, 2007).

between patient and medical practitioner for diagnostic analysis and decision support. Intelligent and collaborative environments will then become available, allowing greatly increased efficiency in providing patient care. Personalised, web-based portals will provide ubiquitous, secure access to healthcare IT systems. Personalised and optimised diagnosis and therapy using genetic technologies, like the DNA chip, and drug-on-demand technologies will lead to a more personalised medicine with greater patient empowerment. Various smart systems like smart pacemakers and smart robots will lead to a totally digital hospital with fully automated functions like digital imaging and networking (Fig. 2.13).

Smart implants are the first step towards *full functional substitution*. The integration of smart sensor systems is a prerequisite to improve functionality. Trends for the future are neural prosthesis for visual, motor, deafness, respiratory and bladder control. These applications will require the development of specific neural technologies such as artificial retinas, optical nerve stimulation, functional electro stimulation (FES) for walking and standing, vestibular implants or bladder stimulators. Also neuro-modulation technologies which address back pain (spinal stimulation), Parkinson's disease essential tremors, epilepsy, depression (DBS), migraine (occipital nerve stimulation) are targeted subjects.

Assisted living for *full rehabilitation* at home or in hospital can be supported by smart robotics in homecare units or in home-support devices for the elderly, e.g. machine-assisted physical rehabilitation. Obesity, diabetes and hypertension (VNS) are chronic diseases which will need assisted devices. Trends for the intra-cardiac market are preventive and predictive medicine to assess risk factors and the monitoring

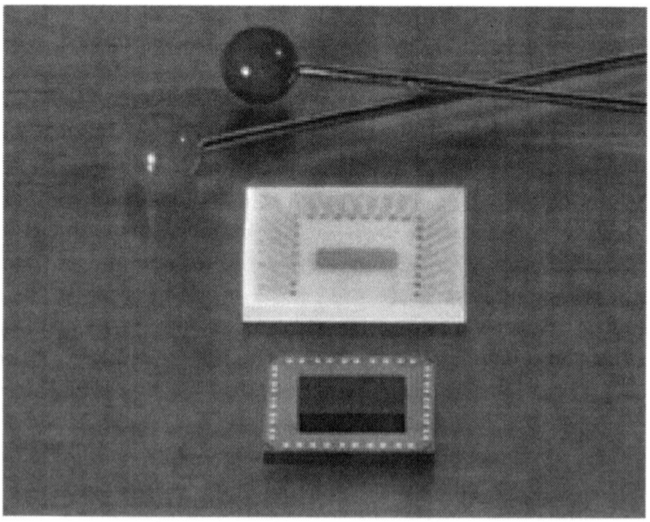

Fig. 2.13 Breath analysis based on multi-functional sensors will become an important non-invasive-tool in differential diagnosis, therapy control and home heath care. The GasFET sensing platform is an integration of several gas sensors and electronics into one Si-chip. It is suitable to detect a wide range of marker gases on-line in human breath (by courtesy of Siemens AG)

of the at-risk population. Reliable and biocompatible cardiac resynchronisation therapy will be supported by implantable cardiac rhythm management devices. The longer-term perspective is complete functional substitution, including automated surgery by smart sensor systems (bio robots), smart prosthesis and implantable sensors.

Monitoring and control of external influences on human health is important for the maintenance of individual health status and will grow in importance owing to increased longevity. Small analytical systems, e.g. MEMS gas sensors based on CMOS technology or smart diagnostic tools like lab-on-chip systems, will support control of the ambient environment. Automated data management systems and software-based signal evaluations in combination with intelligent sensor devices will be needed for an automatic health check. Automatic food storage technology with reliable quality management can improve storage quality, authenticity, consistency of ingredients and product safety. Future applications of smart sensor systems will lead towards a complete environment assistant for the home. These technological developments will especially benefit Europe's ageing population.

2.3.5 RFID

RFID systems are complex ICT systems. A mix of advanced technologies is needed to design and to realise RFID-based systems for ambient intelligence. Moreover, it is essential that R&D projects are designed to work towards real-world applications and towards long-term benefits for the RFID industry as well as for both industrial and private users of RFID in Europe.

The present technological challenges for RFID define research priorities in ten technology fields from microelectronics, microsystems, software engineering and high frequency technology (HF) where mid-term and long-term research is required. Some of the main application specific technology fields are described below.

To a great extent the *design of integrated circuits* is not specific to RFID applications. However, tags for smart objects in particular require future research. In order to reduce the size of tag sensors, energy supply and HF components must be integrated into monolithic silicon chips, e.g. system on chip. Energy supply is a great challenge for smart RFID tags with additional sensors and computing capabilities as well as for mobile readers. Tags and mobile readers are not connected to a constant power supply and in general cannot be recharged on time. Research topics are integrated foil batteries; energy saving algorithms especially for cryptographic functions; energy harvesting using effects like piezo or others to gain energy from the environment and an energy saving power management of all tag components.

Due to the usually low power level of RFID tags, *high frequency or radio technology* is an important issue. There is a need to further improve the antenna design of both tags and readers in order to gain larger reading ranges and predictable reading rates. Today, the metallic antenna is by far the largest component of a RFID tag. Research tasks are printed antennas which can be easily integrated into paper packages and antennas which are integrated into the chip itself called coil-on-chip.

In the case of smart objects which interact autonomously in sensor networks, there is a great need to adapt existing ad hoc network approaches to the restricted hardware and energy resources of RFID systems. In the long run, ultra-wide-band communication will gain more importance due to its low power requirements and to its more efficient use of given frequency ranges.

Silicon-based tag *packaging*, which comprises the application of the chip to a substrate, the antenna and the connection of antenna and chip, is responsible for more than 50% of all cost. It determines how a tag can be integrated into goods and packages and how it must be disposed of after use. There is a strong need to integrate mobile readers into small objects in industrial and end-user environments. Thus packaging topics with respect to RFID are flexible and multi-layer substrates; the integration of chips and antennas into non-standard substrates like textiles and paper; and the development of substrates, conducting paths and bonding materials adequate for harsh environments and for ecologically immaculate disposal.

There is still great potential to enhance the *manufacturing process* of RFID tags. Print-like reel-to-reel manufacturing is a standard process for simple tags but still a challenge for high-quality tags. This includes the handling of thinned and thus flexible chips and the assembly of tags consisting of multiple chips or multiple discrete components. Even more advanced are self-assembly manufacturing processes where the explicit pick-and-place of tag components is replaced by mechanical or electromagnetic key-to-lock placement methods. Another big issue is the long-term convergence of traditional print processes and RFID manufacturing.

The manufacturing of silicon chips is a complex and costly process. The replacement of silicon by *polymer* as base material for integrated circuits and electronic components promises a substantial cost cut.

2.3.5.1 Application Example: Smart Active ID Label for Transport Monitoring

by D. Reuter and A. Bertz, Chemnitz University of Technology and F. Kriebel, KSW Microtec AG, Germany

Active labels integrating inertial sensors and electronics show the potential of RFID for monitoring of shock, inclination and temperature during transportation. The requirements on the inertial sensors, measuring shock and inclination, are different to common applications. Therefore, the technology, the design and the packaging of the sensors as well as the labels have to be modified or developed, respectively. Some specific challenges are the limited energy supply, the very thin and flexible complete system and the high dynamic range of the sensor at low manufacturing costs. The extremely low height of the sensor system, required for integration in a flexible label, could be achieved by two approaches: bonding of thin substrates as well as by applying a thin-film encapsulation technology.

The Smart Active ID Label includes the RF-chip with antenna and battery for the energy supply, as well as the sensor system, consisting of the micro-mechanic transducer and the signal processing electronics. For the monitoring of transportation processes, the system has to detect and record inclination and mechanical

Fig. 2.14 On flexible interposer hybrid integrated sensors and electronics for smart active RFID labels (by courtesy of Fraunhofer ENAS)

shocks. In order to reduce the complexity of the system, it is reasonable to measure both parameters with the same microstructure (Fig. 2.14).

2.3.5.2 Future Trends

RFID research needs to span a range of different technologies in order to meet the two main application challenges around logistic applications: Short-term and mid-term challenge would be the low cost high quality tag but in long-term RFID should be established as a core technology for ambient intelligence where tagged goods and commodities become smart objects able to sense their environment and to communicate to other objects and IT systems.

The application objectives put forward here are those required to ensure the technological breakthrough of RFID. It is equally important, however, to acknowledge the future impact of RFID in gradually creating ambient intelligence and ubiquitous computing. The interplay and smart system integration that is required will not only affect logistics and retail but also, to a greater or lesser degree, most technological applications in everyday life. Ambient living, smart homes, new mobile services and the "Internet of Things" all depend on the successful integration of RFID technology in industry sectors.

2.4 Cross-Cutting Issues

Smart systems will be embedded in an increasing number of applications in the future. Their pervasive nature leads to specific requirements for the development and the exploitation of the components of which they are made and for the know-how required to design and manufacture them. Five drivers seem to emerge as major characteristics for research and technology trends in the smart system integration field.

Smart Systems are characterised by their high level of heterogeneity: They are made of very different materials which have to co-exist despite their different behaviours, e.g. thermal expansion, biochemical interactions, etc. These building blocks will be prepared using various processes and will be exposed to a broad range of environmental constraints. Scenarios of use are also leading to heterogeneous requirements for the communication links which will build on various wireless standards.

Complexity is another driving characteristic for future developments in the field. Most of these systems are designed and built to embed intelligence and to enable the products in which they are used, with the ability to react to their environment and to provide relevant and ergonomic information to their users. The amount of multi-modal data to be processed by the system is very large. The user interfaces have to cope with complex and variable environments while taking into account the context of use and rapidly changing user behaviours.

Most of the applications require the development of *autonomous* solutions and thus face major energy management challenges. Also most of the basic functions embedded in the systems have to be designed according to strong power requirements exploiting the most relevant energy sources. Therefore, the development of generic solutions will have to follow two tracks: low power functions and improvement of energy resources.

The typical characteristics of smart systems are their various constituting elements covering a very large scale of geometric features. Basic blocks are made of layers and structures at the nanometre scale which are embedded in micro-devices. Together they are assembled at the macro-scale in order to meet environmental and user interfaces requirements. The *multi-scale nature* of smart systems will have a very strong impact on the design tools and manufacturing techniques required in the industry.

Finally the four above-mentioned characteristics are reinforcing a trend already observed in the scientific community of teams involved in smart systems developments: *multi-disciplinarity*. Process and product innovation requires a large body of know-how, ranging from chemistry to electronics, mechanical engineering to biology and sociology to software engineering. The ability to manage this "melting pot" and to set up effective and efficient project management methodologies will be a strong challenge as well as a cultural revolution.

These five characteristics will shape the technological environment for smart systems in the next two decades and will have a strong impact on the scientific community by initiating new research fields and providing new opportunities for product innovation.

In the field of materials science and associated processes, the main issues for system integration are the new solutions required to meet the low-cost and high-complexity specifications of smart system applications. Alternatives to semiconductors for planar substrates and processes must be explored to enable the emergence of low-cost large-area intelligent sub-systems. Polymer substrates, for example, are needed for so-called polytronics or flextronics, which will be applied in human-machine interfaces, smart surfaces and labels and more generally in components that exhibit extended areas. Ceramic substrates, capable of withstanding high temperatures and inert to chemical attack, are necessary for harsh-environment operation. Also micro-reactors, micro-combustion-chambers, micro-thrusters and micro-turbines need on ceramic structures. The technologies or active materials based on polymers, ceramics or metals and their composites require further development: their generation and the optimisation of their properties, the adaptation of low-cost processes like printing techniques for the fabrication and integration of electrically active materials and their characterisation and control of effects such as ageing of the active phase, the host material and their interfaces.

Smart surfaces possess sensing/actuation functions and exhibit engineered catalytic, electrical, thermal, mechanical and magnetic properties. The use of nano-scale structures, scaffolds, nano-particles is going to increase in many devices. Functionalisation of such structures will enable new applications in the life sciences and sensing fields.

Thin-film deposition technology must be improved for active materials exhibiting piezoelectric, magnetoresistive, magnetostrictive or shape memory effects. The incorporation of such materials in MEMS-like devices can impart new functionality. Shape-memory alloys, which in addition show super-elastic behaviour, could be used in imaginative new ways like for instance overload protection and self-repair in shock-sensitive structures. Dedicated slurries, feedstock, inks and pastes must be designed and developed to allow printing, casting, injecting, dispensing and spraying of the materials to be deposited on the substrates described above. Many of these deposition and replication technologies presently exist, but the rheology and composition of the materials must be adapted to the critical dimensions of the new micro-devices targeted. The addressing of biological applications and the incorporation of biological structures into systems will call not only for the exploration of new materials in micro- and nano-fabrication but also new fabrication and integration methodologies. Special packaging materials will need to be developed to answer the new demands posed by the widened application fields. Their function will go beyond traditional encapsulation for protection against the environment. Heat dissipation is only a first step. Smart systems interact with their environment and the package will be the first point of contact. Rather than sealing the device, the package will filter and concentrate the relevant interactions with the system's environment. The boundary between active device and package will disappear as the package takes on more and more active functionality. In the end, there will be no more system-in-a-package or system-on-a-package: the system will be the package.

Considering that the *Printed Wiring Board* (PWB) has been the main platform for integrating components to a module or system level in all of the fields of application for the past decades, the further development of this technology is seen as critical to the success of smart integrated systems of the future. The ongoing developments target further miniaturisation of systems, using approaches like 3D integration, including hetero-integration and integration of active and passive components. The path towards integrating the different functions of a complete system, like logic, storage, sensing and actuating are key elements of smart systems using a PWB-based integration platform.

The second challenge comes from the trends and opportunities offered by *miniaturisation techniques*. The trend towards reducing current microstructures in order to reach nanostructures has appeared over the last 3 years. New structures are unveiling completely new scientific problems including nanocharacterisation, new manifestations of basic forces such as van der Waals force[12] and Casimir effects.[13]

Together with the trends towards the development of nanodevices based on new transistor structures or new concepts such as carbon nanotubes, these new devices are also opening the door to new electronic interfaces, signal processing and metrology techniques.

Packaging is one of the most important integration issues today. The proposed system-in-package solutions enable a high degree of miniaturisation by applying 3D techniques on the wafer, substrate, component and package levels, e.g. integration of sensor devices and stacking of several functional layers with integrated components. The functionality of new systems in the different application areas can be considerably enlarged by the implementation of mechanical, optical and/or biological functions. The realisation of such systems is based on new architectures, components and heterogeneous system integration technologies of various devices, including electrical and non-electrical devices, and has to be carried out more cost efficiently, with a high degree of miniaturisation and flexibility in adaptation to different applications. The focus on heterogeneous integration allows designers to integrate devices for sensing, electrical signal- and data-processing, wireless communication, power conversion and storage. The heterogeneous integration concept offers the possibility to integrate system-on-chip solutions (SoC) for sub-systems and results in a higher overall functionality of the systems, a shorter time-to-market cycle and a high degree of flexibility with respect to application requirements.

Application developers willing to integrate complex microsystems or smart subsystems are facing several challenges due to current limitations of *product design tools*. Current CAD software suites are in general specialised on specific functions: finite element simulation for mechanical, chemical or EM phenomena (Ansys, Coventor, Intellissuite, HFSS, etc.), analytical or functional models for high level system simulation (UML, Matlab/Simulink, System-C), electrical behaviour of the devices for integrated circuit design and simulation (Cadance, Matlab/Simulink, VHDL-AMS, Verilog-AMS, System-C). In general, computing power limitations lead engineers to use an increasing level of complexity as long as the scale of the devices to model and simulate is decreasing. This means for complex and heterogeneous systems that specific tools and tasks deal with the adaptation and the transfer of simulation data from one level of model to the next. This is time consuming and one challenge in the future will be to develop methodologies and

[12] van der Waals dispersion forces are intermolecular attractions caused by electrical distortions of the molecules.

[13] Casimir effect is given by the attraction force between two flat surfaces.

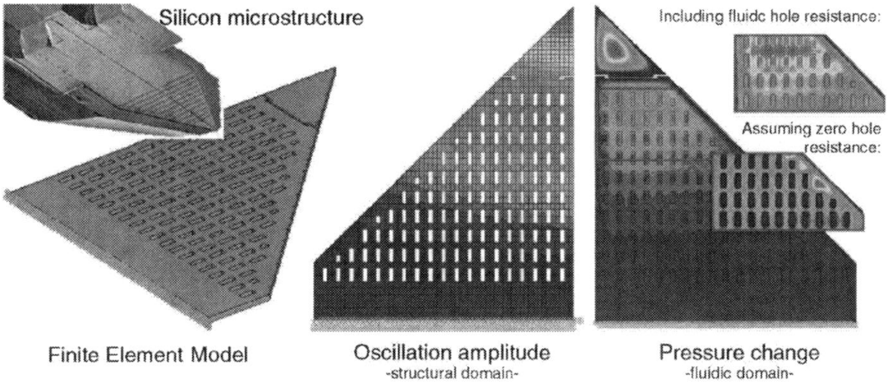

Fig. 2.15 Multi-domain design tools for micro- and nano-machined devices (by courtesy of Chemnitz University of Technology, Germany)

tools that enhance the productivity of system designers. A big challenge for system designers in the next 10 years will be to specify and contribute to the development of methodologies and tools leading to a better unification of software tools for the design of macro-scale systems composed of micro- and nano-scaled devices. This will also have to include knowledge management issues to enable industrial actors to build on former projects and benefit from learning curves (Fig. 2.15).

2.5 Conclusions

Continuous technological innovation is essential for the economic progress of major industries like automotives, aeronautics, telecommunication, medical technology and logistics. Advanced "smart" microsystem technologies are the key drivers of this process. At the level of components they can help to radically improve existing products bottom-up. As standalone devices, they frequently even lead to technology breakthroughs.

Intelligence, which is the ability to be aware of its environment, to make predictions and to reach and execute decisions, is of utmost importance for a smart system. Therefore, advanced solutions are needed for cognition, actuation and control; for autonomous operation and energy efficiency and for compatibility with the harshness or sensitivity of the operation environment.

These functionalities will be enabled by the development of novel components based on advanced materials, and by their multi-level integration into a miniature package. Therefore, nanotechnology, surface engineering, MEMS processes, materials science and packaging technologies will deserve particular attention in the future. Furthermore, aspects like reliability, robustness and safety will have to be taken into account.

Starting the definition of new functionalities for smart systems from megatrends and major global challenges, R&D in smart systems integration is aimed at far more than maintaining global competitiveness of the European industry. Strengthening sustainability, energy efficiency, ambient intelligence and assisted living, smart systems will provide great benefit for the society as a whole.

References

1. Global MEMS/Microsystems Markets and Opportunities, SEMI and Yole Developpement (2007).
2. G. Fischer, Next-generation base station radio frequency architecture. Bell Labs Technical Journal 12, 2007, 3–18.
3. S. Leidich, S. Kurth, T. Gessner, Continuously tunable RF-MEMS varactor for high power applications. *ISMOT, Rom (I)*, 2007 Dec 17–21; Proceedings, pp. 433–436.
4. K. Hiller, J. Nestler, T. Gessner, et al., Integration aspects of a polymer based SPR biosensor with active micro optical and micro fluidic elements. *Smart Systems Integration*, 2007, Paris (F), 2007 Mar 27–28; Proceedings, VDE VERLAG GMBH, Berlin und Offenbach (2007), pp. 295–302 (ISBN 978-3-8007-3009-4).

Chapter 3
RF Technologies and Systems

W. De Raedt, G. Carchon, H.A.C. Tilmans, and E. Beyne

Abstract This chapter gives an overview of assembly and packaging technologies that allow heterogeneous system integration for RF applications

Keywords RF • RF integrated passive device (RF-IPD) • Microwave systems • Millimeter-wave systems

3.1 Introduction

Today's wireless communication systems are showing an explosive growth of emerging consumer and military applications of radio frequency (RF), microwave, and millimeter-wave circuits and systems. Applications include wireless personal (hand-held) communication systems such as mobile phones, wireless local area networks (WLAN), satellite communications, and automotive radar. Future hand-held and ground communications systems as well as communications satellites will require very low weight, volume, and power consumption in addition to higher data rates and increased functionality. Convergence means that many products become more and more multiband and multimode now, using RF communications not only for voice and data transmission which are based on many different standards and frequencies, but also for localization and navigation. As a result, many of them require multiple air interfaces and advanced antennae systems.

These versatile systems will be increasingly enabled by high-density system integration which is becoming more and more important. High-density integration of RF-radio devices not only requires the integration of the active devices (RF systems-on-chip, RF-SOC), but also requires the integration of a large number of passive devices, such as transmission lines, resistors, capacitors, and inductors, as well as functional blocs such as filters and baluns. In order to reduce the system

W. De Raedt (✉), G. Carchon, H.A.C. Tilmans, and E. Beyne
IMEC, Kapeldreef 75, B3001, Leuven, Belgium
e-mail: deradft@imec.be

G.Q. Zhang and A.J. van Roosmalen (eds.), *More than Moore: Creating High Value Micro/Nanoelectronics Systems*,
DOI 10.1007/978-0-387-75593-9_3,

size, a higher degree of miniaturization is required. These passive components do not scale like active CMOS-based IC-technology, making it difficult to integrate all these devices on-chip. Therefore a proper partitioning of the RF system is required. The active devices may be integrated in one or two SOC devices and the external passive devices should be integrated in the SOC package, effectively realizing an RF-System-in-Package (RF-SIP).

An interesting enabling technology for the realization of RF-SIP with integrated passives are multilayer thin-film [or integrated passives device (IPD)] technology [1, 2] and low temperature cofired ceramic (LTCC) technology [1, 3, 4], in which the passives and interconnects are incorporated into the carrier substrate and offer in many ways an attractive solution [5, 6]. The multilayer thin-film technology as used for wafer-level-packaging (WLP) of device wafers (redistribution and bumping) offers very attractive possibilities. A key feature of this technology is the use of photolithographic technology for the definition of the various passive circuit components, resulting in a high degree in miniaturization and high patterning accuracy, with tolerances in the mm and sub-mm ranges. This results in an excellent circuit repeatability and predictability, key ingredients for the realization of first-time right and high manufacturing yield devices.

As transistor dimensions scale down and CMOS- and Si-based semiconductors are increasingly replacing GaAs for microwave and millimeter-wave applications, circuit performance becomes more and more determined by the on-chip passive component quality. However, in the attempt to pace up with this evolution, thinner on-chip metals and dielectrics have a troubling effect on the Q factor of on-chip passives. A cost-effective and attractive solution is to realize on-chip inductors using thin-film WLP techniques, similar to those used for realizing the RF-SIP interposer substrates.

More added advantage is expected from an expansion of the existing library of fixed passives with innovative variable reconfigurable passives as switches, varicaps, and tunable filters. In addition, there is a need to integrate miniaturized high Q resonators and filters to replace the bulky SAW and ceramic (dielectric resonator) filters and conceivably also the quartz crystal reference resonators. Perhaps the only technology at present with the potential to enable the integration of all these passives is micromachining or Micro-Electro-Mechanical System or Structure (MEMS) technology [7]. When applying MEMS to high-frequency circuits [radio frequency (RF), microwave or millimeter wave], the technology is commonly referred to as RF-MEMS [7]. RF-MEMS are currently under development in laboratories around the world and offer the potential to build a multitude of miniaturized components such as switches, voltage-tunable capacitors, high-Q inductors, film bulk acoustic resonators (FBAR), dielectric resonators, transmission line resonators and filters, and mechanical resonators and filters. Research on RF-MEMS continues on both the device level and the system level. The use of MEMS technology opens new perspectives for various wireless applications to achieve especially the required convergence between radio standards. MEMS-technology may revolutionize the choices made in the architecture of transceiver systems and/or radar antennas. In light of the foregoing, the potential and future for RF-MEMS is likely to be situated in the integration of RF-MEMS components, yielding so-called integrated RF-MEMS systems. A related incentive for the use of RF-MEMS components is therefore the integration capability with other passives and active components either in a monolithic or in a cost-effective hybrid way.

The evolution towards very compact 3D system integration is also going to drive heterogeneous integration of RF functionality with the digital systems. This poses tremendous challenges both on technologies that enable the vertical stacking of thin devices and on design (digital and RF) strategies that fully exploit the opportunities that appear with these technology developments to built extremely compact and high performance systems.

3.2 Multilayer Thin-Film Technologies

A technology used to integrate passive components for RF-applications above 1 GHz must allow for the realization of passive components with values relevant to those applications and with a high degree of precision and repeatability. Also, to allow for the integration, a high degree of scaling is required to fit the complex circuits in a small area.

These requirements strongly favor the use of a photolithographic-defined technology, where a large number of devices are collectively realized on large substrates or wafers. We have proposed [8–10] the use of multilayer thin film for this purpose. The infrastructure for this technology was developed for the WLP or silicon back-end-of-line processing. High volume manufacturing equipment with automatic handling is now available for the common Si-wafer sizes.

3.2.1 Thin-Film Technology for Integrated Passives Devices

On a low loss RF-substrate (glass or high resistivity silicon) integrated resistors, capacitors, and inductors are integrated. The basic elements of this technology are a thin-film, high-density metallization technology and a thin-film, dielectric deposition technique, capable of realizing very small via holes in the isolation layers to allow for high-density interconnects between the different layers in the structure. In this way, low loss (high Q) passives such as inductors, capacitors, resistors, and transmission lines for use up to the higher millimeter waves are achievable. An example of an RF-SIP build-up topology, developed at IMEC, is shown in Fig. 3.1

Thin-film technology is well suited for the integration and miniaturization of passive components. Complex materials can be deposited with high repeatability to form the highest quality resistor or capacitor layers. Thin-film lithography assures a high-dimensional accuracy, enabling small tolerances and increased miniaturization and, therefore, avoiding the need for "trimming" of resistor or capacitor values.

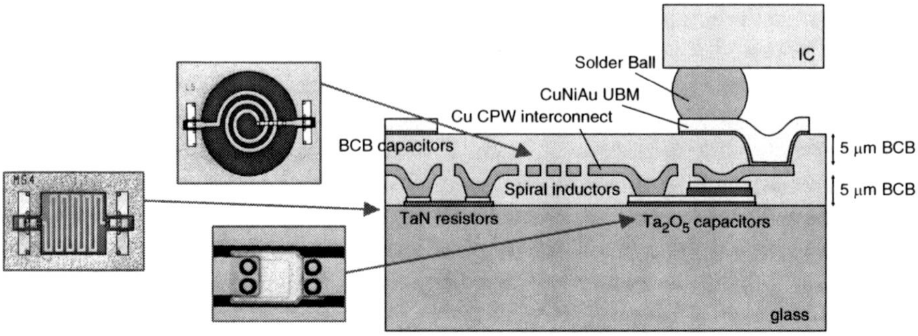

Fig. 3.1 Schematic cross section of IMEC's integrated passives platform technology

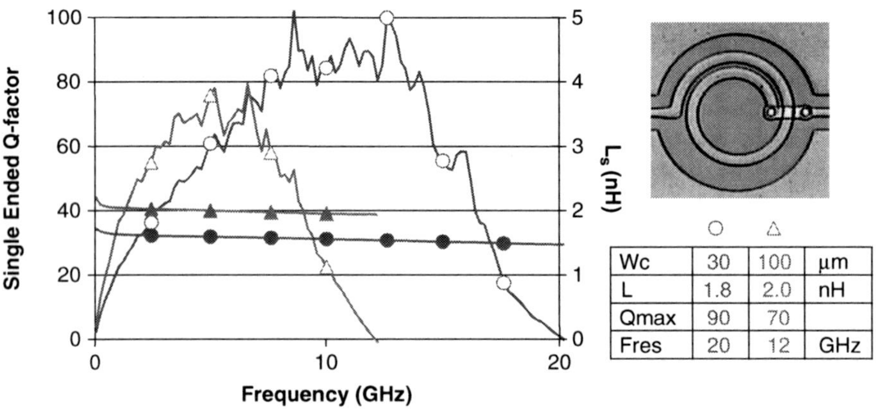

Fig. 3.2 Measured inductance and quality factor vs. frequency of two 1.5 turn inductors: realized on high resistivity Si substrates, with a line spacing of 10 μm and a gap of 150 μm between the outer trace and the ground plane. The inner diameter of the 30-μm trace width (Wc) inductor is 250 μm and that of the 100-μm trace width (Wc) inductor is 300 μm

The electroplated copper lines, described above, are ideally suited for realizing high-quality inductors, particularly those required for high-frequency applications. In order to allow for the practical use of this technology, a design library was developed. This library consists of electrically equivalent circuit models for all the relevant RF-passive circuits, as well as models for the interconnect lines, discontinuities, and wire bond or flip-chip connections. These models are parametric with the main geometric dimensions, allowing for a flexible optimization of the RF-design. The library also automatically generates the mask layout for the circuit, further improving the predictability of the designed circuit. As an example a typical inductor is depicted in Fig. 3.2

Complex filters, coupling structures, and filter functions can be realized effectively using this approach. Some application examples are shown in Fig. 3.3.

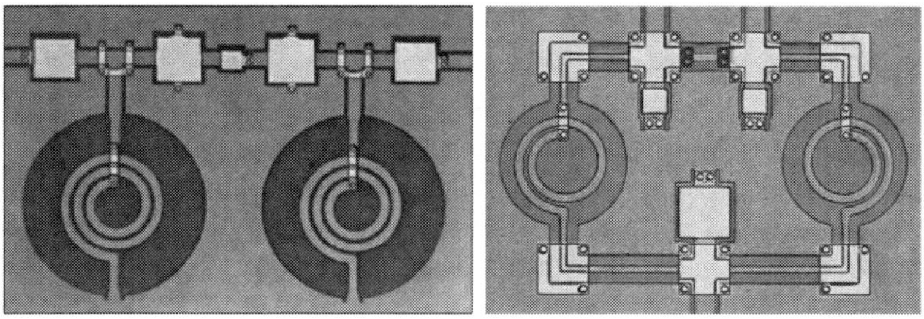

Fig. 3.3 Examples of integrated RF functions in the RF-interposer technology. *Left*: 2.45-GHz band-pass filter. *Right*: 7-GHz Wilkinson power splitter integrated on a high resistance Si substrate, RF-SIP. Characteristics: Return loss < −20 dB, Isolation < −20 dB, Loss 0.5 dB, Size: 1.65 × 1.05 mm^2

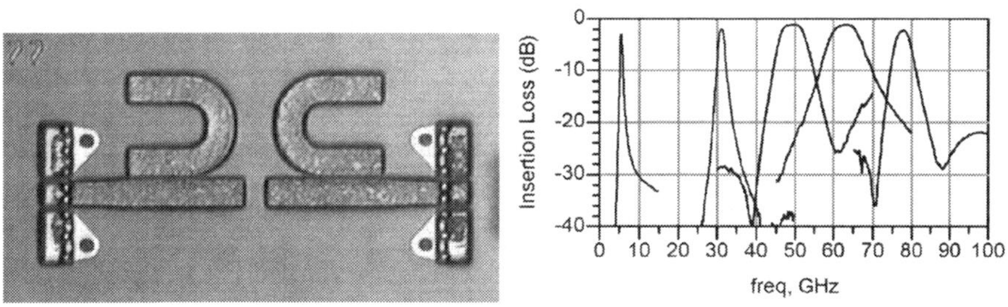

Fig. 3.4 *Left*: Picture of a 77-GHz microstrip band-pass filter integrated in IPD technology; *right*: measured responses of various filters realized in this IPD technology through the full microwave and millimeter-wave band

Table 3.1 Measured center frequency (f_0), insertion loss, bandwidth, and dimensions of the different filters

f_0 (GHz)	Insertion loss (dB)	Bandwidth (%)	Dimensions (mm^2)
5.2	2.95	10.3	2.44 × 1.55
31	2	4.9	3.1 × 0.76
49	1.2	14.3	1.4 × 0.54
63	1.1	12	1.6 × 0.6
77	2.25	5.5	1 × 0.47

This type of technology allows to build a wide range of functions up to the higher millimeter-wave frequency range (see Fig. 3.4) with good performance. The performance of such filters is shown in Table 3.1.

3.2.2 "Above-IC" RF-SOC

A characteristic feature of RF front-end integrated circuits is the relatively large area occupied by on-chip inductors. The large size of these inductors is due to the physical limitations of scaling inductors while maintaining performance (Q-factor). Realizing these inductors on the RF-SIP interposer substrate, as described earlier, is generally not an option as the interconnect parasitics, even a small flip-chip bump, would degrade the performance improvement gained from using high Q off-chip inductors. However, the integrated passives technologies can also be applied directly on the device wafer, "above-IC" [11].

For many high-frequency RF-ICs, the poor quality factors of regular on-chip inductors are a limiting factor. This is due to the relatively high sheet resistance of the on-chip metallization and the losses in the semiconducting silicon substrate. By placing the spiral inductor in the thin-film layer, "above-IC," the distance between the spiral and the lossy substrate is significantly increased. By using a thicker, electroplated Cu conductor, a much lower track resistance is obtained. A FIB cross section of such an inductor process is shown in Fig. 3.5. In this case, a 10-mm thick copper layer and a 12-mm thick dielectric are used. Inductors with Q-factors above 30 up to 5 GHz were obtained over 10-mW cm Si CMOS wafers. The Q factor can even be increased by applying a ground shield on the silicon substrate. Also differential inductors with high-quality factors and very high resonance frequencies can be realized, as shown in Fig. 3.6.

The postprocessing is compatible with both Cu and Al back-end. The technology is cost-effective and consumes no additional Si real estate. Measurements performed on MOS transistors and back-end interconnects show no important performance shifts after postprocessing. The WLP inductors have increased performance and resonance frequency as compared to back-end versions enabling the design of high-performance low-power circuits such as VCOs.

IMEC applied this technology to realize - among other circuits - a 5-GHz and 15-GHz low-power VCO in 90-nm CMOS [12]. The 5-GHz and 15-GHz VCOs

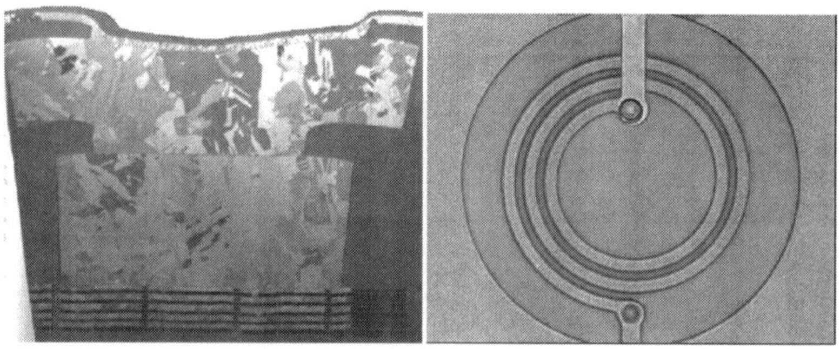

Fig. 3.5 High-Q, 10-μm thick Cu inductor processed on top of a 10-μΩ cm CMOS wafer. *Left*: cross-section contact inductor, *Right*: top view

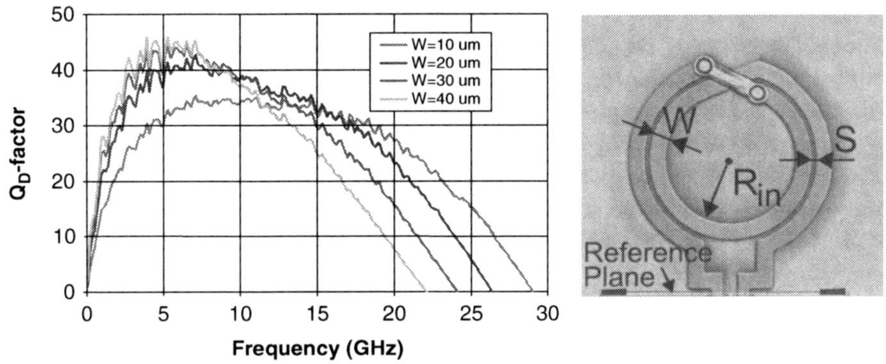

Fig. 3.6 Differential 1.6 nH "above-IC" inductors integrated on a 10 μΩ cm CMOS wafer. Inductor parameters: 2 turns, line thickness 10 μm, line spacing of 10 μm, line widths ranging from 10 to 40 μm and inner diameter of 250 μm

use, respectively, a 3-nH and a 0.6-nH WLP inductor without ground shielding resulting in a differential Q factor of, respectively, 40 and 55. The 5-GHz and 15-GHz VCOs show a low core power consumption of, respectively, 0.33 mW and 2.76 mW, a phase-noise of −115 and −105 dBc/Hz (at 1 MHz offset), and a tuning range of 148 MHz and 469 MHz. For comparison, a 6.3-GHz, 90-nm VCO using a back-end inductor with patterned ground shield has a core power consumption of 5.9 mW with a phase noise of −118 dBc/Hz at 1 MHz.

3.3 RF-MEMS Technology

RF-MEMS technology refers to micromachining of materials [7]. Dimensions may vary from around 1 mm to several millimeters. MEMS for wireless applications have been demonstrated to operate from around 1 GHz to frequencies approaching 100 GHz [13]. RF-MEMS, most in particular switches and resonators, have originated at several (industrial) research labs in the United States and have experienced an exponential growth since the early 1990s. Hughes Research Labs has pioneered rotating- and cantilever-type electrostatically actuated micromachined structures for RF and microwave switching in the late 1980s, with a first publication by Larson et al. in 1991 [14]. Around the same time, Westinghouse has engineered cantilever-type capacitive MEMS relays for use as RF switching elements, with patents dating back to the early 1990s [15]. Following this, Nguyen and Howe from the University of California-Berkeley demonstrated the successful development of a CMOS oscillator implementing a MEMS resonator in 1993 [16]. Hewlett-Packard (HP) has filed a patent on a tunable thin-film acoustic resonator as early as 1993 [17]. Goldsmith et al. from Texas Instruments (now Raytheon) [18] and Yao et al. from Rockwell Science Center [19] published on micromachined electrostatically actuated ohmic (resistive) relays for RF and microwave switching in 1995.

A micromachined microwave switch relying on electrothermal actuation was patented in 1995 by HP [20]. A upsurge of publications on a variety of RF-MEMS devices and systems as switches and switching arrays, switched-line phase shifters, tunable capacitors, microresonators, tunable filters, and reconfigurable antenna arrays from research groups around the world has appeared since the late 1990s [21–24]. Among the large amount of RF-MEMS devices (micromachined inductors, switches, variable capacitors, mechanical and acoustic resonators, FBAR,...), let us focus a bit on switches and FBAR devices.

3.3.1 RF-MEMS Switches

RF-MEMS devices are extremely attractive devices for the realization of reconfigurable RF hardware. High-frequency switches are proposed in wireless communications and radar systems for switching between the transmit (Tx) and receive (Rx) paths, for routing signals to the different blocks in multiband/standard phones, for RF signal routing in phase shifters used in phased array antennas, and much more. RF-MEMS switches offer great potential benefits over semiconductor switches in terms of a high isolation, more particular at high frequencies (>30 GHz), and a low loss over a wide frequency range (in particular compared to FETs at the higher frequencies), extremely low standby power consumption (in particular compared to PIN diodes), and the excellent linearity characteristics [25, 34, 37, 41, 42]. Other advantages that may prove valuable are the integration capability with other high-quality passives (high-Q inductors, varicaps, filters), the flexible choice of the substrate, and the ability to keep the signal circuit separate from the control circuit (relay configuration), which greatly simplifies the bias circuitry. Drawbacks remain the rather high driving voltage (of an electrostatic device) and the relatively slow response. Largely unknown are the packaging complexity, the reliability, and the power handling capability. In short, RF-MEMS switches are not the "Holy Grail" for every problem and every application, but these switches definitely offer great potential benefits for a wide range of applications.

Although these switches can be built using various actuation mechanisms, the electrostatic actuated switch is still most popular. A typical RF-MEMS switch cross section is depicted in Fig. 3.7 and consists of a free standing metal beam which can be pulled down by applying a voltage between the signal line and the bridge (called a switch) or between a separate actuation electrode and the beam (called Relay-style).

Figure 3.8 gives an example of a relay type shunt switch - a shunt switch provides an RF short circuit in the actuated down position - with insertion loss and isolation, respectively, better than 0.5 dB and 20 dB in a wide frequency band spanning from 22 GHz to more than 40 GHz [25]

As a first step toward the realization of tunable fully integrated band-pass filters in MCM-D/RF-MEMS technologies, one can realize a hybrid filter by flip-chipping

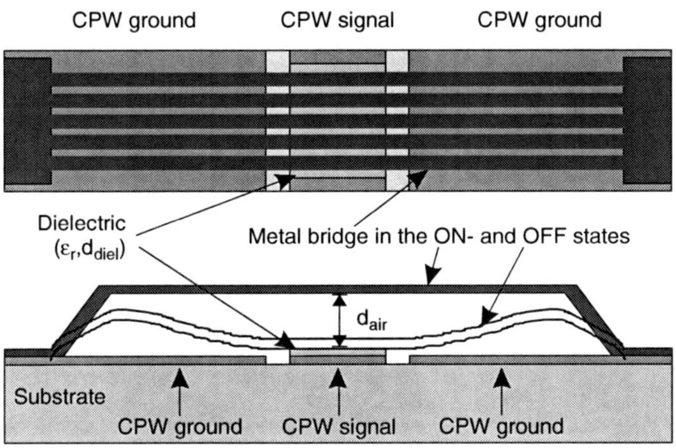

Fig. 3.7 Typical RF-MEMS capacitive shunt switch

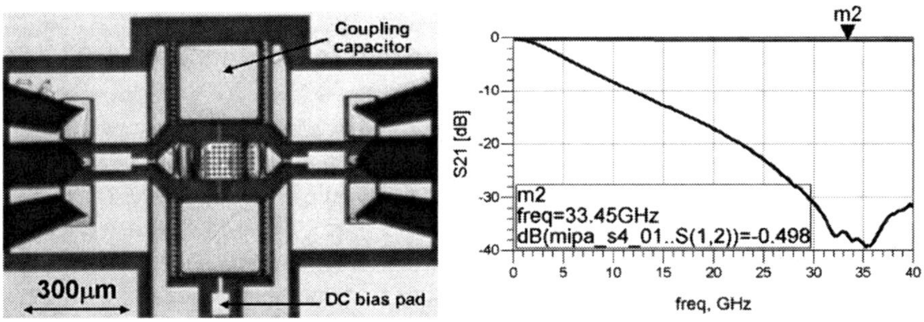

Fig. 3.8 *Left*: Picture of a shunt capacitive relay; *right*: the associated measurement in up and down state

a RF-MEMS chip, constituted of one shunt and two series MEMS capacitors, on a MCM-D substrate [26], incorporating high-Q inductors.

Figure 3.9 presents a close-up picture of the assembly for thin-film IPD GPS-Galileo switchable band-pass filter integration. The thin-film IPD inductors are visible through the glass substrate of the RF-MEMS chip. This example shows how more complex RF functions can be integrated to build new - heterogeneous modules.

3.3.2 FBAR Resonators

Thin FBAR, in short FBARs, can be considered as the micromachined or MEMS version of conventional bulk acoustic wave resonator, e.g., a quartz crystal. Bulk acoustic wave resonators in essence consist of a parallel plate capacitor with a piezo-

Fig. 3.9 Hybrid RF-MEMS on thin-film IPD GPS-Galileo switchable band-pass filter integration

electric layer used as the dielectric. By applying an ac electric signal to the electrodes, a longitudinal acoustic wave is excited in the bulk of the piezoelectric film. This wave is trapped by the reflecting electrode surfaces thus forming an acoustic resonator. In order to attain a high-Q, the acoustic losses into the supporting substrate must be made as small as possible. One way is to isolate the structure from the substrate by (locally) removing the substrate underneath [27]. In FBARs, thin films of aluminum nitride (AlN) or zinc oxide (ZnO) are commonly employed for the piezoelectric layer. For these materials, resonances in the low GHz regime result for a piezoelectric layer thickness on the order of 1 mm and are thus well within reach for thin-film technologies. The resonators are made as small as a few tens to a few hundreds of micrometers on a side, typical for a MEMS design. Membrane-supported FBARs using AlN films have been demonstrated to operate at resonant frequencies in the low-GHz range with loaded Qs typically of a few hundreds but Qs over 1,000 have also been measured [27].

3.3.3 RF-MEMS Packaging and Reliability

RF-MEMS switches, like many other MEMS devices, contain movable fragile parts that must be packaged in a clean and stable environment. A specific low-cost packaging approach is required offering protection during fabrication as well as during operation. The so-called 0-level packaging is preferably carried out on the wafer during wafer processing, prior to die singulation. One approach for the 0-level package is to bond a recessed sealing cap chip onto the MEMS device wafer as illustrated in Fig. 3.10

The bonding must be performed at sufficiently low temperatures (typically below 400°C) such that the metallization and other materials of the RF-MEMS switching device are not adversely affected. In packaging an RF-MEMS switch (or any other RF-MEMS device), the package itself should have minimal effect on the

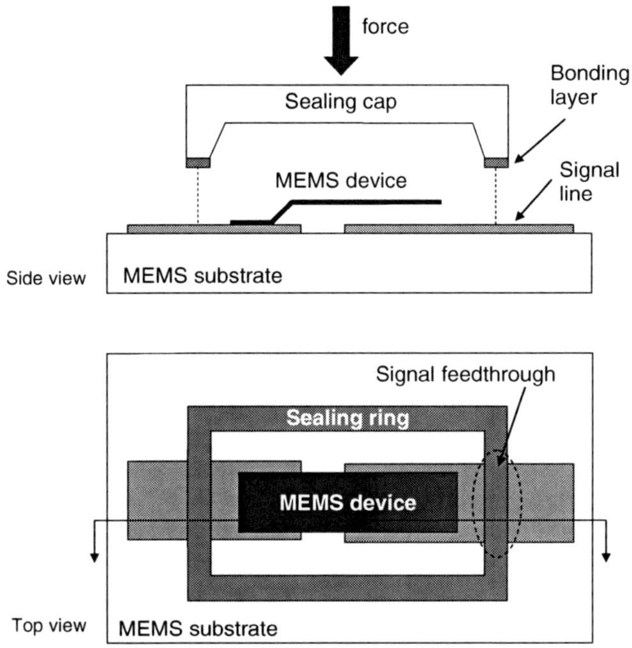

Fig. 3.10 Illustration of the 0-level packaging of a MEMS device based on a chip stacking technique

device performance. In an ideal package, the RF characteristics before and after capping should be the same. This requires low-loss RF transitions and minimal induced loss and detuning of the transmission lines due to proximity coupling to the cap. Several possible implementations for the RF transitions can be envisioned as described in [24]. Other examples of 0-level packaged RF-MEMS switches according the basic scheme of Fig. 3.10 but using different implementations for the RF feedthroughs and the bond and sealing layer, are described by Jourdain et al. [28], Margomenos et al. [29], Fujii et al. [30], and Park et al. [31]. Based on this, it can be said that the RF transitions are not considered the major problem in RF-MEMS 0-level packaging, but the process integration, the cost, and the maximum attainable hermeticity are crucial issues that have to be solved in the near future.

3.4 Using the Third Dimension

Interconnection and packaging of electronic systems is mainly done in a planar 2D way. For further miniaturization and enhancement of performance, the use of 3D interconnections is very attractive and required. Key technologies for realizing these 3D vertical connections are the use of through-die VIA connections for die stacking and multilayer interconnects with embedded die.

3.4.1 Why Moving Toward the Third Dimension

There are basically three important reasons to go into the third dimension:

- The first reason is system size reduction. Traditional assembly technologies are based on 2D planar architectures. Die are individually packaged and interconnected on a planar interconnect substrate, mainly printed circuit boards. The area-packaging efficiency (ratio of die to package area) of individually packaged die is generally rather low (e.g., 5×5 mm die in 7×7 mm package: 50% area efficiency) and an additional spacing between components on the board is typically required, further reducing the area efficiency (e.g., above 1 mm clearance: 30% area efficiency). If we consider the volumetric packaging density, the packaging efficiency drops to very low levels. If in the previous example, we consider the active area of a die to be about 10-μm thick, and the combined package and board thickness to be 2 mm, the volumetric packaging density is only 0.15%. There is clearly room for improvement of the packaging density.
- A second reason for looking at 3D integration is performance driven. Interconnects in a 3D assembly are potentially much shorter than in a 2D configuration, allowing for a higher operating speed and smaller power consumption. This is of particular interest for advanced computing applications. Due to the rising on-chip clock speeds, only a limited distance may be traveled by a signal in a synchronous operating mode. Using 3D-IC stacking techniques, more circuits may be packed in a single synchronous region. This requires a technology with 3D interconnects with low parasitics; in particular low capacitance and inductance are needed to avoid additional signal delay and the same is true for RF applications.
- The third and most interesting reason to consider 3D integration is so-called heterointegration. As silicon semiconductor technologies continue to scale (vertical scaling), the realization of true SOC devices with a large variety of functional blocks (digital, analog, RF) becomes very difficult to achieve. Technologies need specific optimization for logic, analog, memory, microwave, etc. to reach the desired performance levels and circuit density. Furthermore, the substrates used to build active devices may vary significantly between technologies, including nonsilicon substrates, e.g., compound semiconductors. Also systems may contain other planar components, such as MEMS and integrated passive devices. Besides the "vertical" scaling we are also experiencing a "horizontal" scaling. Realizing the full system on a single SOC die is becoming increasingly difficult and often not economically justified. If, however, a high-density 3D technology is available, a "3D-SOC" device could be manufactured, consisting of a stack of heterogeneous devices. This device would be smaller, lower power, and higher performance than a monolithical SOC approach.

3.4.2 3D Systems in a Package for Highly Miniaturized Systems

Especially this last argument is especially true in the evolution of "More than Moore" systems, such as wireless autonomous sensor nodes realizing so-called "ambient intelligence" systems. These are sometimes referred to as smart-dust, e-grains, or e-cubes. This type of systems is from its nature heterogenous. As shown in Fig. 3.11, such systems can be divided into clear subsystems: the radio (antenna, RF-front-end, and base band), the main application (processor, sensors, and actuators), and the power management (regulation, storage, and generation). Each of these functions can be realized as a SIP-subsystem. These may be very small (a few mm to a few cm), enabling its realization using wafer level processing technologies. The 2D subsystems can be stacked on top of each other, realizing a dense 3D-SiP system.

An example of such 3D-SIP integration scheme, realized at IMEC, is a fully integrated low-power RF transceiver shown in Fig. 3.12. This device measures only 7×7 mm. It consists of two CSP-type devices (CSP = "Chip-Scale-Package").

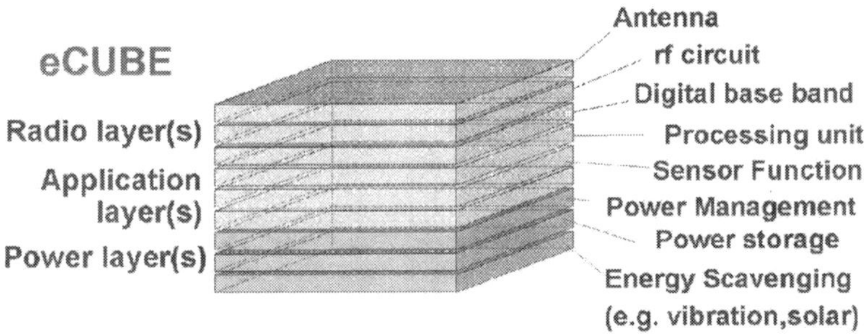

Fig. 3.11 Schematic representation of a 3D-SIP concept "eCube," for the realization of distributed, fully autonomous "ambient intelligent" systems. Each layer in the stack is a fully integrated SIP subsystem

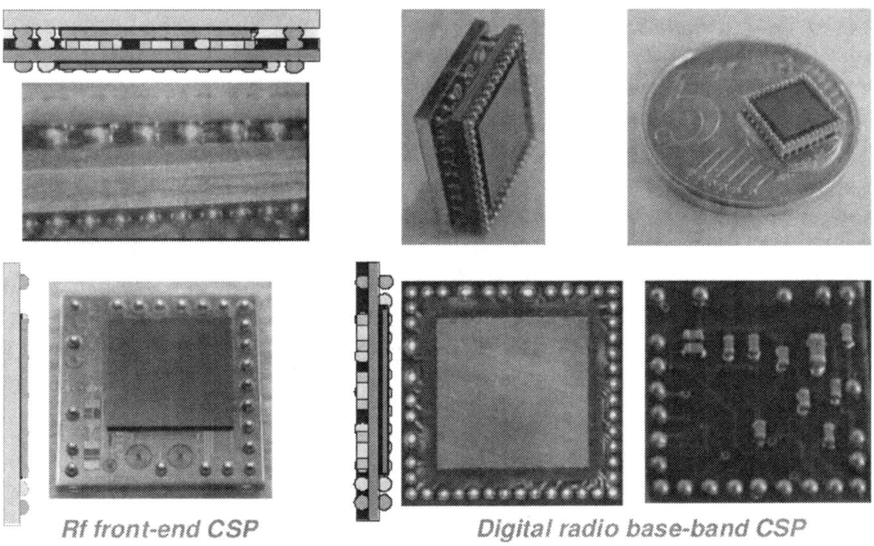

Fig. 3.12 Fully integrated low-power RF radio, measuring 7 × 7 × 2.5 mm, realized by 3D stacking of CSP packages. 3D joining using microbumps

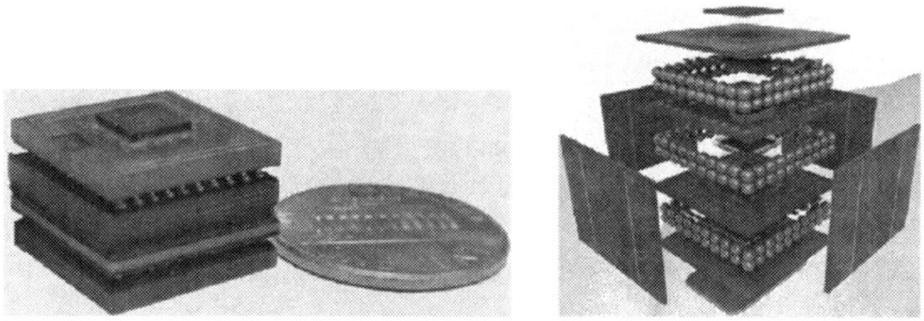

Fig. 3.13 Photograph (*left*) and schematic view (*right*) of 1 cm^3 eCube (14 × 14 mm), developed at IMEC for a medical application. Small solar cells power scavengers, to provide for long time autonomy, may be added on the module sidewalls

The top CSP is realized using IMEC's RF-IPD technology with integrated passives [9, 10]. The bottom CSP is a double-sided high-density printed circuit board with a high density flip chip die on the bottom side and several discrete passive components mounted on the topside. The connection of this bottom part to the top part is obtained by using solder balls on the topside of the bottom laminate and encapsulation of the topside devices.

A more integrated complete system realization of such e-cube is depicted in Fig. 3.13 The 3D-SIP consists of a RF-SIP, with integrated antenna, RF transceiver chip, a low-power DSP SIP, a 19 channel EEG/ECG sensor die, and a power SIP [32, 33].

3.4.3 Packaging Technologies up to Millimeter-Wave Integration

It is very challenging to explore the RF-SiP approach and extend it into the millimeter-wave frequency range. Since more traditional PCB technologies suffer from performance degradation (losses, tolerances, and cost) when moving to these higher frequencies, IPD technologies, such as described earlier (see Sect. 2.1), become also very attractive, especially in the combination with 3D concepts that allow through wafer VIAs and using the third dimension to add functionality. As an example of this approach, a cavity filter constructed using a high resistivity silicon wafer with VIA wall - delineating the cavity - can be built with very good performance while at the same time the filter can act as the vertical 3D millimeter-wave interconnection between two functional RF layers (Fig. 3.14). This is just one example to demonstrate the creative use of this type of technologies and techniques for the realization of novel integrated solutions. In this case, the insertion loss of the filter was lower than −1.6 dB at 60 GHz.

3.5 Further Trends

It is clear that heterogeneous integration of components originating from different technologies is driven by a strong demand of upcoming "More than Moore" system requirements, especially for systems containing RF functionality. In order to further

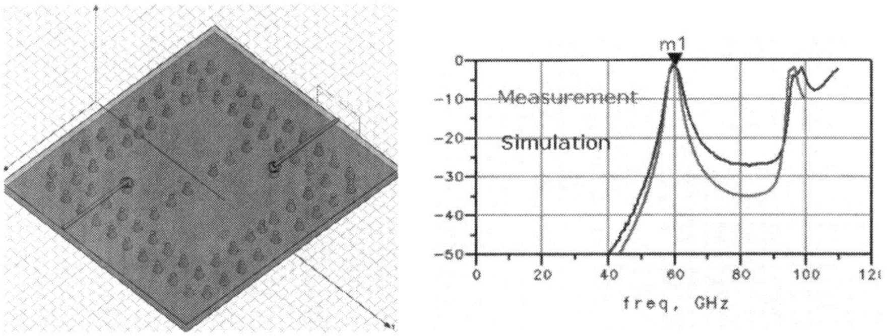

Fig. 3.14 *Left*: Design drawing of a cavity filter at 60 GHz (size); *right*: simulated and measured insertion loss of a cavity filter realization

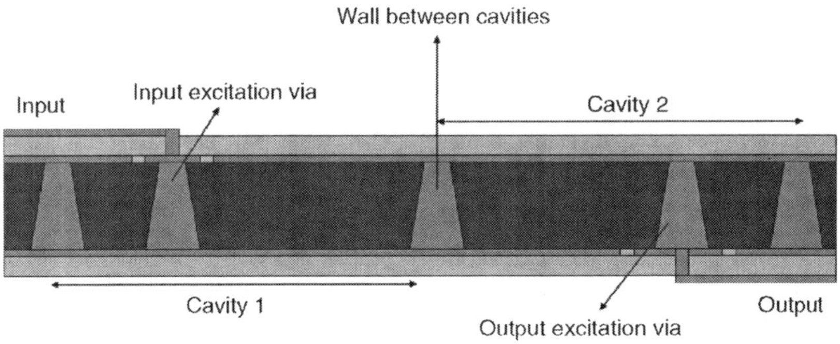

Fig. 3.15 Filter cross section showing the principle for a 3D vertical interconnect in the filter example of Fig. 3.14

decrease component and module sizes - even in 3D-SiP systems - the embedding of bare die components becomes a must. To achieve this, many routes are possible. A first solution is to embed bare die components into a host substrate. This substrate can be flex, a micromachined silicon substrate or even a mold compound based material, shaped as a new wafer. As an example of this last method, the wafer reconstitution technique is described below in Sect. 3.5.1. Another even more aggressive solution is thinning the active die to a very thin layer (5–30 mm thick) followed by positioning the active die onto a cheap host substrate containing the interconnections and high performance RF passive components (see Sect. 3.5.2). Both methods allow to reduce the active RF chip area considerably since - usually - the large high Q passive components take otherwise also a large - expensive chip area.

3.5.1 Wafer Reconstitution Techniques

A new innovative way to perform heterogeneous integration using bare die components is called "embedded Wafer Level Ball grid array (eWLB)" technology [38–40].Compared to standard WLP, where chip size is equal to the package size,

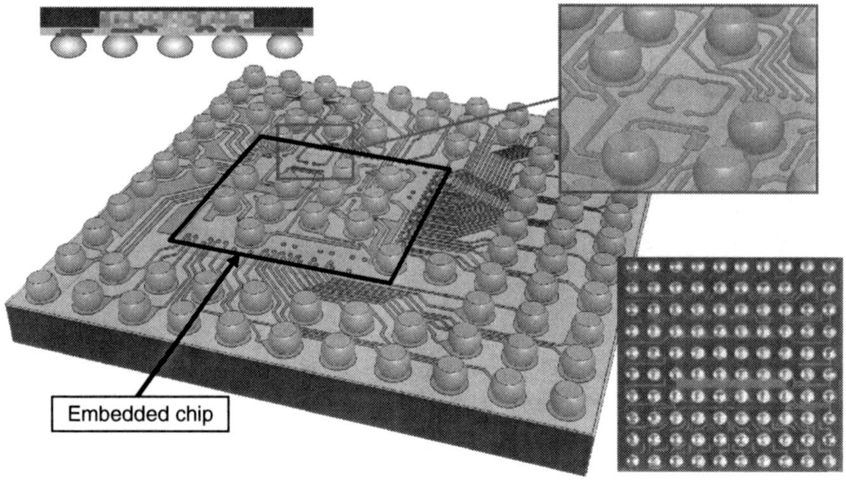

Fig. 3.16 Conceptual drawing of the eWLB heterogeneous integration technology (Courtesy of Infineon). The bare chip device is the gray area on the module while the black area is the mould compound material of the reconstituted wafer

this technology allows more contacts to the board. In the eWLB approach, the bare die (gray area in Fig. 3.16) is embedded in a mold compound material that forms the reconstituted carrier substrate (black area). After the embedding, IPD or above-IC process techniques (see 3.2.1 or 3.2.2) can be used to add low loss passive components such as inductors and transmission lines for building low cost but high performance RF modules.

The schematic process flow of this concept is shown in Fig. 3.17. After the die embedding process, a redistribution layer (consisting of metallization and dielectric layers) is added which can incorporate typical low loss microwave or millimeter-wave passives. Electron microscope (SEM) cross sections of an embedded die are depicted in Fig. 3.18. This solution is a cost-effective way to build novel complex modules – even using commercial standard bare die components - that reduces function and module sizes very considerably. This approach is very attractive toward RF system integration since bare die components – without large RF package parasitics are used and the small bond pad pitches of, e.g., advanced CMOS processes can be addressed in a natural way: nowadays redistribution layer technologies are routinely able to use 30-μm pitch metallization.

3.5.2 3D Integration Using Ultra-Thin Chip Stacking

A different road to 3D stacking consists in stacking thin die on active device wafers and using multilayer thin-film technology [34] to interconnect the thin die with the host wafer. Such approach allows for a high level of system level flexibility and

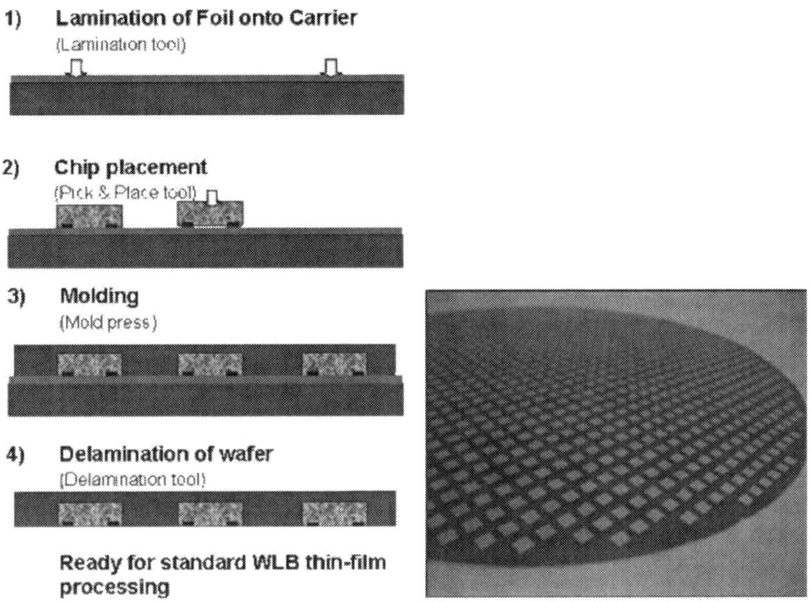

Fig. 3.17 *Left*: schematic fabrication flow of a reconstituted wafer: first a foil is laminated on a temporary carrier, next bare die chips are placed face down on the laminated carrier followed by the molding process. Finally the carrier is removed from the molded reconstituted wafer. The bare dies with free top metallization are then ready for thin-film processing. *Right*: picture of such a reconstituted wafer, ready for further processing with passives and interconnection metallization and dielectrics (courtesy of Infineon)

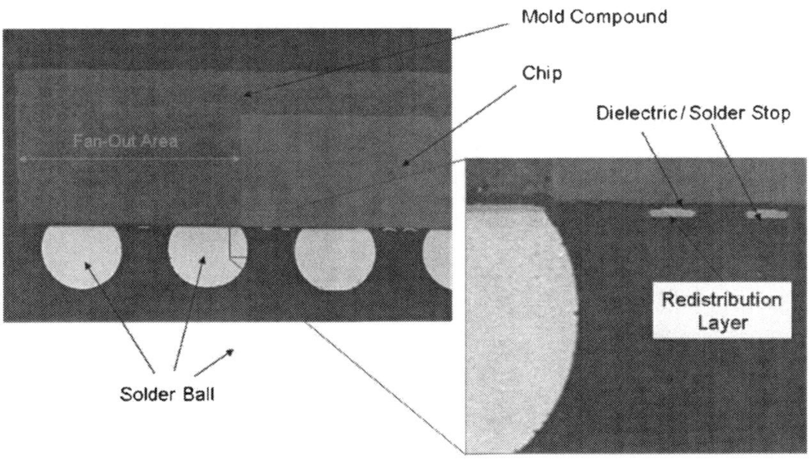

Fig. 3.18 SEM cross section of a molded bare die chip, together with an additional interconnect layer (redistribution layer) and attached solder balls (courtesy of Infineon)

integration density. This technique also allows for stacking of die with largely varying dimensions, as well as the integration of thin-film passive components in a 3D interconnect stack.

Several approaches of this type of stacking are under investigation [35]. IMEC's ultra-thin-chip-stacking (UTCS) approach [36] uses very thin (10–20 μm) Si die, embedded in a redistribution technology. In order to realize a multilayer die UTCS stack, four basic processes are used:

- Ultra-thin chip-on-die (UTCD) process
- Ultra-thin chip embedding (UTCE) process
- Ultra-thin chip-in-flex (UTCF)
- UTCF-stacking process for more than two die layer UTCS structures

The UTCD process is shown schematically in Fig. 3.19. A wafer with active, tested die is bonded to a temporary silicon carrier wafer. Using a combination of coarse and fine grinding, the active wafer is thinned to a thickness of 15–20 mm. Plasma etching is used to remove any remaining Si damage and to obtain the desired final thickness. Plasma etching is then used to etch the scribe lanes of the die. The next step consists in dicing the carrier wafer to obtain the UTCD chips for further processing.

This ultra-thin die is then placed on the integration substrate and can be embedded using the IPD technology based on Cu-interconnects and BCB dielectrics as described in Sect. 3.2.1.

In contrast to the traditional die-stacking technologies, the dies that are stacked can have arbitrary sizes at any layer and do not need any particular process or design adaptation. This approach is also very much compatible with RF and microwave applications where there is a strong need for low-loss, high-Q passive components such as inductors, capacitors, transmission lines, etc. Using the above-IC

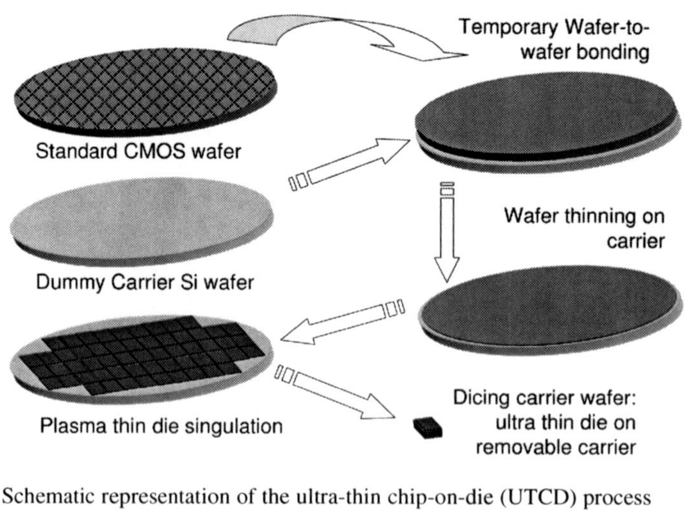

Fig. 3.19 Schematic representation of the ultra-thin chip-on-die (UTCD) process

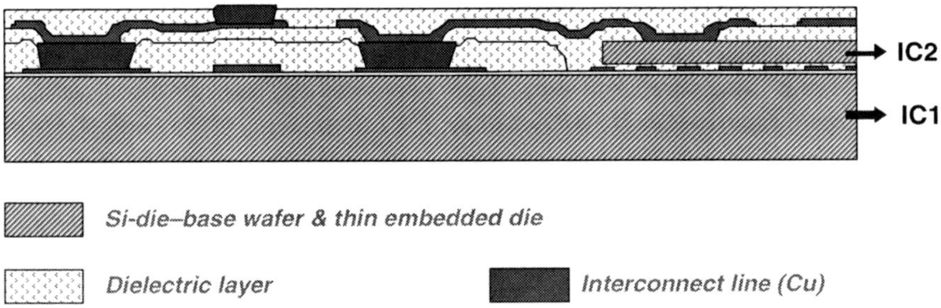

Fig. 3.20 Schematic cross section of a UTCE stack

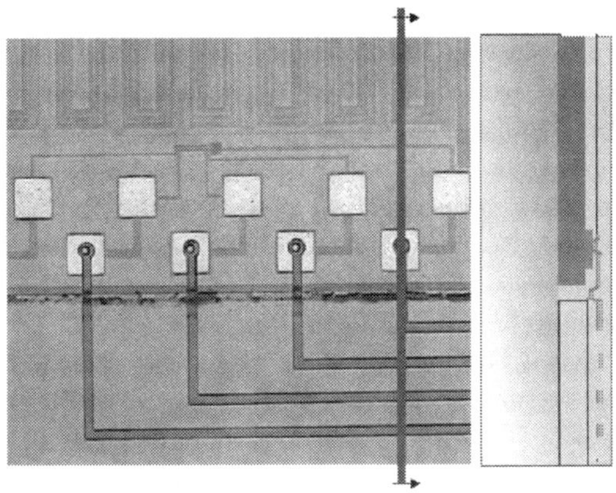

Fig. 3.21 Example of a 15-μm thin Si-die, transferred to a host substrate and electrically connected to that substrate using the UTCE die embedding technique

technology (see Sect. 3.2.2), we showed that it is possible to fabricate these low loss passives on top of active silicon. However, in many cases, the area used by these passives cannot be completely reused for active components due to possible parasitic coupling effects between the active circuits underneath and the passives. Using the UTCS embedding techniques, the active RF chip area can be minimized (thus reducing cost) while the passives can be built on the embedding substrate keeping the parasitics (capacitances, inductances at a very low level, and this is mandatory at increasing frequencies (Figs. 3.20 and 3.21).

3.6 Concluding Remarks

Many technologies for heterogeneous systems are currently being developed since these technologies may fill the gap between pure SOC (systems on chip) and board level systems. Especially for systems containing RF functionality, dedicated technologies are still required to realize the needed level of performance. The most prominent example is found in today's ubiquitous wireless communication technology, requiring low cost, high performance, low power or efficient power, and multiband adaptive RF systems. The components and technologies that can achieve part of these requirements do exist (RF-MEMS devices in many flavors to allow reconfigurability, wide bandgap materials, advanced CMOS integration technologies,...) but none of them can fulfill them all. Therefore the most efficient and cost-effective way is to use heterogeneous integration concepts. There is also a strong push toward the reuse and modularity of RF building blocks and circuits for cost reduction and time to market reduction reasons. At this moment, the era of RF systems is developing exponentially and still new applications arise at the horizon: emerging applications include highly integrated low-cost, short-range collision avoidance and road-safety systems in vehicles (24, 77, 110 GHz) radars, imaging devices for aircraft take-off and landing safety systems, and imaging systems for access control, the detection of metallic and nonmetallic materials (explosives, ceramic weapons). This will involve the development of high-speed spatial resolution, short and medium range antenna sensors, radars, and imaging systems operating at frequencies up to 1 THz. Also in the area of sensing, e.g., millimeter-wave spectroscopy is still in its infancy but first applications in chemical analysis are showing up. The main challenges are achieving the required performance regarding frequency and bandwidth and reducing sensitivity to effects such as humidity, temperature, and mechanical impact.

We are living only at the beginning of an exciting era for RF system heterointegration, the "More than Moore" era for RF is developing fast!

Acknowledgment The authors also are particularly indebted to Klaus Pressel (Infineon) for helpful discussions and additional information.

References

1. Brown W D 1999 Introduction and overview of microelectronic packaging in *Advanced Electronic Packaging – with Emphasis on Multichip Modules* edited by W D Brown (New York, NY: IEEE).
2. Carchon G, Nauwelaers B, Pieters P, Vaesen K, De Raedt W and Beyne E 2001 Multi-layer thin-film MCM-D for the integration of high performance wireless front-end systems. *Microwave J.* **44**:96–110.
3. Brooks R, Guilford M and Hendricks D 2001 0201 Component assembly: A tough test. *OnBoard Technol.* 2001 50–53.
4. Scrantom C Q and Lawson J C 1999 *LTCC Technology: Where We Are and Where We're Going To*, IEEE MTT-S International Topical Symposium for Wireless Applications, Feb. 1999, pp. 193–200.
5. Ulrich R and Schaper L 2003 *Integrated Passives Technology* (New York, NY: IEEE).
6. Tummala RR, White GE, Sundaram V and Bhattacharyam S 2000 SoP: The microelectronics for the 21st century with integral passive integration. *Adv. Microelectron.* **27**:13–19.
7. Maluf N 2000 *An Introduction to Microelectromechanical Systems Engineering* (Boston, MA: Artech House).

8. Beyne E, Van Hoof R and Achen A 1995 *The Use of BCB and Photo-BCB Dielectrics for High Speed Digital and Microwave Applications*, Proceedings of the 1995 International Conference on Multichip Modules, Denver, CO, April 19–21, 1995, pp. 513–518.
9. Pieters Ph, Brebels S and Beyne E 1996 *Integrated Microwave Filters in MCM-D*, Proceedings IEEE-MultiChip Module Conference MCMC-96, Santa Cruz, CA, February 6–7, 1996.
10. Carchon G, Vaesen K, Brebels S, De Raedt W, Beyne E and Nauwelaers B 2001 Multilayer thin-film MCM-D for the integration of high-performance RF and microwave circuits. *IEEE Trans. Compon. Packag. Tech.* **24**(3):510–519.
11. Carchon G, et al. 2003 *High-Q RF Inductors on Standard Silicon Realised Using Wafer-level Packaging Techniques*, MMTS Conference, Philadelphia, PA, June 9–12, 2003.
12. Sun X, Dupuis O, Linten D, Carchon G, Soussan P, Decoutere S and Beyne E 2006 High-Q above-IC inductors using thin film wafer-level packaging, demonstrated on 90 nm rf-CMOS 5 GHz VCO and 24 GHz LNA. *IEEE Trans. Adv. Packag. Technol., Part B* **29**(4):810–817.
13. De Los Santos H 1999 *Introduction to Microelectromechanical (MEM) Microwave Systems* (Boston, MA: Artech House).
14. Larson L E, Hackett R H and Lohr R F 1991 *Microactuators for GaAs-Based Microwave Integrated Circuits,* Transducers'91, San Francisco, CA, June 1991, pp. 743–746.
15. Buck D C 1994 *Non-Contact Two Position Microelectronic Cantilever Switch*, US Patent 5,367,136.
16. Nguyen C T-C and Howe R T 1993 *CMOS Micromechanical Resonator Oscillator*, Technical Digest, International Electron Devices Meeting, pp. 199–202.
17. Ruby R C and Merchant P P 1996 *Tunable Thin Film Acoustic Resonators and Method for Making the Same*, US Patent 5,587,620.
18. Goldsmith C, Lin T H, Powers B, Wu W-R and Norvell B 1995 *Micromechanical Membrane Switches for Microwave Applications*, 1995 IEEE MTT-S Digest, pp. 91–94.
19. Yao J J and Frank-Chang M 1995 *A Surface Micromachined Miniature Switch for Telecommunications Applications with Signal Frequencies from DC up to 4 GHz,* Transducers'95, Stockholm, Sweden, pp. 384–387.
20. Field L A and Ruby R C 1995 *Thermally Actuated Micromachined Microwave Switch*, US Patent 5,467,067.
21. De Los Santos H J and Richards R J 2001 MEMS for RF/Microwave wireless applications: The next wave - Part II. *Microwave J.* **44**(7):142–152.
22. Rebeiz G M and Muldavin J B 2001 RF MEMS switches and switch circuits. *IEEE Microwave Mag.* **2**:59–71.
23. Rebeiz G M, Tan G-L and Hayden J S 2002 RF MEMS phase shifters: Design and applications. *IEEE Microwave Mag.* **3**:72–81.
24. Tilmans H A C 2002 MEMS components for wireless communications, *EUROSENSORS XVI Prague Czech Republic* Sept. 15–18 2002, pp. 1–34 (on CD-ROM only).
25. Rottenberg X, Soussan P, Stoukatch S, Czarnecki P, Nauwelaers B, Carchon G, De Wolf I, De Raedt W and Tilmans H 2006 *RF-MEMS technology platform for agile mobile and satellite communications*, in: *Proc. EuMC*, Sept. 10–15 2006, Manchester, UK, pp. 1723–1726.
26. Oya Y, Okubora A, Van Spengen M, Soussan P, Stoukatch S, Rottenberg X, Ratchev P, Tilmans H, De Raedt W, Beyne E, De Moor P, De Wolf I and Baert K 2004 *A Reliable and Compact Polymer-Based Package for Capacitive RF-MEMS Switches*, International Electron Devices Meeting, pp. 31–34.
27. Lutsky JJ, Naik RS, Reif R and Sodini CG 1996 *A Sealed Cavity TFR process for RF Bandpass Filters*, International Electron Devices Meeting, San Francisco, CA, USA, Dec. 8–11 1996, pp. 95–98.
28. Jourdain A, Brebels S, De Raedt W and Tilmans H A C 2001 *The Influence of 0-Level Packaging on the Performance of RF-MEMS Devices*, 31st European Microwave Conference, London, UK, 24–28 Sept. 2001, pp. 403–406, 419.
29. Maragomenos A, Peroulis D, Herrick K J and Katehi L P B 2001 *Silicon Micromachined Packages for RF MEMS Switches*, 31st European Microwave Conference, London, UK, Sept. 25–27 2001, pp. 271–274, vol. 1.

30. Fujii M, Sato S, Hosoya K and Imanaka K 2001 *Micro Machined Relay with Vertical Feed Through and Wirebond-Less Package*, 51st relay (NARM) Conference, pp. 16.1–16.4.
31. Park Y-K, Park H-W, Lee D-J, Park J-H, Song I-S, Kim S-W, Song C-M, Lee Y-H, Kim C-J and Ju B-K 2002 *A Novel Low-Loss Wafer-Level RF-MEMS Packaging of the RF-MEMS Devices,* MEMS 2002 Las Vegas, NV, USA, Jan. 20–24 2002, pp. 681–684.
32. Stoukatch S, Winters C, Beyne E, De Raedt W and Van Hoof C 2006 *3D-SIP Integration for Autonomous Sensor Nodes*, Proceedings of the 56th IEEE Electronic Components and Technology Conference, pp. 404–408.
33. Baert K, Gyselinckx B, Torfs T, Leonov V, Yazicioglu R, Brebels S, Donnay S, Vanfleteren J, Beyne E and Van Hoof C 2006 Technologies for highly miniaturized autonomous sensor networks. *Microelectron. J.* **37**(12):1563–1568.
34. Beyne E, *Multilayer Thin Film Technology as an Enabling Technology for System in Package (SiP) and "above-IC" Processing*, IEEE/SEMI 29th International Electronics Manufacturing Technology Symposium, July 2004, pp. 91–99.
35. Proceedings of the First and Second conference on "3D Architectures for Semiconductor Integration and packaging", *RTI International*, Burlingame, California, April 13–15, 2004 and Tempe, Arizona, June 13–15, 2005.
36. Beyne E 2001 *Technologies for very High Bandwidth Electrical Interconnects Between Next Generation VLSI Circuits*, IEEE-International Electron Devices Meeting 2001 Technical Digest, December 2–5, 2001, Washington, DC, S23–p3.
37. Rebeiz G M and Muldavin J B 2001 RF MEMS switches and switch circuits *IEEE Microwave Mag*. **2**, 59–71.
38. Brunnbauer M, Fürgut E, Beer G, Meyer T, Hedler H, Belonio J, Nomura E, Kiuchi K, and Kobayashi K 2006 *An Embedded Device Technology Based on a Molded Reconfigured Wafer*, Proceeding of the 56th Electronics Components and Technology Conference ECTC 2006, p. 547.
39. Wojnowksi M, Engl M, Brunnbauer M, Pressel K, Sommer G, and Weigel R 2007 *High Frequency Characterization of Thin Film Redistribution Layers for Embedded Wafer Level BGA*, Ninth Electronic Packaging Technology Conference EPTC 2007, Singapore, December 2007.
40. Brunnbauer M and Meyer T 2008 *Embedded Wafer Level Ball Grid Array (eWLB)*, IMAPS, Fourth International Conference on Device Packaging, March 19th, 2008, Scottsdale, AZ, USA.
41. Goldsmith C, Randall J, Eshelman S, Lin T H, Denniston D, Chen S and Norvell B 1996 Characteristics of micromachined switches at microwave frequencies 1996 *IEEE MTT-S Digest* 1141–44.
42. Mihailovich R E, Kim M, Hacker J B, Sovero E A, Studer J, Higgins J A and DeNatale J F 2001 MEMS relay for reconfigurable RF circuits *IEEE Microwave and Wireless Components* Lett. **11**, 53–55.

Chapter 4
High Voltage and Power

G. Ricotti and G. Maggioni

Abstract High voltage and power are essential for successful MtM product creation. This chapter gives an introduction for the basics of high-voltage and power technologies. Application examples of solid-state lighting and hybrid cars are presented.

Keywords High voltage • Power • Solid-state lighting • Hybrid car

4.1 Introduction

High voltage (HV) and power can be defined as any voltage higher than that used in the classical digital I/Os within the state-of-the-art semiconductor processes, i.e., starting at 3.3 or 5 V. HV interfaces and functions are important parts of most (small) systems, usually as part of an input/output (IO) system that interfaces the system to the real world.

They are usually needed when the I/O device requires a high-power or high-voltage drive (e.g., electromechanical actuators or LCD displays) or when high-voltage capability is required in protection circuitry that allows sensitive electronic circuits to be used in harsh environments. High-voltage capabilities are also required in power management, power conversion, and power distribution circuits. Power management solutions are necessary to drive low-voltage CMOS circuitry from battery or AC-line power sources in a wide range of consumer products. Automotive systems need to drive electromechanical actuators such as fuel injection systems, solenoids, start motors, and electric windows but are also increasingly typified by

G. Ricotti (✉) and G. Maggioni
STMicroelectronics, Milan, Italy

G.Q. Zhang and A.J. van Roosmalen (eds.), *More than Moore: Creating High Value Micro/Nanoelectronics Systems*,
DOI 10.1007/978-0-387-75593-9_4, © Springer Science+Business Media, LLC 2009

low-voltage electronic systems in body control and safety systems that need protection circuit from voltage spikes.

The high-voltage and power technologies can help to solve some grand social and economic problems, such as energy, healthcare, and environment protection. Examples are energy saving by solid-state lighting (SSL) systems, health assistance by electro-biomedical systems, and pollution reduction by hybrid/electric cars or energy harvesting. Examples of existing and emerging applications of HV include the following:

- Power management solutions to drive low-voltage CMOS from a variety of sources. This includes a large range of applications, from battery-powered mobile gadgets to mains-powered consumer products.
- Car systems:
  - Conventional electric systems based on 42/14-V car batteries, driving mechanical actuators such as in fuel injection systems, solenoid drivers, starter engines, and electric windows.
  - Emerging technologies such as x-by-wire, EVT.
  - Hybrid car electronics, involving 200-V (and above) electric motor driver circuits.
- Solid-state lighting
- Power supplies for PCs
- Climate control
- Industrial applications
- Electrical power generation and distribution:
  - Battery chargers
  - New portable energy sources, e.g., microfuel cells, microbatteries
  - High-power DC-to-AC converters (e.g., for solar cells)
  - Decentralized and regenerative energy supply (wind power plant, fuel cell, microturbine, photovoltaic, etc.)

  The application parameters are as follows:

- On-state resistance (Rds, on) and breakdown voltage (BVds)
- Switching performance (Qgd)
- Robustness
- Operation temperature range
- Integration density when combined within a CMOS environment
  - Logic density (gate/mm^2, Mbit/mm^2)
  - Current density in metallization layers
- Parasitic effect immunity

All these parameters contribute on device level to an overall system performance that can be classified with respect to energy conversion efficiency and cost efficiency.

4.2 SSL Technology

4.2.1 General Lighting Technology

During the last 10 years the enormous business of portable systems such as mobile phones, PDAs, portable computers, IPods, etc. has focused the research on finding solutions that could provide a longer battery life.

Obviously big improvements have been obtained in battery chemistry and in the efficiency of energy management, but especially on the utilization devices such as microprocessors, displays, speaker drivers, and mainly on the lighting.

The utilization of portable systems in dark environments was and continues to be the most critical condition of power consumption for many reasons. One among these is the difficulty in matching between the source's electrical parameters (voltage and current) and the lighter's parameters.

The classical incandescent lamp, based on an electrical resistor heated by the joule effect generated by current flow, generally is the easiest load that can well be matched to the source's characteristics. It is easy to design a lamp with the optimum electrical resistivity adaptable to the voltage level of the main supply networks of portable systems. It can cover a very wide range of supply voltages, from hundreds of millivolts up to hundreds of volts.

Other advantages are low cost and the simple production technology. Unfortunately, many drawbacks affect this object, starting from high temperature generated and, as immediate consequence, the extremely low light efficiency, the limited lifetime, and the poor white light quality.

If the lamp is designed for high voltage and medium-low power, fragility becomes an issue. In fact the very thin resistive wire required can be easily broken by a simple shock, especially when incandescent. Because of the high-temperature range the "package" is generally glass and this is another source of fragility.

It is easy to understand that the extremely low cost of incandescent light systems has been the main reason of their enormous diffusion in all application fields, for sure where the energy source is unlimited but in the past also in portable equipment due to missing alternative technology.

The strong need to light small watches and mobile phones has encouraged the research to work and develop new lighting technology with the key target of light efficiency and quality.

4.2.2 New Technology

Many new lighting methodologies based on different physics phenomena have been developed: everybody remembers the first generation of digital wristwatches based on a red LED (light-emitting diode), shown in Fig. 4.1. The LED typical supply voltage is from 1.2 to 2.1 V.

Fig. 4.1 Digital wristwatch implementing red LED display

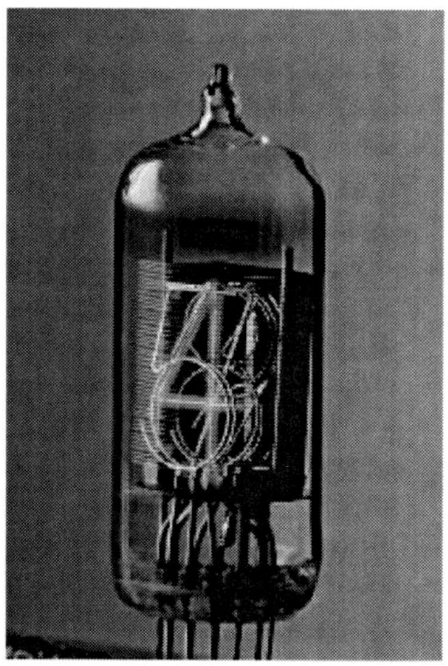

Fig. 4.2 An example of NIXIE tube with numerical digits

Before the LCD technology based on a liquid crystal display that does not emit light, the alphanumerical displays were based on a NIXIE technique (Figs. 4.2 and 4.3) widely used in watches, pocket calculators, electronic balances, and typically in flippers.

A nixie tube [1] is an electronic device used to information. The glass tube contains a wire-mesh anode and multiple number-shaped cathodes. The tube is filled with a gas at low pressure, usually neon with a little percentage of mercury and/or argon.

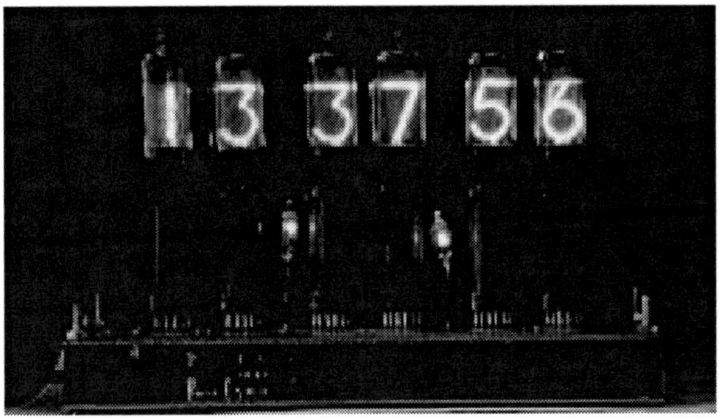

Fig. 4.3 Watch based on nixie display

Fig. 4.4 Electroluminescent display

Average longevity of nixie tubes varied from about 5,000 h for the earliest devices to 200,000 h or more for some of the last introduced ones. It should be noted that there is no formal definition for what constitutes the "end of life" of nixies (mechanical failure excepted) but a continuous reduction of light intensity.

The nixie tube requires a main supply voltage in the range of 120–200 V plus some other bias and control voltage lines.

Electroluminescence (EL) [2] is an optical and electrical phenomenon of light emission in response to an electric current or a strong electric field applied to a particular material. This is different from light emission resulting from heat (incandescence) or from the action of chemicals (chemoluminescence). Electroluminescence (Fig. 4.4) is the result of irradiative recombination of electrons and holes in a material (usually a semiconductor). The excited electrons release their energy as photons - light. Before recombination, electrons and holes are separated either as a result of doping of the material to form a p-n junction (in semiconductor electroluminescent devices such as LEDs) or being excited by the impact of high-energy electrons accelerated by a strong electric field (as with the phosphors in electroluminescent displays).

4.2.3 Fluorescent Lamps and Driving Methodology

We can find an important example of battery-supplied object on space satellites with humans on board, where light is required to have a good comfort. Since the first satellites were realized in sixties, lighting has been generated by fluorescent lamps (Figs. 4.5 and 4.6).

Fig. 4.5 Driverless fluorescent lamps

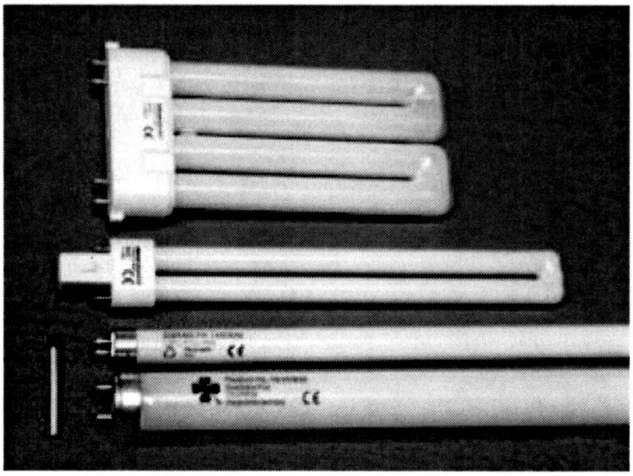

Fig. 4.6 Fluorescent lamp with modern electronic driver

Nowadays these kinds of lamps are widely diffused in private houses and especially in public structures or offices – thanks to their relatively low cost and their improved light efficiency, higher than classical incandescent lamps.

Unfortunately, the main AC (alternate current) supply networks cannot directly drive these lamps because they need a series of nondissipative current limiters and dedicated start-up circuits.

Where the space around the lamp is not a problem the driver is very simple and composed of two main passive devices; a simple coil usually called reactor realizes the current limiter. The start-up is performed with a nonlinear component called "starter" that generates with the reactor an overvoltage pulse able to trigger the discharging through the low-pressure gas inside the tube. Once discharging has been triggered, the tube behaves as if it has low impedance, due to the good conductivity of the ionized gas, and so a current limiter is mandatory and it is realized by the coil.

A power factor corrector (PFC), simply implemented by a normal capacitor, is also needed to rephase the voltage and the current.

A typical driving circuit is shown in Fig. 4.7.

About 20 years ago electronically driven fluorescent lamps, represented in Fig. 4.6, have been introduced in the consumer market with a big success, mainly due to the possibility to replace classical incandescent lamps through the same standard connector.

This has been made possible - thanks to the electronic driving that allows to place in a small volume the reactor and the start-up circuitry. The base of the lamp contains an integrated circuit (IC), and few discrete components such as a small coil and a capacitor are also present. The IC drives the coil with high-frequency voltage pulses to generate the extra voltage necessary to trigger the gas discharge. With increase in the frequency, the dimensions of the coil and the capacitor automatically scale down to few cubic centimeters.

The IC makes also possible all the optimizations such as the immediate ignition without flickers and the adaptive power factor correction. These lamps are 2–3 times more efficient than incandescent ones and have longer lifetime.

To support this business some silicon companies have invested in high-voltage technology, allowing the design of ICs directly supplied by the network power line (off line) without any scaling down of transformers or preregulators.

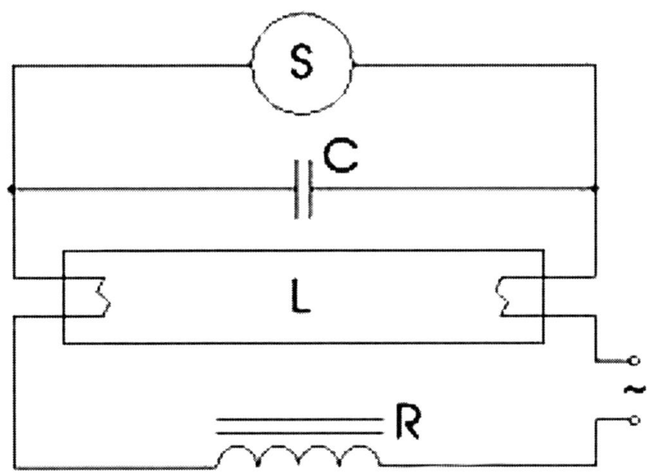

Fig. 4.7 Full passive fluorescent tube driving circuit

Obviously, these technologies do not follow the Moore's law, because these are not driven by low lithography but by all the isolation techniques that allow high electrical field in micrometric dimensions. A deep analysis of the silicon technology for fluorescent and LED lamps will be presented in the following section.

4.2.4 IC for the Fluorescent Lamps

Figure 4.8 shows an example of IC for lighting applications that is integrated in one small package with all the functions required for the correct operation of the electronic ballast and the PFC function.

The PFC provides the rephasing of the supply power line, assuring no noise injection on the domestic network. That means no more interference on television or radio when fluorescent lamps are powered. Moreover, it contributes in keeping high overall efficiency.

The electronic ballast is an adaptive current limiter able to optimize lamp's lighting, its fast ignition, and mainly the lamp's life. A half-bridge drives and controls the management of the preheating and the ignition pulse for the lamp as well.

The silicon technology, needed to design these drivers, integrates mainly CMOS plus high voltage (up to 700 V) DMOS or IGBT.

These kinds of ICs are able to guarantee flexibility, protection, and optimum internal signal management between the PFC section and the current controller during the ignition sequence, delivering the right current level and shape according to the lamp's characteristics (aging) and ambient environment (temperature).

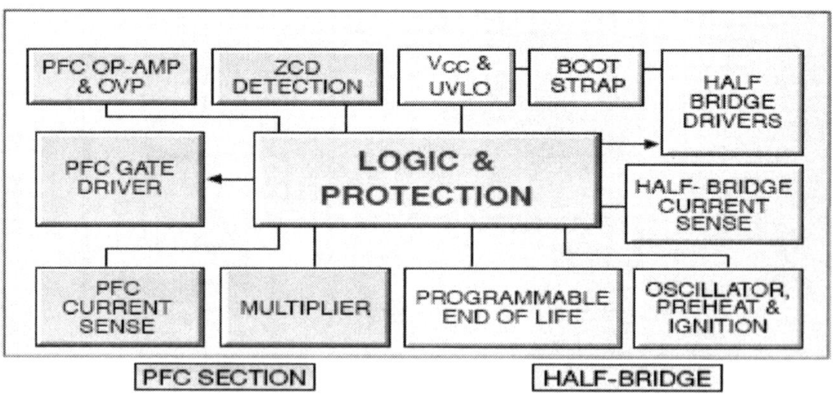

Fig. 4.8 A fluorescent lamp driver IC for OFF line connection

These driving devices are not simply a sum of two different ICs (a PFC and a half-bridge controller), but include functionalities that increase the system reliability, protecting the ballast in case of bad design or external component failures. As an example, the saturation of the PFC coil is prevented and its correct operation is constantly monitored; this is an important feature for general safety of high-voltage systems.

Many patents protect the integrated circuitry that controls the aging of the lamp installed in different configurations.

During lamp's aging, the end-of-life detection is performed by monitoring the current variation needed to keep the lamp switched on at the optimum light level. When this current becomes too high, causing a huge drop in efficiency, the IC controller can be programmed to switch off the lamp.

The main features and characteristics of the fluorescent offline lamp drivers can be summarized as follows:

- Since preheating and ignition phases are independently programmable, a unique IC is suitable for different lamp types/sizes.
- The controlled lamp voltage/current during ignition and the overcurrent protection during running mode increase the ballast reliability.
- The programmable end-of-life detection (Fig. 4.9), compliant to the two standard ballast configurations, allows flexible design (lamp to ground or block capacitor to ground).
- Overvoltage and feedback disconnection detection on the PFC stop the IC, avoiding ballast damage.
- PFC current sense and inductor saturation detection considerably increase the system safety.

These complete high-voltage ICs make easy to build high-performance electronic ballasts for fluorescent lamps.

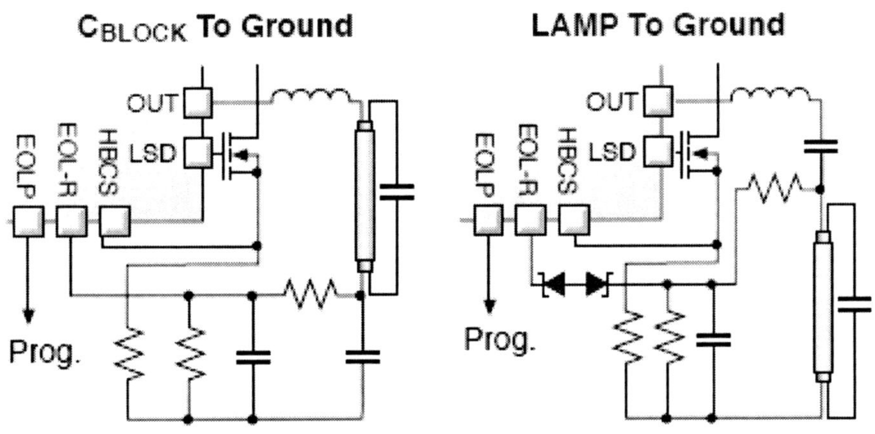

Fig. 4.9 Ballast and PFC configuration for end-of-life detection

4.2.5 LED Lighting

The LEDs perfectly respect the definition of SSL. Many are the uses of LEDs, organic LED (OLED), or polymer light-emitting diodes (PLED) as sources of illumination rather than that of electrical filaments or gas.

The term "solid state" refers to the fact that the light in a LED is emitted by a solid object - a block of semiconductor – rather than a vacuum or gas tube, as in the case of traditional incandescent light bulbs and fluorescent lamps.

Unlike traditional lighting, however, SSL creates a visible light with reduced heat generation or parasitic energy dissipation. The light emitted by the traditional incandescent lamp is a consequence of the high temperature reached by the filament, so 90% of the power is wasted in thermal effect; in other words, it is lost.

Commercial white LEDs provide more than 40 lumens/W compared with 8–15 lumens/W provided by incandescent lighting. This can be considered the most important parameter that has been encouraging the continued research investment on these kinds of light sources.

This high efficiency obviously means a longer battery life in portable applications; this is a very important parameter but not the only one. If a light source generates high heating power, a large heat sinker is also needed to dissipate the thermal energy in an adequate volume.

The LED lamps can be very compact and useful in portable equipment by placing many LEDs to distribute light on the wide display or on the keyboard.

In the house lighting system, today this concept seems not to be relevant, because there is enough room and especially because the LED lamps can replace the traditional solutions (Fig. 4.10).

For sure, architects have already thought about new forms of lighting art using these very compact light sources distributed in an ambience to generate different atmospheres or to capture eyes' attention on some particular objects.

Fig. 4.10 An example of commercial LED lamp compatible with the existing lamp connectors

Solid state also means very high reliability and strong mechanical robustness; LEDs are going to replace stop lights of the car, not only because of the insensitiveness to vibrations, but, most of all, for their quickness (order of microseconds) in turning on. An incandescent lamp takes 150 ms to turn on: during this time the following car, at 80 Km/h, has covered another 3.5 m before starting to brake.

Moreover, LED's lifetime typically lasts for 100,000 h or more, reducing maintenance costs. In comparison, an incandescent light bulb lasts approximately for 1,000 h. In other words, considering a daily operative duty of 3–4 h, LEDs would be replaced every 30 years while an incandescent lamp every year.

LED lighting systems have already proved to be very effective in indicator applications where brightness, visibility, and long life are important, such as in traffic signals. Becoming the high voltage and power technologies with the white LED more powerful and effective, LEDs are going to be used in more general applications such as the home lighting.

The modern illumination systems have to do more than just switch light on and off: they have to enable programmed lighting situations integrating settings depending on human presence and on daylight intensity.

HV and power technologies will thus allow not only a substantial energy saving but also comfort and flexibility, smaller dimensions and weights. To illuminate an ambience also the "light color," measured by the correlated color temperature (CCT), is important in achieving the right comfort.

While filament lamps provide different CCT from warm white to bluish light, today the research on LEDs is widening the color range, extending from classical bluish toward warm white.

Another actual and important aspect of this lighting system is the dimming, that is, the possibility to program or adjust the light intensity according to the situation. For example, in the dining room and when watching the television, a soft light is preferable, but to read a newspaper during the frequent advertisement breaks, the user would need to increase light density.

Incandescent LED lighting can be dimmed easily. Dimming of fluorescent lighting is more expensive because of the cost of the dimming components required.

In the following, the driving methodology and electronic technology needed to perform the system key parameters will be described.

4.2.6 IC for LED Driving

There are two main categories of LEDs: the white-blue LEDs, with a typical voltage drop of 3–4 V, and the green-red-yellow LEDs, with a typical voltage drop of 2 V or less.

It is possible to make another distinction based on the forward current, that is, the low-current LEDs, from 15 to 50 mA and the high-current LEDs, from 350 to 1,000 mA.

When LEDs do not come from a special selection based on their voltage drop, if connected in parallel, they have different brightnesses depending on the individual voltage drop. The result could be not acceptable in many applications.

The solution is to connect LEDs in series. In this way, they are driven by the same current and so no difference in brightness is present. The disadvantage of this solution is that a high voltage is required.

A typical way to realize a constant current source to keep constant the brightness is to use a DC-DC converter (Fig. 4.11). The current loop can be obtained simply connecting a resistor between the voltage feedback pin and GND. The DC-DC converter will adjust the output voltage (V_{out}) in order to control and keep constant the V_{FB}, so the LED's current will be automatically controlled by the Ohm's law. Some tricks can be used to reduce the voltage drop across the sense resistor in order to increase the whole efficiency. The output voltage can be in the range of tens of volts depending on the number of LEDs in series and on the current level.

To modulate the brightness it is possible to act in two ways: by varying the DC value of I_{LED} or by modulating the constant current level in PWM mode.

According to the level of the main supply voltage and the number of LEDs in series, the DC–DC converter topology can be a boost converter, a buck-boost converter, or a simple buck converter. This choice has an important impact not only on the converter topology but mainly on the silicon technology.

When batteries supply the LEDs, the driver IC is designed in a low-voltage mix mode technology, integrating high-quality power DMOS able to guarantee high efficiency and driving capability with low supply. Considering a supply based on a double NiMH (nickel metal hydride) battery, the input voltage range is 1.8–2.4 V; it means that the DMOS used to drive the external coil has to guarantee low $R_{ds,on}$ with just 1.8 V applied to its gate source.

The generated output voltage depends on the number of LEDs in series and can reach 20–30 V; this fixes the technology voltage class that is, in any case, in the range of few tens of volts. The previous example is referred to as boost topology.

An interesting and actual application of LEDs is the wireless optic mouse supplied by one or two NiMH cells or one LiIon (lithium Ion) cell, while the main supply voltage range is between 0.9 and 5.2 V and the load is composed by just one

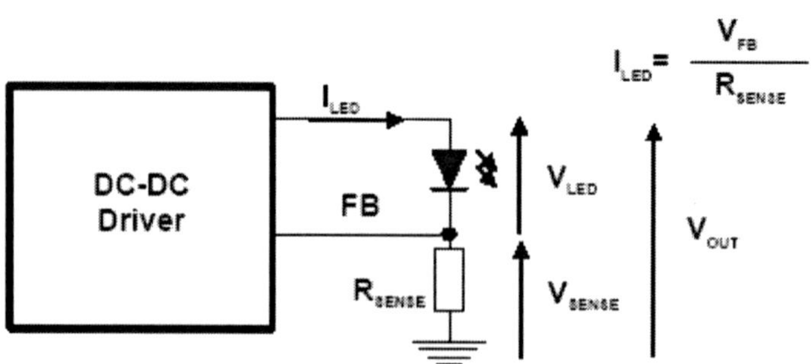

Fig. 4.11 Current mode driver for LEDs

LED with a typical voltage of 1.6–2.1 V. Here, a 7-V high-performance mix mode technology is the best choice.

The most interesting emerging application is the LEDs' driver supplied directly by the main domestic electric line in the range of 175-V AC or 230-V AC (depending on country standards) that becomes 175-V DC and 326-V DC, respectively, after rectification.

In the field of MtM application this is the most interesting LED driving methodology, and the circuitry will be a quasi-conventional step-down AC/DC current mode converter, but the silicon technology must be of high voltage (generally up to 700 V) to have enough capability to manage and survive to the noise spikes and overvoltages present on the "dirty" domestic power lines.

In Figs. 4.12 and 4.13, two examples of HV ICs are shown, designed on a junction-isolated technology based on a 2-μm minimum lithography. It is easily

Fig. 4.12 An example of a HV integrated circuit on a junction-isolated technology

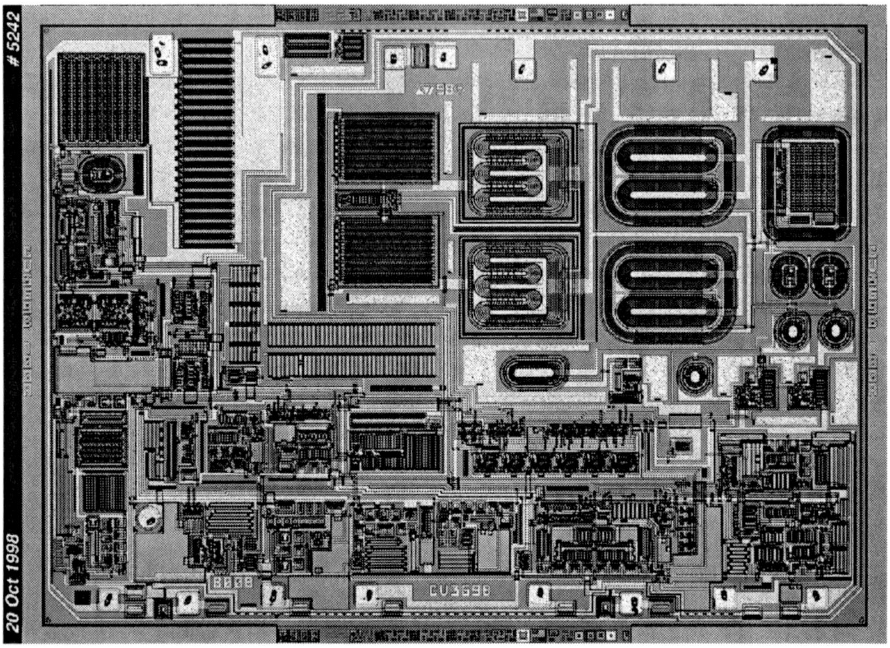

Fig. 4.13 An example of a HV integrated circuit on a junction-isolated technology

visible that the rounded geometry of the HV transistors is adopted to avoid the concentration of electric field like the "tip effect."

The right technology could be the SoI (silicon on insulator) that can allow high-voltage components in a very dense layout, because an oxide guarantees the isolation between the active devices in a deep trench instead of classical reverse-biased diodes. This will allow to integrate also a large-sized logic to interface a power line modem able to receive and respond to the commands delivered through the power line by a central box of the future (and actual) domotic systems.

4.3 Automotive IC Technology

4.3.1 Introduction and General Automotive Survey

Generally speaking, car systems nowadays are not so different from the ones put into production more than a century ago with a thermal engine working with crude oil derivatives. Although big improvements have been done in general performances, internal combustion engines still suffer from relative low efficiency and high-pollution production.

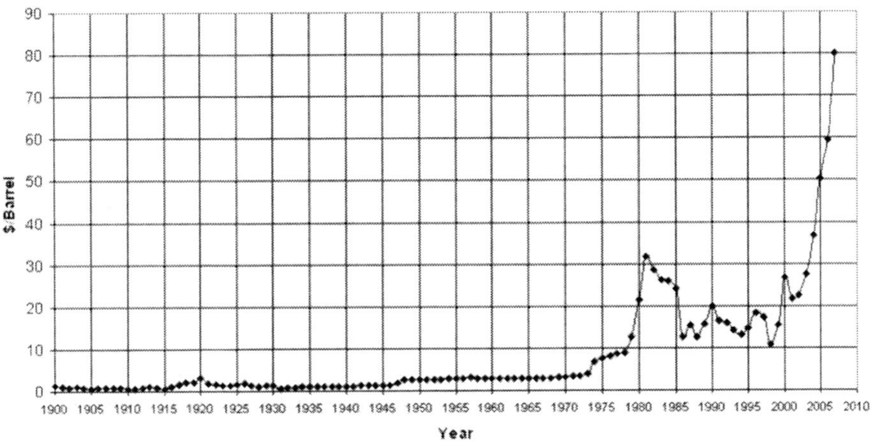

Fig. 4.14 Crude oil price history

Modern engines have an average efficiency of only 25–40%; this means that 60–75% of energy is dissipated in heat or friction losses. Global efficiency is also related to the use range of the car: idle, urban driving, and highway conditions have big difference in terms of engine performance.

Another source of inefficiency in car system is the amount of auxiliaries such as pumps, fans, and actuators normally driven by mechanical link.

As far as production of pollutant gases is concerned a big discussion is now in place on global worming and transportation contribution, despite the fact that more than 95% of greenhouse gasses is due to natural water vapor; carbon dioxide (CO_2) is the another major contributor (70% excluded water vapor).

Transportation system plays a significant role in CO_2 pollution (15% of total anthropogenic CO_2 production). For example, when burning 1 L of gasoline 2.3 Kg of CO_2 is produced. This becomes 2.7 Kg for diesel engines. When you drive a distance of 500 Km with your gasoline car, assuming an average fuel consumption of 7 L/100 Km the CO_2 production will be $7 \times 5 \times 2.3 = 80.5$ Kg. This figure is one of the key drivers for future engine efficiency improvement.

This tough situation is becoming difficult to sustain also due to the price of crude oil that was historically few dollars per barrel till the 1970s but starting from that point, driven by political and production strategy reasons, it is projected near $100 per barrel at the end of this decade (Fig. 4.14).

For the discussed reasons the car system is in a breakpoint; the automotive application driving forces are related to pollution and fuel consumption reduction.

A lot of research activity is in the power train arena, which is following the engine efficiency evolution. Figure 4.15 describes the evolution of the engine typology where the most important trend is the replacing of standard indirect diesel and

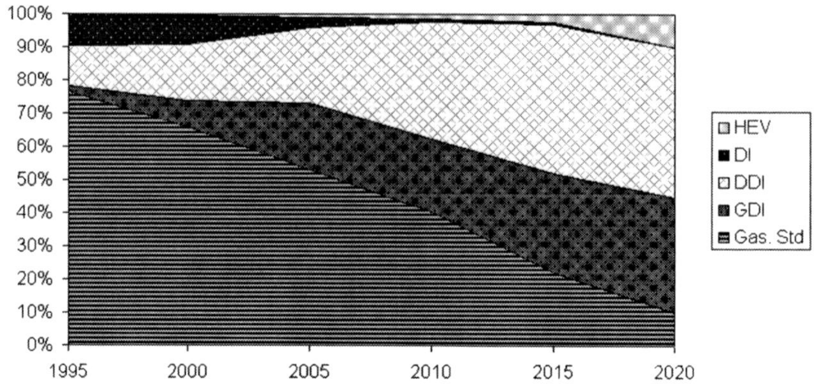

Fig. 4.15 Engine evolution

gasoline systems with direct injection in the combustion chamber and the increasing use of electrical and hybrid cars.

The key drivers of this evolution are the environmental care, the strict pollution regulations, the high fuel price, and the public/private financial incentives.

4.3.2 *Engine Evolution: Major Trends*

4.3.2.1 Gasoline Engine

The spark-ignited engine toward meeting the Euro6 pollution standard will be downsized, with a turbocharger and direct injection fuel loading.

This big change will force an evolution of the solenoid injectors that will be capable to inject fuel in the cylinder in high-pressure condition; the standard 14-V power supply will be increased to 80–150 V trough on-board DC/DC converter.

4.3.2.2 Diesel Engine

As regards air pollution caused by diesel engines, the main problems arise from the presence, in the exhaust gas, of nitric oxides (NO*x*), particulate matter, carbon monoxides (CO), and hydrocarbons (HC).

The direct injection with solenoid injectors in diesel engine is nowadays a common standard but is facing a technical limitation when fast multiple injections are necessary in very low emission conditions because of the slowness of the injectors themselves.

The pressure of common rail system is in the range of 1,800–2,000 bar; a key feature of high-pressure systems is that they can provide good fuel spray formation, which is needed for low particulate and hydrocarbon emissions.

To fulfill latest emission legislation, multiple injections in the same combustion cycle and very short actuation timing are necessary; in high-performance platform up to seven multiple injections with minimum timing of less than 100 μs have been achieved.

A possible solution is the use of faster piezoelectric injectors; also in this case the needed supply voltage will be higher than the standard 14-V battery line and will be in the range of 300 V.

4.3.2.3 Hybrid Cars

This is the most promising engine technology in terms of pollution content and driving comfort and is the combination of an internal combustion engine and an electrical motor controlled by an electronic system able to share the two power contributions, taking care of the driving conditions so as to minimize the fuel consumption.

This technology allows functionalities not possible in a conventional system, such as the start/stop driving, the energy recovery during the braking phase, and the improvement of standing-start performance and torque management (smooth or boost capability).

The standard hybrid drive train system includes a high-voltage battery pack (150–300 V), the main inverter for traction (400–600 V), and the buck and boost DC/DC converters for auxiliary loads (lamps, entertainment, etc.) and for the inverter supply.

In all the new power train systems, an updated HV power silicon technology will be necessary allowing higher integration to minimize the cost and the wiring, thereby increasing the reliability and the electromagnetic compatibility.

4.3.2.4 Engine Auxiliaries

The service system in a car is becoming important in overall car efficiency due to the increased numbers of nodes following the drivability, safety, and convenience request.

There are tens of actuations in a modern car (pumps, fans, compressors, etc.), most of them mechanically linked to the engine with pinions or belts.

Mechanical actuation suffers from low reliability, weight, and poor regulation performance; the actual trend in car systems is the replacement of mechanical and hydraulic components by electrical ones.

The most important advantage of this solution is the possibility to put the auxiliary loads in the complete regulation loop to minimize fuel consumption acting "on-demand" actuation instead of constant delivery.

In Fig. 4.16 there are some examples of new systems that have migrated in the electrical or regulated domain and their direct impact on the fuel consumption efficiency. This big change will involve the usage of electrical motors that, for permanent

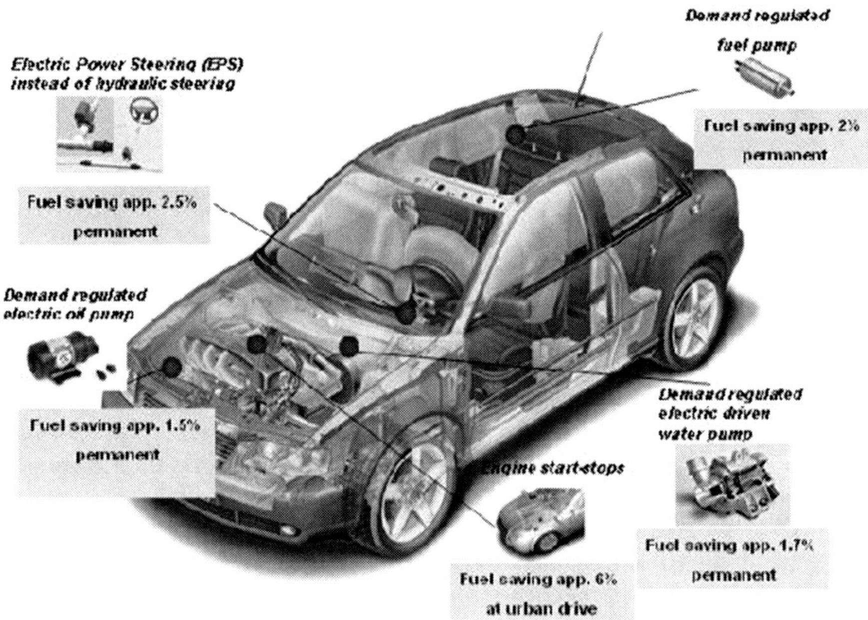

Fig. 4.16 Auxiliaries evolution

use and safety applications, will be brushless DC (BLDC) for achieving better long-term reliability than the standard one.

Future service systems ask for high-power IC technology capable of driving these actuators that are ranging between 10 and 30 A and can easily integrate high-voltage gate driver for high-current motors of 100–200 A.

4.3.3 High-Voltage and High-Power Silicon Technology

Following the big change in the car system IC technologies for automotive are following a different road map compared with other applications where CMOS-based scalability is well established. The new requirement in terms of power and voltage is setting a specific trend; new driver families require a combined technology platform where digital content, energy capability, and high voltage and high current will survive in a single chip.

In Fig. 4.17 the automotive standard voltages are depicted: on the left side there is the battery voltage range for standard single battery and double battery systems, and on the right side there are the common voltage ranges required for the output stages. The output working voltage range (in the past limited to 70 V) is now extended up to 300 V and is more due to the beginning of a new age in the engine concept.

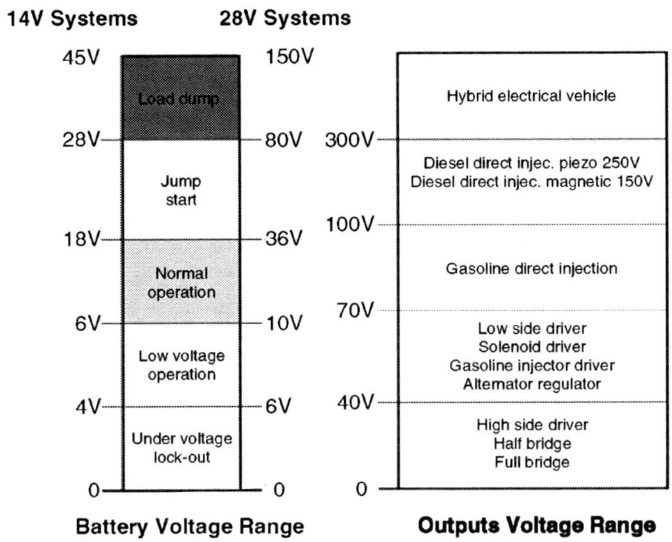

Fig. 4.17 Automotive voltages

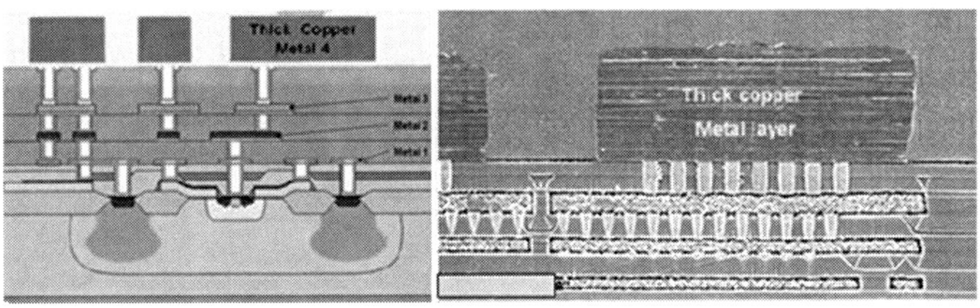

Fig. 4.18 Thick copper metal

The extended voltage range is not the only new requirement for the newcomer automotive smart power devices; the increase of current content with the rise of energy clamp capability requires a dedicated approach to realize robust power elements in a way to guarantee long-term reliability performances.

One of the possible solutions for devices capable of 10–20 A is the introduction of thick copper for the final routing level; the advantage of this process module is the lower resistance of copper compared with aluminum and the possibility to reach relatively high thickness in the range of 5–10 μm, allowing more easy chip floor plan and facilitating bonding on active area.

Figure 4.18 shows a 0.35-μm BCD process metals scheme and the relative microscope picture; the latest layer is a 6-μm copper trace.

The extended voltage range capability of automotive devices is the latest request from the market following the demand of injector predriver for diesel and gasoline direct injection systems.

The BCD roadmap normally includes a family of processes capable of 700 V, used mainly for industrial and consumer purpose. A typical IC realized with this technology is the electronic ballast for lamp application. This type of device is quite simple; a very limited analog CMOS portion and a pulse level shifter with a HV "resurf" lateral DMOS (Fig. 4.19) will form the main building blocks.

For this reason the technology used was not updated during the years for these kinds of devices – being a submicron solution not economically justified.

This solution in any case is not suitable for modern automotive predriver because the complexity needed for this device is much higher; the pulse level shifter suffers from poor common mode transient immunity and low latch-up performances.

Another drawback of pulse level shifter is the lack of real isolation between signal low-voltage circuitry and high-voltage floating epitaxial pocket, since the quasi-isolation is made by the two DMOSs loaded by two resistors in the pulse generator schematic.

The new challenge in smart power BCD process is the ability to integrate a coreless transformer (Fig. 4.20) that allows to overcome the mentioned weakness of pulse level shifter.

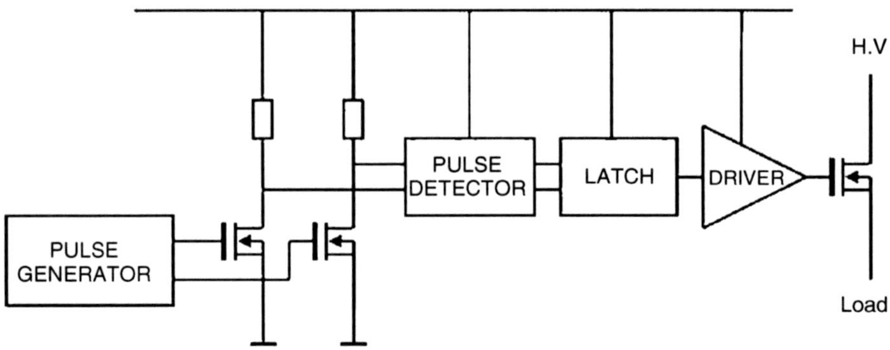

Fig. 4.19 BCD-offline level shifter

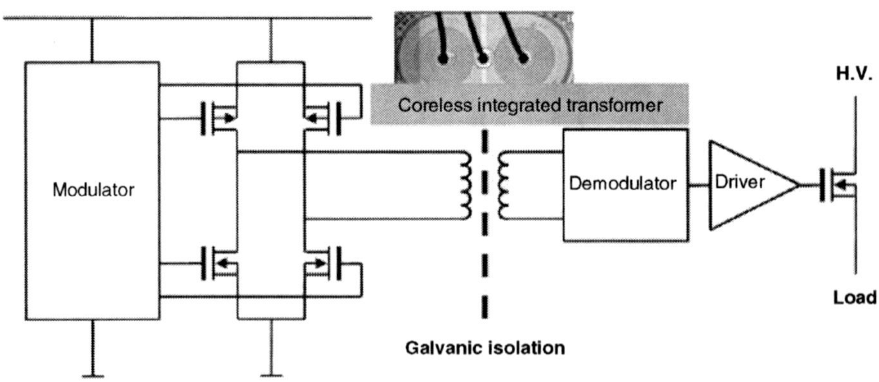

Fig. 4.20 Transformer level shifter

In fact in the transformer topology, the transient immunity and the latch-up performance are much higher, allowing a very high data rate and a real galvanic isolation.

This kind of structure can be imported in a submicron BCD platform using standard process steps and material and can be an interesting module for the future automotive process.

4.4 Research Subjects

The following subjects related to HV and power deserve more attention from R&D community:

- Reduction of transistor on-state resistance to reduce power losses and breakdown voltage.
- Improve switching performance for higher switching efficiency. Improve electrical robustness (e.g., reverse voltage, ESD, etc.) and operating temperature range.
- Optimise R_{on} versus voltage.
- Density of HV components.
- Surface or vertical device.
- Type of isolation (component pitch).
- Current density in metals.
- Parasitic effect immunity.
- Develop high-voltage/high-current interconnect architecture with thick Cu metallization.
- Develop new isolation technologies, such as selective SOI, to allow more flexible integration of HV devices with CMOS, lower leakage, and higher operating frequency.
- Develop energy scavenging systems for autonomous systems.

4.5 Energy and Efficiency [3]

4.5.1 Energy Generation

The specificity of the energy at the microscopic scale lies in the spread aspect of the energy needs. This spreading is linked to the tremendous increase of autonomous and portable systems (telephones, computers, robots, etc.), and soon to the generalization in the Ambient Intelligence environment of autonomous microsystems, microsensors, and microactuators. The development of batteries and fuel cells is a very important research area not only for chemistry but also for micro/nanoelectronics industry because the use of planar and collective processes (flexible supports, or compatible with integration on silicon) can provide cost-effective solutions. Thermal micromachines and microgenerators are also possible solutions even if these remain rather complex to be integrated. Among the low-power sources, the photovoltaic generation still raises many practical problems (packaging, manufacturing) or problems connected to the low conversion yield and to the very weak level of energy storage. The research on thermoelectric generation should be mainly oriented on materials and their integration in actual microsystems. Finally, the recovery of mechanical energy (vibrations, flow, or compression of fluids), with electrostatic, magnetic, piezoelectric, or magnetostrictive devices is a research subject with increasing attention. But since this domain is extremely large in terms of possibility of power sources and recovery principles, the best orientations remain unclear.

4.5.2 Energy Efficiency

The rising demand for energy in all forms and the decreased traditional energy resources have made it evident that energy must be used more efficiently. With the dramatic increase of leakage currents, design for high-power efficiency becomes an even more challenging task, due to the introduction of Cu/low-*k* CMOS technologies. In idle and shut-off mode, remedies such as MTCMOS, back biasing, etc. can be leveraged to reduce the impact of leakage power dissipation. However, most of these strategies are not applicable in active mode and in challenging future applications, e.g., in 10 GBASE-T Ethernet-over-Copper echo cancellation, leakage contribution to total active power becomes dominating in worst case leakage process and application corner. In such designs the low-power design space features many dimensions (e.g., multi/adaptive VDD, multi-VT including random modulation on extra stack level, device sizing/stacking including long Le, clock frequency optimization by parallelism/pipelining on architecture level, etc.). Navigating in this design space to minimize total active power dissipation is challenging due to strong interactions with timing, yield, and especially with noise margin. This consequently affects reliability. It has already been shown here in a 90-nm design featuring nearly 100 millions of transistors and operating at 800-MHz sample rate that careful optimization can yield huge savings in total power dissipation. Nevertheless, there is still a need for comprehensive optimization strategy taking all these interactions into account. The associated research subjects are as follows:

- Development of a framework on challenging applications in 65- and 45-nm CMOS technologies as well as in applying predictive technology models for future CMOS generations
- Dynamic control of digital supplies
- PMU adapting to analog dynamic range
- Low power or batteryless devices
- Ultralow-voltage/ultralow-power digital circuits
- New transceiver techniques to deal with lower voltage of new CMOS technologies while maintaining required dynamic ranges

4.5.3 *Power for Mobile Communication*

Future mobile communication systems require low-power implementation of complex functionality with highly efficient and highly linear CMOS solutions well beyond state of the art. The overall trust is dedicated to significantly improve the power/performance as well as the cost efficiency balance compared with current architectures. Total power efficiency, standby management, and deep-sleep current consumption effect battery size, bill-of-material cost, and product acceptance especially once portable electronics equipment, audio players, mobile phones, or digital cameras are envisaged. Mobile devices beyond third-generation (3G) products are all multistandard. They are expected to become more power hungry and they will require highly integrated power management solutions within separate power domains. One needs to meet the overall design requirements within the thermal envelopes for maximum battery lifetime and smallest chip and PCB area.

Major innovations to create more efficient overall platforms are expected to come from separation of power domains, unorthodox voltage regulators, use of sleep transistors, or running multiple power supply rails. Innovations for power management units are expected to be the integration of power circuits in deep-submicron technologies and the integration of external components on chip. Application and architectural savings are quantified using the product of the energy and the performance denoted (energy × performance). Work will concentrate on the feasibility, design, and development of power-efficient information technology devices and systems using approaches that extend beyond traditional CMOS scaling to achieve low-cost, reliable and fast systems. There are at least two aspects of power efficiency to enhance: back-off efficiency and peak efficiency. The associated research subjects are as follows:

- (Non)scaling effects calling for variation-tolerant design or soft-error mitigation
- Embedding power management and power output functions while obeying analog dynamic range restrictions
- Decreasing analog chip area and power consumption while concurrently removing OPAmp structures
- Inhibiting Vdd reduction with scaling, leakage (Vt), interconnect, and costs
- Implementing adaptive body biasing - reverse versus forward, supply gating, and active management methods.

References

1. http://en.wikipedia.org/wiki/Nixie_tube
2. http://en.wikipedia.org/wiki/Electroluminescence
3. ENIAC Strategic Research Agenda, 2006 update, Monte Carlo, November 30, 2006

Chapter 5
Sensors and Actuators on CMOS Platforms

Mark Lantz, Christoph Hagleitner, Michel Despont, Peter Vettiger, Mario Cortese, and Benedetto Vigna

Abstract The monolithic or hybrid integration of sensors and actuators (S&A) onto CMOS platforms is of great importance to reduce the device size and cost as well as to facilitate new functionalities and better performance. Hence, S&As on CMOS platforms are important and convincing examples of "more than Moore." This chapter focuses on a broad range of aspects of S&As on CMOS platforms. First, the basic concept is introduced, followed by a detailed description of the fundamentals of current S&A concepts. The current market situation is then described, demonstrating the great opportunities for S&As. A substantial part of the chapter is devoted to the various techniques and fabrication processes for monolithic and hybrid integration of S&As on CMOS. Successful examples of very large-scale integration (VLSI) of S&As on CMOS platforms are also described. Current research into the use nanostructures for new or better S&A functionalities or improved performance is described in an additional subsection. Finally, the equally important back-end-of-the-line aspects in large-volume production, packaging, and testing are presented. The chapter closes with the authors' outlook and view on the challenges and opportunities in research, development, production, and marketing of new and better S&As, VLSI-type S&As on single-chip CMOS platforms, and S&A networks.

Keywords Sensors • Actuators • CMOS • MEMS • NEMS • Nanotechnology

5.1 Introduction

Today sensors and actuators (S&As) are used almost everywhere to sense and monitor parameters and to control actions of importance and interest in our daily environment accordingly. They play an essential role as an interface between the

M. Lantz (✉), C. Hagleitner, M. Despont and P. Vettiger
IBM Zurich Research Laboratory, 8803, Rüschlikon, Switzerland
e-mail: mla@zurich.ibm.com

M. Cortese and B. Vigna
STMicroelectronics, MEMS & Healthcare Product Division, 20010 Cornaredo (Milano), Italy

G.Q. Zhang and A.J. van Roosmalen (eds.), *More than Moore: Creating High Value Micro/Nanoelectronics Systems*,
DOI 10.1007/978-0-387-75593-9_5,

electronic systems and the user or environment. Because a large number and variety of devices are required for performing different functionalities, the S&A market is huge. However, the majority of today's devices are stand-alone, and are controlled and supported by vendor electronics. Electronics and S&A devices are commonly assembled on a printed circuit board. In addition, the many different types of sensor and actuator functionalities also require different device structures and fabrication processes. As a result, there is not a single fabrication technique capable of producing all S&A structures required for the large variety of applications. To illustrate the broad spectrum of S&As and the large number required, for example, in today's cars, Fig. 5.1 shows the S&A devices installed in a high-end car. More than 30 devices monitor and control engine functions, safety, navigation, and passenger comfort. Typically, such devices are introduced into main stream products within a few years. A similarly great demand exists for use in traffic control, building management, environmental control, ID sensors, etc.

Because every S&A requires control electronics to support its functionalities, the interfacing, interconnection, and integration of an S&A device with its control complementary metal oxide semiconductor (CMOS) electronics (mostly application-specific integrated circuits or ASICs for short) are of great interest and importance. Common to almost every application is a CMOS-based electronics platform. The S&A devices, which perform the actual monitoring and control function, need to be interconnected to or even monolithically integrated onto the CMOS platform. Driven by the IT industry, CMOS miniaturization and scaling are progressing very rapidly, reaching integration levels of about 10^{11} transistors per chip, with minimum

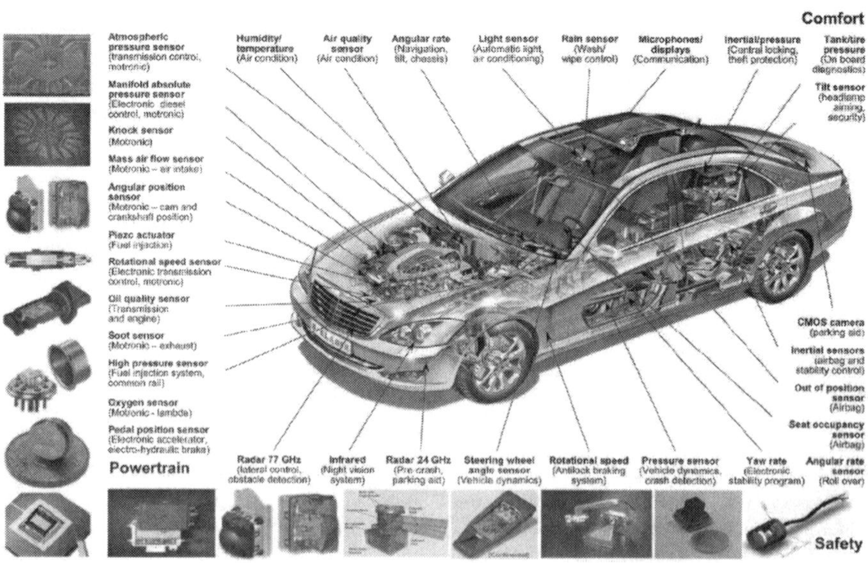

Fig. 5.1 Sensors and actuators as installed in a typical car today (Source: Mercedes Benz, Daimler AG, Germany)

dimensions well below 100 nm. The great advances in micro and nanofabrication techniques for CMOS also stimulated considerable research and development efforts, which began more than 30 years ago, into silicon-based S&A devices and fabrication techniques such as silicon micromaching. Today the majority of S&As are made of silicon. By applying modern micromaching techniques, full-wafer, high-throughput, and high-yield fabrication have been successfully established for large-volume S&A production. This technology is more generally referred to as microelectromechanical systems or MEMS technology.

Although silicon-based S&As have reached a high degree of maturity and are used successfully in large volumes, the challenge of high-yield and low-cost hybrid or monolithic integration of the S&A devices onto a CMOS platform is still an emerging area of great interest and importance. Considerable R&D efforts are focused worldwide on this challenge. Moreover, both of these integration approaches are suited and will be applied and explored for very large scale integration (VLSI) of S&A devices onto a CMOS platform. The famous Texas Instruments (TI) micromirror chip is an excellent example of a successful VLSI S&A product integrated on a CMOS platform. In this device, millions of aluminum micromirrors are electrostatically actuated by the circuits of the CMOS platform onto which the mirrors have been integrated by post-COMS fabrication processes. In addition, there is growing research interest into new S&A principles or devices based on nanometer-scale structures and materials such as carbon nanotubes (CNT) or nanowires to achieve better performance.

This chapter will provide an overview of the fundamentals of S&As, the state of the art in products and applications, hybrid and monolithic integration of S&As onto CMOS platforms, the use of nanostructures for advanced performance as well as a near- and long-term outlook.

5.2 Sensing Fundamentals

An electronic sensor, often referred to as simply a sensor, is a device that detects a change in a physical parameter and converts this change into an electronic signal, i.e., a current, a voltage, or a charge. More generally, electronic sensors can be defined as a form of transducer that convert various forms of energy (e.g., thermal, mechanical, and magnetic) into electrical energy. This electrical signal is typically processed in some way, e.g., amplified, and/or filtered, and transmitted, stored, processed further, or converted into a human-readable form, e.g., used to drive an optical display.

Today, a huge variety of sensors are available that exploit many different physical phenomena for converting the physical parameter of interest into an electrical signal. A convenient way to classify the many types of sensors is in terms of the type of the energy being converted. Here, we adopt the classification scheme of Lion [1] and distinguish between six basic energy types: electrical, mechanical, thermal, optical/electro-magnetic radiation, magnetic, and chemical. Examples of

Table 5.1 Electronic sensor principles and example sensor types

Energy	Parameter	Physical effect	Example sensors
Mechanical	Position, velocity, acceleration, force, stress, pressure, strain, mass, torque, stiffness	Piezo-electric, piezo-resistive, capacitive, inductive, (optical)	Accelerometer, strain gauge, load cell, switch, motion sensor, touch sensor, gas or liquid flow sensor, mass flow sensor
Thermal	Temperature, power	Seebeck effect, thermo-resistance, pyroelectricity, thermal noise	Thermocouple, thermal resistance
Optical (EM radiation)	Light intensity, temperature, wave length, refractive index	Photo-electric, photo-voltaic, photo-electric	Photo diode, photo transistor, photo multiplier tube, position sensitive detector, CCD,
Magnetic	Magnetic field, magnetization	Magneto-resistance, inductance, Hall effect	Hard disk head, tape head
Chemical	pH, humidity, chemical/ bio agents, gas	Potentiometry, conductometry, amperometry	pH sensor, gas sensor, humidity sensor

Adopted in part from [2] and [3]

some of the physical parameters from each category are listed in Table 5.1, which has been adapted in part from [2] and chapter VII of [3]. In addition, it lists typical examples of sensors and the physical effect utilized for the conversion of energy to the electrical domain.

A complete list of sensor types and the physical effects they utilize is beyond the scope of this work. For a more thorough discussion, the reader is referred to [4].

Another way to classify electronic sensors is in relation to the conversion process itself, from which two categories arise: direct or active sensors and indirect or modulating sensors [2]. A direct sensor converts one form of energy directly into an electrical signal. For example, a photo detector converts photons (optical energy) into a current. As a consequence, direct sensors do not require a source of electrical energy to produce a signal. In contrast, in an indirect sensor, the physical parameter of interest modulates an electrical signal. A typical example is that of a resistive sensor, such as piezo-resistor, magneto-resistor (giant magneto resistance (GMR) sensor), or thermo-resistor. Such resistive sensors are typically biased with a constant voltage, so that the change in resistance resulting from a change in the parameter of interest (e.g., strain, magnetic field, temperature) results in a modulation of the current flowing through the sensor.

Many sensors do not convert energy directly to the electrical domain, but first convert the energy into another form of energy that is more convenient to convert to the electrical domain. For example, in an optical displacement sensor the displacement of an object (a form of mechanical energy) is converted into an optical intensity, which is then converted to the electrical domain by a photo detector.

A recent trend in sensor technology is miniaturization through the use of micro and nanofabrication technologies. In itself, miniaturization typically results in improvements in the response time and the sensitivity of a device and in a reduction in the power consumption. Additional benefits of using microfabricated sensors are size and cost reductions and the possibility to directly integrate electronics with the devices, enabling even further reductions in size and cost, thereby making it feasible to add redundant sensors to a system to improve its reliability. Another important benefit of microfabricated sensors is the possibility to integrate several different sensors on the same platform so that physical quantities can be measured simultaneously or the accuracy of the measurement of a specific physical quantity can be increased. For example by simultaneously measuring the property of primary interest and the local temperature, any temperature dependence of the physical effect used for the transduction of the first quantity can be compensated for.

Perhaps one of the most powerful concepts facilitated by microfabricated sensor technology and more generally by CMOS-based sensors technology is that of massive parallelization. In the past, data describing the spatial variation of a parameter of interest were typically acquired by taking a single sensor and raster scanning it over the area of interest. The main drawbacks of this technique are the complexity involved in positioning the sensor and the time required to acquire the data. An alternative approach is to build an array of identical sensors and use them to simultaneously acquire an image of the spatial variation of the parameter of interest. This concept has been exploited very effectively in optical imaging devices, such as the image sensor arrays used in digital cameras. For a detailed discussion of imaging, the reader is referred to Chap. 8. Another application of parallelism is to build an array of similar sensors in which each is functionalized to respond to a different stimulus. This concept has tremendous potential for biological sensing applications. The concept of massive parallelism of microfabricated S&As will be discussed in more detail in the later sections of this chapter.

Examples of microfabricated sensors include accelerometers, pressure sensors, magnetic field sensors, temperature sensors, radiation sensors, chemical sensors, and biosensors [4, 5]. A description of all of these technologies is clearly beyond the scope of this work. However, many of these sensors employ displacement sensing either to measure a mechanical signal of interest directly, such as pressure or acceleration, or to measure another parameter of interest that was first converted into a displacement signal such as the bending of a cantilever beam in response to the absorption of a chemical or biological compound on its surface. In fact, displacement sensing is one of the most highly developed areas in the field of microfabricated sensors because it can be applied so widely and is relatively straightforward to implement. Hence, in the remainder of this section, we will focus on common techniques for displacement measurement in microfabricated sensors: piezoelectric, piezoresistive, capacitive, and thermal. A brief description of each of these techniques is given below, together with a few application examples.

5.2.1 *Piezoelectric Sensing*

Piezoelectric materials generate an electrical moment in response to a mechanical stress. This is known as the direct piezoelectric effect. Moreover, the piezoelectric effect is reversible in that the application of an electric field to the material induces a strain, which can be used for actuation. This is known as the indirect piezoelectric effect. In both sensing and actuation, piezoelectric materials are used for building a capacitor-like structure in which a layer of piezoelectric material is sandwiched between two metal electrodes, as illustrated in Fig. 5.2. Application of a strain to the piezoelectric material results in the generation of charge at the electrodes. The magnitude of this charge is directly proportional to the applied strain. Piezoelectric sensors cannot generally be used for static sensing as a fixed strain will result in an ideally constant charge on the electrodes, which because of inevitable imperfections in the system will leak away, resulting in a decreasing signal with time [4].

Piezoelectric sensing is typically used to detect the strain that results from the displacement of a spring, i.e., a displacement is converted into mechanical strain energy, and the strain is then converted into a charge/voltage. To maximize the signal produced in this way, the piezoelectric material should be positioned at the location of maximum strain, which for a cantilever-based sensor is near the base of the cantilever.

Piezoelectric materials can be divided into three general classes: single crystals, polycrystalline/ceramic materials, and polymetric materials. Quartz is an example of a single-crystal piezoelectric material and is one of the most commonly used piezoelectrics, with excellent mechanical properties. However, it is not typically used in microfabricated or CMOS-based sensors because of integration difficulties. Polymeteric materials, such as polyvinylidene fluoride (PVDF) or polyimide, are potentially compatible with microfabrication techniques, but exhibit a rather weak piezoelectric effect [6]. Ceramic materials exhibit the strongest piezoelectric effect and are more straightforward to integrate into a microfabrication process. Lead zirconate titanate ($Pb(ZrTi)O_3$, also known as PZT) is the most commonly used ceramic piezoelectric material. It exhibits a strong piezoelectric effect and has been used in numerous microfabricated-sensor applications. Recently, there have also

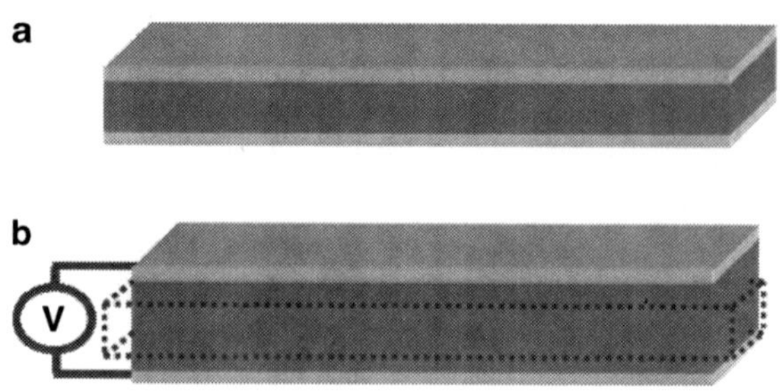

Fig. 5.2 Capacitor structure for piezoelectric sensor/actuator. The piezoelectric material is shown in *blue* and the metal electrodes in *yellow*

been efforts to use lead-free piezoelectric materials such as AlN to comply with new or anticipated regulations to reduce the use of lead. Piezoelectric sensing is less commonly used than piezoresistive or capacitive sensing, but its applications are similar and include accelerometers [7], atomic force microscope (AFM) cantilever displacement sensing [8], and infrared sensors [9].

5.2.2 *Piezoresistive Sensing*

Piezoresistivity is an effect in which the electrical resistance of a material changes in response to an applied stress. In metals, this change in resistance results predominantly from geometric effects. In semiconductors, the resistivity of the material also changes in response to stress. To illustrate these effects, we consider a rectangular conductor of length l, width w, thickness t, and resistivity ρ_0. The electrical resistance along the length is given by

$$R_0 = \rho_0 l / wt. \tag{5.1}$$

Stretching this resistor along its length results in a change in resistance given by [10]

$$\frac{\Delta R}{R_0} = \frac{\Delta l}{l} - \frac{\Delta w}{w} - \frac{\Delta t}{t} + \frac{\Delta \rho}{\rho_0}. \tag{5.2}$$

The first three terms in (5.2) result from geometric effects, i.e., the increased length and decreased area of the conductor. The fourth term accounts for any strain-induced change in the electrical properties of the material. Poison's ratio, v, can be used to relate the change in width and thickness resulting from a change in length:

$$\nu = -\frac{\Delta w / w}{\Delta l / l} = -\frac{\Delta t / t}{\Delta l / l}, \tag{5.3}$$

and hence

$$\frac{\Delta R}{R_0} = \frac{\Delta l}{l}(1 + 2\nu) + \frac{\Delta \rho}{\rho_0} \tag{5.4}$$

The gauge factor, G, of a piezoresistor is a measure of the sensitivity of the device to strain and is defined as [10]:

$$G = \frac{\Delta R / R_0}{\varepsilon}, \tag{5.5}$$

where $\varepsilon = \Delta l / l$ is the applied strain. In metals, the resistivity is approximately constant with strain, i.e., $\Delta \rho \approx 0$, leading to $G = (1 + 2v)$ for a rectangular metal resistor. Taking $v \approx 0.3$ as a typical value for metals, we obtain $G \approx 1.6$. In semiconductors, the resistivity of the material changes strongly in response to an applied stress owing to changes in the mobility of carriers in the material [10, 11]. In semiconductors, such as silicon and germanium, the change $\Delta R / R_0$ due to strain-induced resistivity changes is approximately 50 times larger than the geometrical effects, and hence we can neglect the first two terms of (5.4). Accordingly, for a semiconductor, (5.4) and (5.5) can be reduced to

$$\frac{\Delta R}{R_0} = \frac{\Delta \rho}{\rho_0}, \tag{5.6}$$

and

$$G = \frac{1}{\varepsilon}\frac{\Delta \rho}{\rho_0}. \tag{5.7}$$

In microfabricated sensors, the materials most commonly used for piezoresistive sensing are doped single-crystal and polycrystalline silicon. Metal-based piezoresistive strain gauges have also been demonstrated. The applications of piezoresistive sensors are similar to those of piezoelectric sensors, i.e., for sensing the strain resulting from a displacement, and, like piezoelectric sensors, piezoresistors should be fabricated in the region of maximum strain. Compared with piezoelectric sensors, piezoresistive sensors have the advantage of being suitable for static measurements and the disadvantage of higher power consumption. In general, piezoresistive sensing is a rather simple and robust technique to implement in microfabricated sensors, and application examples include accelerometers [12], force [13], fingerprint [14, 15], pressure [16, 17], and flow sensors [18, 19]

5.2.3 Capacitive Sensing

Capacitive displacement sensing utilizes the change in capacitance that occurs when two conducting electrodes, mounted on insulating support structures, are displaced relative to each other, see Fig. 5.3. This change in capacitance is typically sensed using a detection circuit that converts the capacitance change into voltage, frequency shift, or pulse width [20].

In microfabricated capacitive sensors, three basic geometries or electrode configurations are typically used as illustrated in Figs. 5.4 and 5.5. Figure 5.4a illustrates the simplest geometry, namely, that of completely overlapping parallel plates. In this case, if the electrode area, A, is large compared with the gap, d, the capacitance is well approximated by the following equation [20]:

$$C = \frac{\varepsilon_0 \varepsilon_r A}{d}. \tag{5.8}$$

Here, ε_0 and ε_r are the permittivity of free space and the relative permittivity, respectively. This geometry is typically used for sensing displacements in the direction normal to the surface of the electrodes, where the capacitance is inversely proportional

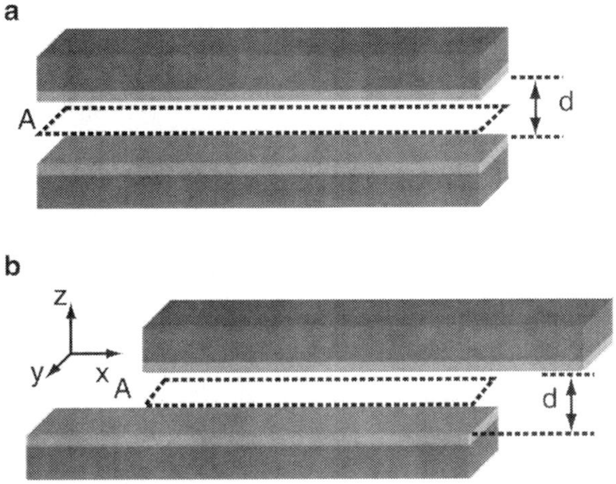

Fig. 5.3 Illustration of a parallel plate capacitor displacement sensor

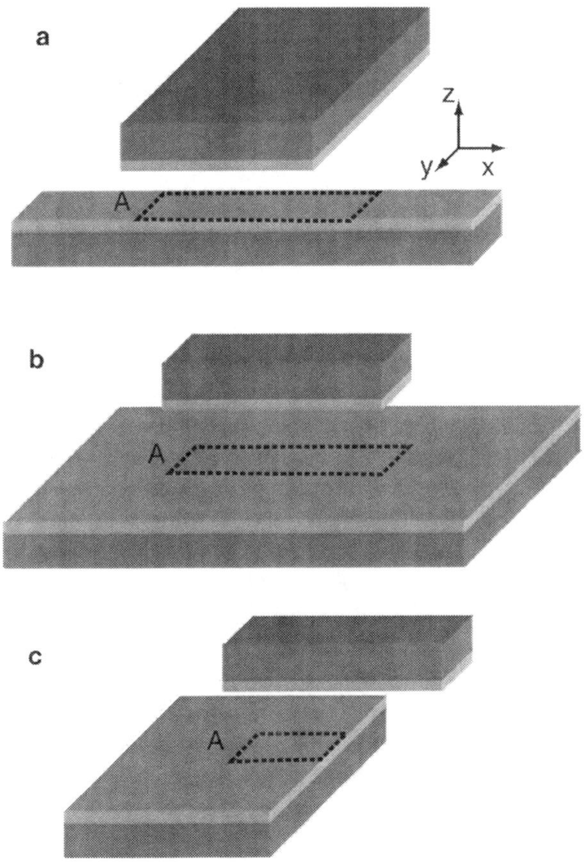

Fig. 5.4 Schemes of overlapping electrodes to reduce sensitivity to motion in the x and/or y directions. The *dotted outline* indicates the area of overlap of the electrodes

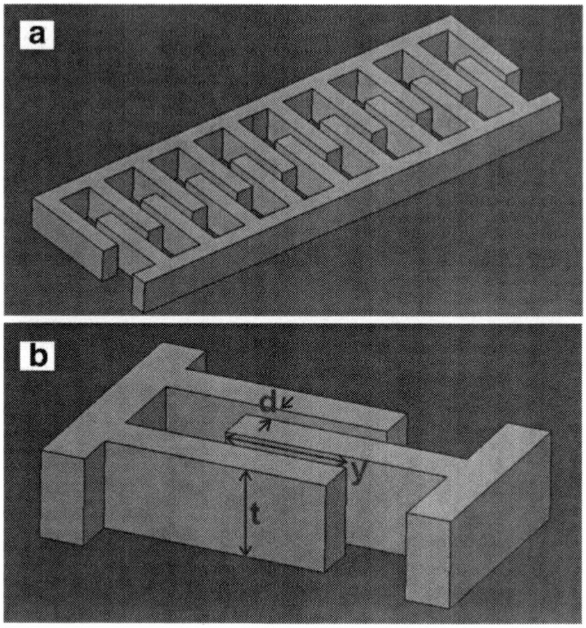

Fig. 5.5 Interdigitated (comb) electrode capacitive sensor

to the displacement, i.e., nonlinear. For small gaps, the capacitance change as a function of the displacement can be very large, making this technique suitable for high-resolution sensing of small displacements. One drawback of this configuration is its sensitivity to motion in the plane of the electrodes. This can be mitigated either by making the device very stiff against motion in these directions or by over- or underlapping the electrodes (Fig. 5.4).

An alternative geometry for sensing displacements in the plane of the electrodes is shown in Fig. 5.3b. Equation (5.8) can also be used to describe the capacitance of this configuration by taking A as the area of electrode overlap as indicated by the dotted line in the figure. With this configuration, the capacitance changes approximately linearly with displacement, making it suitable for applications requiring a larger dynamic range. One drawback of this approach is cross-coupling due to either the strong sensitivity to displacement in the out-of-plane direction or tilting of the electrodes. The geometry illustrated in Fig. 5.4b is sensitive to displacements in both the x- and the y-direction. However, by using unequal electrode dimensions, as illustrated in Fig. 5.4c, this sensitivity can be confined to one axis.

The third electrode geometry commonly used is illustrated in Fig. 5.5. Here the effective area of the capacitor is increased by using interdigitated electrodes. The capacitance of this configuration is given by [21]

$$C = \frac{2n\varepsilon_0\varepsilon_r ty}{d} \tag{5.9}$$

where n is the number of electrode pairs, t is the thickness of the electrodes, d is the gap between electrodes, and y is the overlap between the fingers (Fig. 5.5b). The capacitance of this structure changes linearly with displacement in the y-direction, i.e., as the fingers move in and out.

One drawback of capacitive displacement sensors is that the voltage required to sense the capacitance results in an attractive electrostatic force between the electrodes (see Sect. 5.3). For small electrode spacing, which leads to high sensitivity, the forces can be considerable. A second issue is that of stray capacitance, which can reduce the sensitivity of the device. Stray capacitance effects can, however, be compensated for in the design of the detection circuit or minimized by integrating the detection circuit as close as possible to the sensor. A third issue arises from the change in the dielectric constant of air because of changes in humidity, temperature, and pressure. These effects are relatively small and can also be compensated using a reference capacitor with stationary electrodes [20]. Compared with piezoresistive sensors, capacitive sensors are in general more sensitive, require less power, exhibit less drift, and are less temperature sensitive [22], but entail a more complex design of the detection circuit. It is also interesting to note that capacitive sensors measure displacement directly, whereas piezoresistive and piezoelectric sensors measure displacement via strain, and as such they can be more sensitive to stress due to packaging [22]. Example applications of capacitive sensing include pressure [22], acceleration [22], movement [22], humidity [23, 24], and inclination sensing [25].

5.2.4 Thermal Sensing

Heat conduction through the ambient air between two objects separated by macroscopic distances is a relatively slow process. However, as the distance between the two objects is reduced to the scale of a few micrometers, the speed and efficiency of heat transport through the air increases to a level where it is feasible to use the change in heat flow as a function of distance as a transduction mechanism for displacement sensing. This idea was exploited by Dauderstädt et al., who used a resistive heater and a thermopile to sense displacement in a microfabricated accelerometer [26]. More recently, it has been used as a means of sensing the displacement of an AFM cantilever relative to a surface [27] and to measure the displacement in a two-axis scanning system [28].

There are two basic designs or geometries used for thermal sensing that are, to some extent, analogous to the capacitive sensor designs shown in Fig. 5.4b, c. The first is illustrated in Fig. 5.6. The basic sensing element consists of a thin strip of highly doped silicon, which has a region in the center that has a lower doping level and acts as a resistor, see Fig. 5.6a.

Typical dimensions for such a heater are 4 μm in width, 5 μm in length, and 0.5 μm in thickness, with a similar width and thickness for the highly doped strip but with a total length on the order of several tens of a micrometer. To sense a displacement,

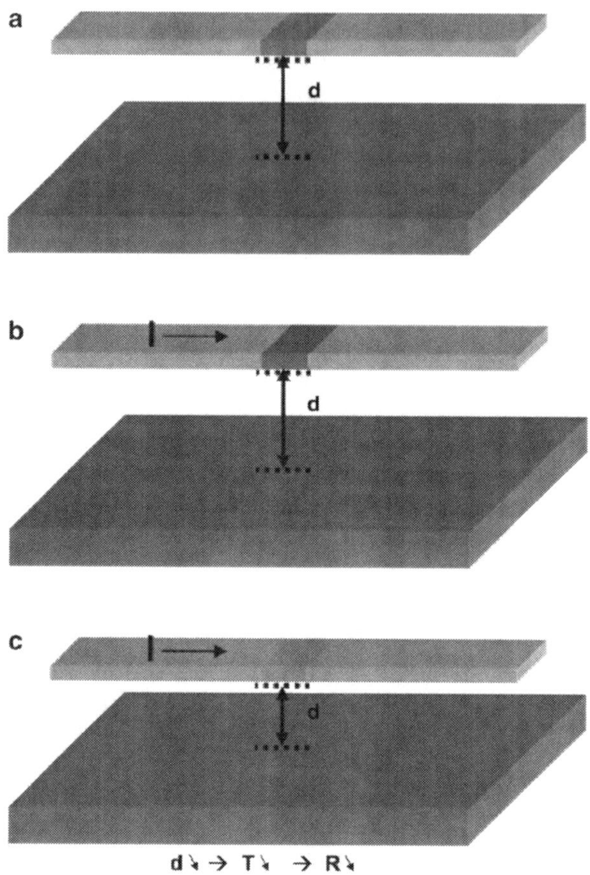

Fig. 5.6 (**a**) Illustration of a thermal sensing scheme. The sensor consists of a thin strip of highly doped silicon (shown in *blue-gray*) with a region of lower doping near the center (shown in *gray*). The sensing strip is positioned a distance *d*, above the object of interest (shown in *gray*). (**b**) Application of a current through the strip results in an increase in temperature of the resistor. (**c**) As the spacing between the resistor and the object of interest, *d*, is reduced, the temperature and resistance of the heater also decrease

the strip is positioned opposite the object of interest, see Fig. 5.6a. When a current is passed through the strip, an increase in the temperature of the resistor occurs, which in turn causes a resistance increase because of increased scattering. If the strip is long and thin and the spacing *d* sufficiently small, a large fraction of the heat generated in the resistor will be conducted through the ambient air and into the object of interest, which acts as a heat sink. The efficiency of this cooling mechanism strongly depends on the distance between resistor and heat sink. Decreasing this distance results in an increase in cooling efficiency and, therefore, in a decrease of the temperature and resistance of the sensor, as illustrated in Fig. 5.6c. For this geometry, the change in resistance vs. distance is highly nonlinear. For high-resolution sensing applications, such as in AFM, the typical spacing *d* is on the order 0.5–1.0 μm, the heating power is on the order of a few milliwatts, and the sensitivity is on the order of $\Delta R/R \sim 10^{-4}$ [27]. The bandwidth is a function of the heat capacity of the sensor and the thermal resistance. For the typical dimensions given earlier, the bandwidth is on the order of tens of kilohertz. The resolution of such a sensor is a rather complicated function of its geometry, doping levels, heating power, sensor noise, and the measurement bandwidth. For a detailed analysis of the scaling of the sensitivity and bandwidth as a function of sensor geometry, the reader is referred to [29].

The second geometry used for thermal sensing is illustrated in Fig. 5.7. Here the length of the resistor is increased to a value that is comparable to the desired dynamic range. The sensor is then positioned above an edge on the object of interest at a distance d that typically is on the order of a few micrometers. The application of current through the device results in an increase in both temperature and resistance of the heater. The displacement of the object of interest in the negative x-direction results in an increase in cooling of the resistor, and hence a decrease in the temperature and the electrical resistance. The larger dimensions of the resistor used in this scheme typically result in an increased power consumption and a reduced bandwidth. The linearity possible with this approach can be improved by a careful design of the structure to achieve uniform heating and by using a pair of sensors operated in a differential configuration.

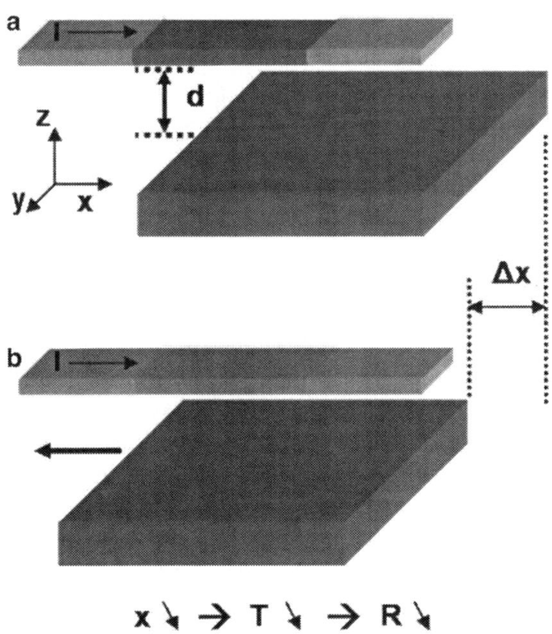

Fig. 5.7 (**a**) Thermal sensing geometry for large linear range. (**b**) Displacement of the object of interest in the negative x-direction results in an increase in cooling and hence a decrease in sensor temperature and resistance

In general, thermal sensing is sensitive to ambient temperature changes that cause a DC drift in the signal. However, such effects are less important for AC sensing applications or can be compensated for using a reference sensor.

5.3 Actuation Fundamentals

Actuators are another type of transducer that converts various forms of energy into mechanical energy. Electrical actuators are probably the most common subclass, and are often referred to simply as actuators. These devices convert electrical energy into mechanical energy, such as a force or a displacement. Examples of macroscopic actuators include rotary electric motors, solenoids, loudspeakers, relays, and piezoelectric actuators. Today, electrical actuators have become extremely common, with applications ranging from cars, watches, consumer electronics, and home appliances to medical diagnostic equipment and robotics.

The advent of microfabrication technology triggered considerable research effort into the building of micro and nanoscale actuators. As electrical energy is the most readily available energy source in microfabricated CMOS and/or silicon devices, actuators built using these technologies tend to be driven almost exclusively by electrical energy. Applications for such microfabricated actuators range from analogues of macroscopic devices, such as switches and relays or valves and pumps, to more novel devices such as inkjet printer heads and actuated arrays of mirrors for optical displays.

One of the challenges of miniaturizing actuators arises from friction and stiction that can increase dramatically on the micro and nanoscale as the real contact area increases or because of meniscus forces that arise from ambient humidity. To avoid such effects, the majority of microfabricated actuators use spring or flexure guiding rather than sliding contacts to direct the motion of the actuator. When contact sliding or repeated contact between surfaces is necessary for an application, a careful design of the surfaces is required to reduce friction and adhesion, but even then it is difficult to achieve a long lifetime for such a device. Techniques to reduce friction and adhesion include a careful choice of the contacting materials, the use of self-assembled monolayers for lubrication, the patterning of the surfaces to reduce the real area of contact, and operation in low-humidity or vacuum conditions.

An in-depth discussion of flexure design or of the techniques to mitigate friction would exceed the scope of this work, and we will focus instead on the basic actuation mechanisms used in microfabricated actuators. In general, there are four well-established techniques for generating forces: electrostatic, piezoelectric, thermal, and electromagnetic. Here we provide a brief description of each of these techniques along with a few application examples. Emerging actuation techniques, such as the use of electroactive polymers or shape-memory materials, are less developed and will not be discussed. Direct manipulation of liquids using capillary forces or electro-osmotic pumping is another interesting emerging field, which, however, is also beyond the scope of this chapter.

5.3.1 Electrostatic Actuators

Electrostatic actuation is probably the simplest form of actuation to implement in a microfabricated device and therefore is also the one most commonly used. There are three basic actuator designs, which are similar in geometry to the three capacitive sensing schemes described in Sect. 5.2: the parallel plate actuator, the linear surface drive, and the comb drive.

5.3.1.1 Parallel Plate Actuator

A parallel plate actuator is essentially an air-filled capacitor in which one of the electrodes is fixed and the other is mounted on a spring suspension and therefore is free to move, as illustrated in Fig. 5.8. The application of a voltage between the electrodes results in an attractive force between the plates. If the area of the electrodes is large compared with the spacing, then the force is well approximated by [30]

$$F = -\frac{\mathrm{d}E}{\mathrm{d}z} = -V^2 \frac{\partial C(z)}{\partial z} = -V^2 \frac{\varepsilon_0 \varepsilon_\mathrm{r} A}{2d^2}, \tag{5.10}$$

where E is the energy, V is the voltage applied across the electrodes, A is the area of the electrodes, and d is the spacing between the electrodes. The force is a function of both the applied voltage and the spacing between the plates such that as the voltage is increased, the force increases and the gap between the two plates decreases, resulting in a further increase in force. When the rate of change of the force vs. spacing equals the stiffness of the spring, the moveable electrode jumps into contact with the fixed electrode. For a linear spring and the idealized parallel plate geometry as shown in Fig. 5.8, this instability occurs at a distance of 2/3 of the zero applied force spacing. The electrodes in a parallel plate actuator are usually covered with a layer of insulation to prevent shorting if they come into contact.

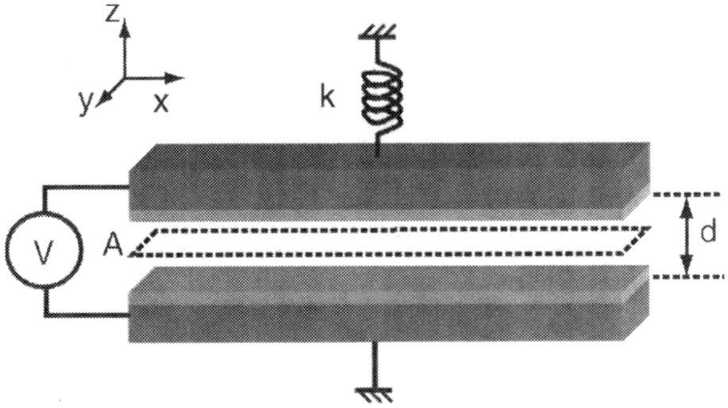

Fig. 5.8 Schematic illustration of a parallel plate actuator. The bottom electrode is mechanically fixed, while the top is mounted on a spring suspension and therefore free to move

The most common application of this type of actuators is in switches and relays. A broader application of parallel plate actuators has been limited by the nonlinear response, instability, and limited actuation range. These shortcomings have stimulated the development of other approaches, such as the linear surface drive and the comb drive.

5.3.1.2 Linear Surface Drive

The linear surface drive exploits the electrostatic force that acts to align a misaligned parallel plate actuator. The basic geometry is illustrated in Fig. 5.9a. In this configuration, the force F_x, acting in the x-direction, is given by [30]:

$$F_x = \frac{\varepsilon_0 \varepsilon_r l V^2}{2d}, \tag{5.11}$$

where V is the applied voltage, l is the length of the overlapping electrodes, and d is the spacing between electrodes as illustrated in Fig. 5.9a. Here again we assume that the spacing is small compared with the area of the electrodes. The force is a linear function of the length of the electrodes, but independent of the overlap distance x, see Fig. 5.9a, as long as the electrodes are misaligned.

The dynamic range of such an actuator can be increased by fabricating a periodic array of electrodes such as that illustrated in Fig. 5.9b. The device is operated

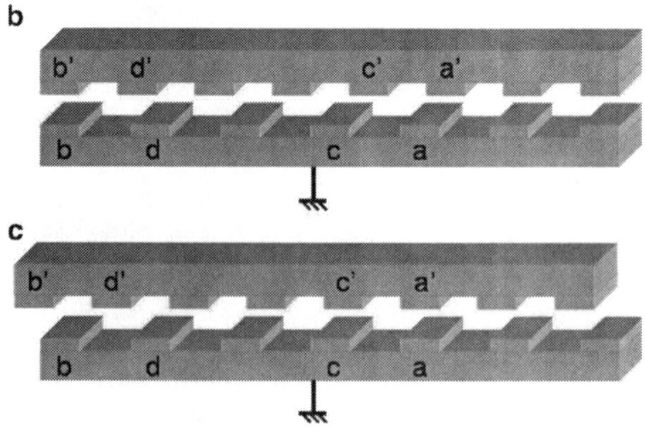

Fig. 5.9 (**a**) Linear surface drive actuator; (**b**) linear surface drive with a large array of periodic electrodes. Application of a voltage between a–a′ results in a displacement in the –x direction as shown in (**c**)

by applying voltages between pairs of electrodes sequentially. For example, by applying a voltage between the electrode pair a–a′ (and all similar pairs), a force is generated in the $-x$ direction, which moves the top electrode to the left as shown Fig. 5.9c. The motion can be sustained by subsequently applying the voltage between electrode pair c–c′. Similarly, starting from the configuration shown in Fig. 5.9b, motion to the right ($+x$) can be achieved by applying a voltage first to electrode pair b–b′ and then d–d′.

Apart from the rather complicated driving electronics required, this approach also suffers from coupling of the motion in the linear direction to the out-of-plane direction because of the large force component generated in this direction. A further challenge arises from the rather small electrode spacing required to achieve significant forces, which is typically on the order of one micrometer. Achieving such a small spacing over large electrode areas can be challenging. Despite these challenges, however, the concept has been developed to a high level by Agilent Laboratories, which used it to develop a prototype scanning system for a data storage system [31].

5.3.1.3 Comb Drive

A comb drive consists of a set of interdigitated electrodes, similar to those used for capacitive sensing, as illustrated in Fig. 5.5. Typically, one set of these electrodes is fixed (the stator), whereas the other (the rotor) is supported by a spring suspension system and is free to move. Ideally, the suspension system is compliant in the direction of desired motion (the y axis in Fig. 5.5) and relatively stiff against motion in other directions. If a voltage V is applied between the rotor and stator, the force F acting on the rotor is given by [21]:

$$F = \frac{n\varepsilon_0\varepsilon_r t}{g}V^2, \tag{5.12}$$

where n is the number of electrode pairs, t is the electrode thickness, and g is the gap between electrodes. The derivation of this equation is based on a one-dimensional parallel plate geometry, in which 3D effects, such as fringing fields, are ignored. The inclusion of such effects results in only a small correction of approximately 5% to the force [32]. Typically a ground plane is fabricated below both rotor and stator, and set to the same potential as the rotor to minimize forces acting on the stator in the out-of-plane direction.

In general, electrostatic actuators require driving voltages in the range of tens of volts to achieve appreciable force. Compared with the linear surface actuator, the forces generated by comb drives are typically smaller, as is the coupling into the out-of-plane direction. Example applications of comb drives range from resonators, electromechanical filters [32], gyroscopes [33], microgrippers [34], micropositioners [35], and relays [36], to positioning systems for scanning tunneling microscopes [37] and probed-based data storage [38, 39].

5.3.2 Piezoelectric Actuators

Piezoelectric actuators use the inverse piezoelectric effect described in Sect. 5.2. In general, the materials used in piezoelectric actuators are the same as those used in piezoelectric sensing, see Sect. 5.2. The basic actuator structure is that of a capacitor with a layer of piezoelectric material sandwiched between two conducting electrodes, as illustrated in Fig. 5.2. If the bottom electrode is kept fixed, application of a voltage across the electrodes results in a displacement of the top electrode. The maximum shape change is relatively small, less than 1% of the volume of the piezoelectric material. The force, however, can be quite large. For actuators of similar volume, piezoelectric materials can generate significantly larger forces than the other actuator types described here. The relatively small displacement resulting from the strain induced in the material can be directly used for actuation, or the strain can be coupled to another structure to amplify the displacement. An illustration of this concept is shown in Fig. 5.10.

In general, piezoelectric actuators exhibit low power consumption, but typically require high voltage and or mechanical amplification to achieve significant displacements. A further drawback of piezoelectric actuation arises from creep and hysteresis effects, which can limit the static positioning performance unless closed-loop feedback is used. Example applications include switches [40, 41], tunable capacitors [42], micropumps [43], actuators for optical components [44], and secondary actuators for hard-disk heads [45].

5.3.3 Thermal Actuators

Thermal actuators use resistive heating and thermal expansion to generate a force and displacement. There are two basic types: monomorphs, typically used for in-plane actuation, and bimorphs, used for out-of-plane actuation.

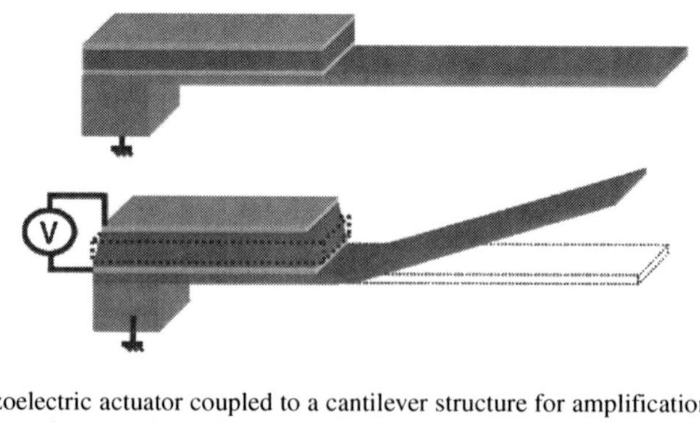

Fig. 5.10 Piezoelectric actuator coupled to a cantilever structure for amplification of the motion. Application of a voltage to electrodes causes the piezoelectric material to expand in the vertical direction and contract in the in-plane directions. The resulting strain causes a deflection of the cantilever in the upward direction

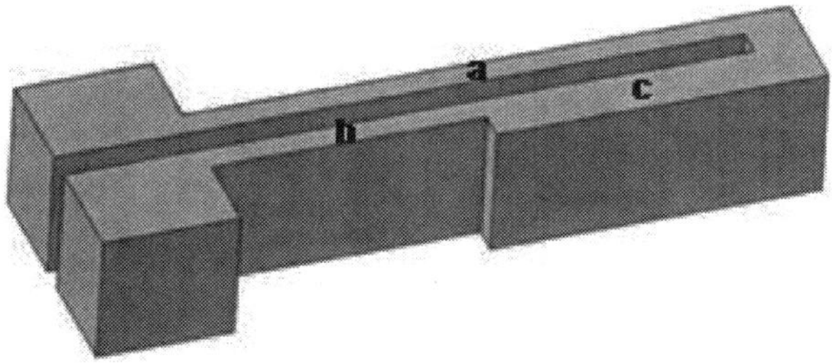

Fig. 5.11 Thermally activated beam (adapted from [46])

Figure 5.11 shows an example of a monomorph thermal actuator. The structure is free-standing and hence thermally isolated. Such a device is typically fabricated from a conducting material, such as doped silicon or polysilicon, and therefore acts as an electrical resistor. Application of a current through the structure results in the dissipation of electrical power and a rise in temperature. The section of the structure labeled 'c' is thicker than the rest of the structure, and hence stiffer, has a lower electrical resistance and a higher thermal conductance. The lower electrical resistance of section 'c' results in less power dissipation in this region and hence less heating. The temperature rise in section 'c' will therefore be lower than in the rest of the structure, giving rise to less thermal expansion than in region 'a.' This difference in thermal expansion in turn results in a clockwise rotation of the free end of the structure upon heating. The displacement of the free end, Δx, is given by [46]:

$$\Delta x = \frac{\alpha \, \Delta T_{\mathrm{bc}} \, l^2}{g\left(0.7707 + 0.3812 t^2 / g^2\right)}, \tag{5.13}$$

where α is the coefficient of thermal expansion, g is the center-to-center beam distance of sections 'a' and 'c', l is the length of sections 'b' and 'c', and t is the thickness of sections 'a' and 'b.'

The second type of thermal actuator, the bimorph, is illustrated in Fig. 5.12. In this approach, a 'u'-shaped cantilever beam is fabricated from a bilayer composed of materials having different thermal expansion coefficients. Many material combinations are possible. Typically, one of the layers is a conductor, such as doped silicon, and acts as a resistive heater when a current is passed through it. Upon heating, the difference in thermal expansion between the two layers results in the generation of stress in the beam and a displacement of the free end, similar to the piezoelectric actuator depicted in Fig. 5.10.

The change in radius of curvature of the beam, ρ, as a function of temperature is given by [47, 48]:

$$T - T_0 = \frac{1}{\rho}\left(\frac{2\chi - Y^{\mathrm{T}} G^{-1} S}{Y^{\mathrm{T}} G^{-1} A}\right), \tag{5.14}$$

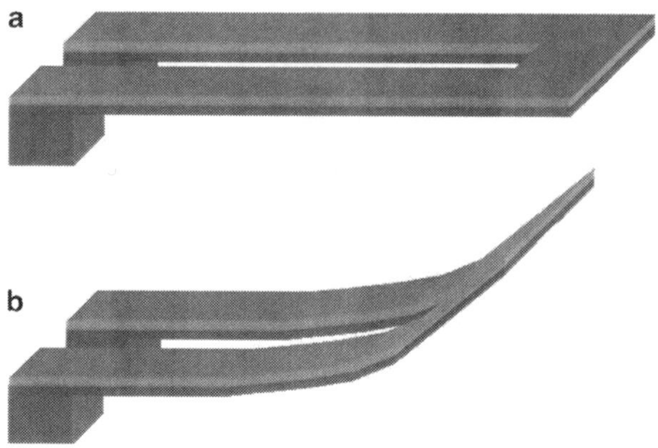

Fig. 5.12 (**a**) Bimorph thermal actuator: here the lower layer is doped silicon and the upper layer silicon dioxide. (**b**) Application of a current through the silicon legs results in heating of the free-standing structure. The temperature rise induces a stress due to the larger thermal expansion of the lower layer, causing the free end to bend upwards

where T is the temperature, T_0 is the reference temperature, ξ is the flexural rigidity of the beam. Matrix G describes elastic moduli and the geometry of the beam, matrices Y and S describe thickness effects, and A describes the effect of the coefficient of thermal expansion. For details, the reader is referred to [47].

In general, thermal actuators are relatively simple to implement, can be operated with low voltages, and generate relatively large forces. Drawbacks of this class of actuator are their high power consumption and, for the bimorph actuator type, a sensitivity to changes in the ambient temperature. In addition, thermal actuators are typically slower than other actuation schemes, as they are limited by the heat capacity of the structure and the thermal time constants of heating and cooling. Application examples include microgrippers [49], actuating mirrors [50], optical switches [51], as well as rotary and linear stepper motors [52].

5.3.4 Electromagnetic Actuators

Electromagnetic actuators use the force exerted on a current-carrying conductor in a magnetic field to generate a displacement. The magnetic field is typically generated either by a permanent magnet or by another current-carrying conductor or coil, i.e., an electromagnet. Although electromagnetic actuation is common in macroscopic applications, it is less common in microfabricated applications because it does not scale favorably to small dimensions [53], i.e., if the volume of an electromagnetic actuator is scaled to smaller dimensions, the magnitude of the maximum force the actuator can generate decreases by a much larger scaling factor. This scaling is particularly unfavorable for devices that rely on the interaction between two microfabricated electromagnets. The case of an electromagnet interacting with a permanent magnet is more favorable, but still does not scale well. This scaling problem arises primarily from the difficulty of generating significant magnetic fields in small microfabricated conductors. To circumvent this problem, most microfabricated electromagnetic actuators are operated in an externally generated magnetic field. Note also that a few electromagnetic actuators that do not rely on external fields for operation have been reported, for example [54, 55].

Here we will focus on microfabricated actuators that rely on an external magnetic field. We can classify such devices into two categories: actuators based on microfabricated permanent magnets and devices based on microfabricated conductors/coils. Each approach is discussed in more detail later. The external magnetic field in such devices is typically generated using a macroscopic electromagnet or a macroscopic permanent magnet. Also, several hybrid devices that use microfabricated coils and miniaturized magnets that were manually assembled into compact systems have been reported [56–59]. These hybrid devices use millimeter-scale microfabricated coils and can therefore still be reasonably efficient. Combining conventionally manufactured miniature coils and magnets with a microfabricated silicon, mechanical system can be even more power efficient [60].

Figure 5.13 shows an example illustration of a microfabricated magnet actuator, similar to that reported in [61]. Here, a microfabricated magnet made from a hard magnetic material is fabricated on a spring suspension. A similar device based on a soft magnetic material has also been reported [62]. The magnetization direction

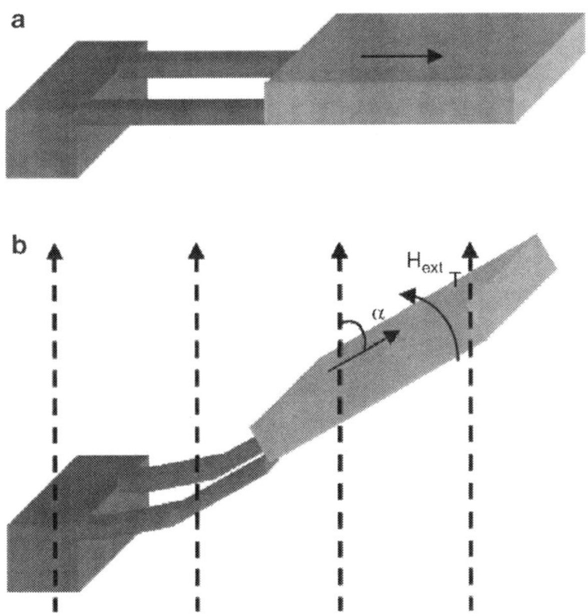

Fig. 5.13 Microfabricated magnet actuator. The *solid arrow* indicates the direction of magnetization, and T is the torque generated by the externally applied magnetic field H_{ext}

is in the plane of the device as indicated by the arrow in Fig. 5.13a. Application of an external magnetic field perpendicular to the plane of the device generates a torque, T_{mag}, on the magnet, which is given by

$$T_{mag} = V_{mag} M H_{ext} \sin(\alpha), \tag{5.15}$$

where V is the volume of the magnet, M is the magnetization, H_{ext} is the intensity of the external applied magnetic field, and α is the angle between the applied field and the magnetization, as shown in Fig. 5.13b. This torque causes the magnet to rotate out of the plane until the magnetic force is balanced by the restoring force on the spring suspension.

Figure 5.14 shows a schematic of a microfabricated coil actuator, demonstrating the actuation principle used in [56, 58, 59]. The actuator consists of an elongated coil positioned above a pair of permanent magnets. The actuation principle is based on the Lorentz force in which the current flowing in a wire in a magnetic field generates a force that is proportional to the cross product of the current and the magnetic flux density times the length of the wire.

Thus, for the arrangement shown in Fig. 5.14, the upper and lower sections of the coil generate a force in the same direction, whereas the left- and right-hand-side sections of the coil generate equal but opposite forces. The net force, F, is thus in the plane of the coil and perpendicular to the long axis of the coil with a magnitude given by [58]

$$F = BIL_{eff}, \tag{5.16}$$

where B is the magnetic flux density, I is the current flowing in the coil, and L_{eff} is the effective length of the coil given by

$$L_{eff} = 2(\text{number of turns})L_{avg}. \tag{5.17}$$

Here L_{avg} is the average length of the long axis of the coil, i.e., the average of the inner and the outer diameter of the long axis. This basic structure can be repeated to generate forces or displacements of a device in more than one direction [56, 58, 59].

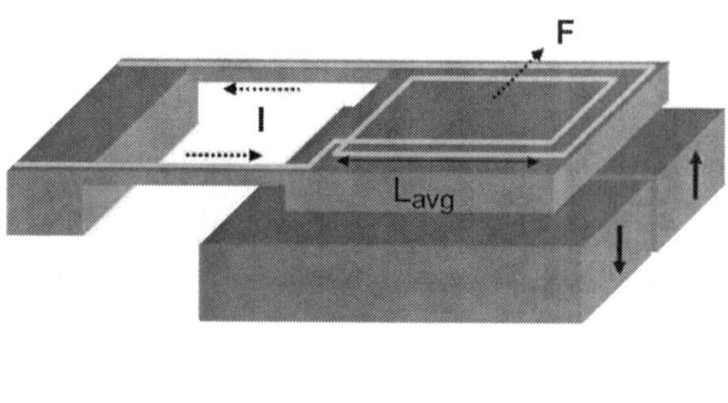

Fig. 5.14 Microfabricated coil actuator

Compared with the other actuation schemes discussed in this chapter, electromagnetic actuation has the advantage of a low operating voltage, a potentially long displacement range, and negligible hysterisis. The disadvantages include relatively high power consumption and poor scaling to the micro or nanoscale. These relative advantages and disadvantages have led to the use of electromagnetic actuation in relatively large devices, i.e., in millimeter-scale devices, where large displacements are required. Relatively few reports of microscale device applications exist.

5.4 S&A Applications and Products

The most well-known microfabricated actuators are thermal inkjets for printers [63] and the tiny micromirrors used in videoprojectors based on digital light processor (DLP™) technology [64], as these products find widespread use in our offices. The most widely used microfabricated sensors include accelerometers, pressure sensors, flow sensors, gyroscopes, and microphones. Microfabricated actuators are mainly used in the consumer market, whereas the largest use of microfabricated sensors occurs in the automotive, industrial, and medical market segments.

All active and passive safety systems in cars, such as Vehicle Dynamic Control™ and airbags, are based on microfabricated accelerometers and gyroscopes. Tiny pressure sensors measure the ambient and the intake manifold pressure to optimize the fuel-to-air ratio and thus reduce gasoline consumption. Pressure sensors in the tires increase driving safety by constantly measuring the tire pressure and alerting the driver in case it falls below a preset threshold. Pressure and flow sensors are also used in the industrial market for factory automation, preventive machine diagnostics, and optimization of operating conditions of the machines. In the medical market, accelerometers are mounted in pacemakers to adapt the pacing signal frequency to the physical activity of the patient. Ultra-miniaturized pressure sensors are used for temporary invasive pressure measurements in several parts of the human body, for example [65].

In the past few years, our unconscious interaction with the microfabricated S&A world has increased dramatically. We are all actors in a large-scale adoption wave that we like to call "the MEMS consumerization wave." Microfabricated S&As are penetrating several large consumer markets at a rapid pace. Laptops, mobile phones, digital cameras and camcorders, personal multimedia players, such as Apple's iPod® touch, personal navigation devices, and many other electronic gadgets start to make use of microfabricated sensors. Free-fall detection, more intuitive user interfaces, and image stabilization are just a few of the many applications of sensors in the consumer market.

The MEMS consumerization wave will make inexpensive and tiny sensors that consume little power available that, when coupled with a wireless module, will enable the deployment of wireless sensor networks. Many technical and business experts believe that after the "nomadic era," in which we are currently living, wireless sensor networks and domestic robots will represent the next big commercial adoption wave for semiconductors and for appropriate, microfabricated S&A products [66].

Many books address in detail the different applications of MEMS, especially in the automotive market. Thus we will only briefly discuss the history of automotive applications here and then focus on the more recent "consumerization wave" driven by the emerging consumer applications of motion and pressure sensors. Lastly, we will provide a brief overview of potential future market applications.

5.4.1 Micromachined Sensors in the Automotive Market

All modern cars use many different micromachined sensors for safety, comfort, and environmental control applications. Almost all these systems require the fusion and concurrent use of signals from different sensors. Pressure sensors, one and two-axis accelerometers, yaw and roll gyroscopes, and airflow sensors are currently the most commonly used microfabricated S&A in cars [67].

The first commercial application of a microfabricated sensor dates back to 1974, when silicon pressure sensors were starting to be used in electronic fuel-injections systems in cars to monitor the manifold absolute pressure (MAP) and mass airflow. The micromachined mass-airflow sensors measure the mass of air that enters the engine. These data, combined with information from a nonmicrofabricated oxygen sensor and a microfabricated MAP sensor, are used to optimize the air-to-fuel ratio and thus reduces gasoline consumption. In the 1980s, there was a proliferation of pressure sensors: from the manifold absolute pressure to barometric pressure, fuel pressure, oil pressure, and others. Barometric-pressure sensors measure the altitude as the atmospheric pressure drops when going from sea level into mountain regions, and this sensor provides important information so that a car's engine can adjust its operating conditions.

In the 1980s, the US government required that all cars produced after April 1, 1989, include a driver's side airbag. In 1993, the US National Highway Transportation Safety Administration (NHTSA) mandated dual front airbags in all passenger vehicles; and a few years later, side-impact airbags also became mandatory. The core of the airbag system is a microfabricated accelerometer that detects any sudden change in the vehicle's acceleration and, in combination with an algorithm, decides whether to trigger an explosive charge that rapidly deploys the airbag. The accelerometers used for airbags are located in the central part of the car and are able to detect accelerations from 20 to 70*g* (*g* being the acceleration of the Earth's gravity). First-generation airbags were lethal for children and dangerous for anyone sitting close to the steering wheel. The latest generation of airbag systems are safer, as they deploy the airbags at a speed that matches the weight of the passenger. Some passenger-occupant detection (POD) systems even use pressure sensors to detect the way in which the passenger is seated immediately before the airbag is being deployed. In the 1990s, the number of accelerometers per car grew exponentially. Some automakers started to use more accelerometers in the periphery of the car (satellite accelerometers with a full scale ranging from 100*g* to 500*g*) to increase the chance of saving lives in case of accidents because in this way the microcontroller had more time to decide whether to deploy the airbag [65].

After 1996, active safety systems such as Vehicle Dynamic Control (VDC™) or Electronic Stability Control (ESC®) started to be deployed. Airbags are a passive safety system because the bag is deployed after the accident event has happened. ESC® is an active safety system as it helps prevent the accident. ESC systems require high-grade low-*g* accelerometers (up to 5*g*), yaw rate gyroscopes, and magnetic speed sensors at each wheel. They help drivers to handle a car on icy or rainy roads better, by preventing the dangerous situation of under-or over-steering. All accelerometers used for these applications are manufactured using silicon, whereas gyroscopes are both silicon and quartz-based. ESC systems are the precursor of anti-rollover systems (ARO). Both ARO and ESC use the same sensor technologies, and, in fact, although they represent slightly separate approaches, they have now been combined into a single system. Modern ESC systems are currently limited to European vehicles. But in 2004, the NHTSA released a study that showed a 35% reduction in single-vehicle crashes for cars with ESC, and up to 67% for so-called sport utility vehicles (SUVs). Thus, in April 2006, the NHTSA mandated the use of ESC active safety systems in all passenger vehicles, beginning in 2012 [67].

After 2000, in which the high failure rate of a specific tire brand was implicated in more than 270 deaths and almost 1,000 injuries, pressure sensors started to gain momentum for measuring the pressure in the tire and to alert the driver in case of anomalies. All vehicles manufactured in 2008 or later will have a tire-pressure-monitoring system. The system consists of a pressure sensor, a simple microcontroller or state machine, and a wireless link. It also contains an accelerometer to wake up the systems when the vehicle starts moving, and some systems have a secondary accelerometer to identify on which side of the car the tire in question is mounted.

Last but not least, car-navigation systems based on Global Position System (GPS) receivers also use accelerometers, gyros, and magnetic sensors for dead reckoning in case urban canyons, tunnels, or blind spots cause a fading of the satellite signals.

5.4.2 Microfabricated Actuators in the Consumer Market

As mentioned, there are currently two examples of highly successful microfabricated actuators in the consumer market: thermal actuators for ink jet printers and electrostatic actuators for video-projectors [68].

Ink-jet print heads have hundreds of microscopic channels connected to ink-filled chambers. Thanks to the thermal conductivity of silicon, each chamber can be heated very rapidly (in a few microseconds) by a resistive heating element so as to vaporize part of the ink and thus propel a small volume of ink toward the paper

Fig. 5.15 Hewlett Packard ink-jet cartridge with a schematic drawing of the fire chamber and the different phases for drop formation and ejection

Fig. 5.16 Drawing of a micromirror from Texas Instruments for video projector

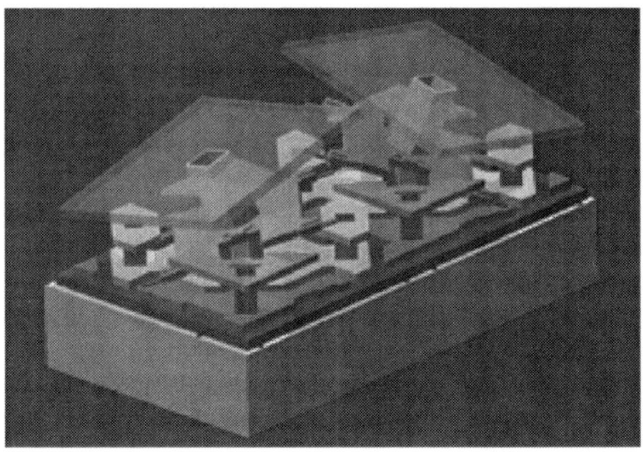

in a tiny droplet, see Fig. 5.15. The microfabricated device is either assembled on the edge of the ink cartridge itself, i.e., disposable with the cartridge, or integrated into the printer, and thus lasting longer. Hewlett Packard, Canon, and Kodak are some of the companies using this technology in their printers. In recent years, there has been an increase in the use of permanent ink jets vs. disposable ones, primarily for cost reasons.

In video projection technology, micromirror-based video projectors are competing successfully with LCD-based devices. In micromirror-based devices, tiny aluminium mirrors are manufactured on top of a CMOS circuit (Fig. 5.16). The mirrors are electrically actuated and thus able to deflect light that comes from a lamp accurately to generate a high-quality image on a screen.

5.4.3 "The MEMS Consumerization Wave"

The adoption of microfabricated S&As in the automotive market is driven by regulations, whereas vision and imagination have been driving the proliferation of microfabricated sensors in the consumer market. In the past few years, a few visionary companies invested heavily to support the development and manufacturing of sensors suitable for the consumer market. This triggered an extraordinary surge in the number of S&A applications in the consumer market, with user-interface applications being the most dominant across several market segments – from mobile phones to portable multimedia players, and from remote game controllers to 3D pointers. Figure 5.17 shows the adoption history of MEMS for user-interface applications in different segments of the consumer market. But S&As are also gaining ground in other consumer applications, for example in hard-disk drives, as described next.

5.4.3.1 Hard-Disk Drive Free-Fall Protection

Free-fall protection in hard-disk drives (HDD) was part of the first installment of "the MEMS consumerization wave." HDD-based devices, such as MP3 and MP4 players, laptops and mobile phones, camcorders and digital cameras use a three-axis accelerometer to protect data stored in the magnetic disk if the device is dropped. For example, if a free-fall-protected computer is being dropped or falls down, an accelerometer senses zero gravity, and a dedicated microcontroller signals

Fig. 5.17 History of adoption of MEMS for user-interface applications in the consumer market across several industries: From game controllers to mobile phones, from remote controllers to portable multimedia players (PMPs)

the read/write head to park away from the sensitive disks, before the head crashes onto the disk, causing the loss of data or possibly damage to the drive. In such an event, the plastic chassis of the computer may break, but the most important part of the personal computer, the data, are likely to survive. Nowadays, all major brands use accelerometers mounted on the motherboard to sense acceleration along all axes. Customer interviews confirm that for safety reasons most users prefer personal computers that are equipped with such data-protection systems [69].

5.4.3.2 Mobile Phones and Portable Multimedia Players

In 2003, some mobile-phone manufacturers approached the MEMS industry for two reasons: On the one hand, they were interested in replacing surface acoustic wave filters with the smaller and more reliable bulk acoustic wave filters and the large electric condenser microphones with small, SMD-compatible silicon microphones. On the other hand, they aimed at embedding a wellness and health dimension in mobile phones by means of a pedometer function. Later on, in 2006, tiny accelerometers were introduced into some mobile phones to act as a simpler user interface. Motion sensors were regarded as the "new mouse" for handheld devices. Initially, this application of microfabricated sensors was only known to a few people until Apple launched the iPhone™ in June 2007. In this button-less phone, the accelerometer is a key building block of the human–machine interface [70, 71].

The combination of microfabricated accelerometers and appropriate application software eliminates the need for conventional switches or buttons and thumb wheels for scrolling, zooming, and panning of Web pages, e-books, and spreadsheets. This is an innovative way to solve the well-known "small button – big finger" problem that bothers many users. In fact, although small cell phones are convenient and easy to carry, their small display screens and limited graphic capabilities reduce the overall user experience. The sensor can detect basic human movements and use them as an input for display orientation, which in turn simplifies how the user views downloaded pages. The user can navigate through Web pages or pan through maps by simply tilting the device in the desired direction. The step from mobile phone to portable multimedia player was a very easy and obvious one. In fact, the Apple iPod® touch MP4 player is currently using a 3× accelerometer for a more intuitive user interface.

5.4.3.3 Microfabricated Accelerometers in Game Consoles

Nintendo pioneered the use of microfabricated accelerometers on a large scale to achieve a novel and disruptive human–machine interface. In May 2006, Nintendo launched the Wii™ game console equipped with two single-hand controllers. Each of them is powered by a very small 3× accelerometer, supplied by STMicroelectronics and Analog Devices, and is able to detect tiny movements of the hand so as to render the user experience more interactive. "Playing is believing" is the value proposition of motion sensors in game consoles. The success of Nintendo's new home game console stems from many factors, with microfabricated sensors being a key contributor [72–74].

Sony's PS3™ also uses a combination of 3× silicon microfabricated accelerometers and 1× piezoelectric yaw-rate sensors to realize a more anthropomorphic wireless game controller. The step from game controller to remote control for TVs, PCs, and set-top boxes has been a very natural one. In July 2007, Logitech announced a controller called "Free Air" that uses a combination of 3× accelerometer from STMicroelectronics and a 2× gyroscope from InvenSense to allow the user to navigate on a display. The combination of gyroscopes and accelerometers enhances equipment performance while enriching the user's experience and device usability.

5.4.3.4 Sensors for Optical Image Stabilization

Currently, gyroscopes are used for image stabilization in camcorders and digital cameras with more than four to five megapixels, to avoid blur effects caused by unconscious muscular contractions in the frequency range of few Hertz. Most products are still using piezoelectric gyroscopes, but silicon gyroscopes are rapidly finding their way into this market, thanks to smaller dimensions and higher performance in terms of sensitivity and stability. Moreover, they can measure angular rates along the pitch and roll axes simultaneously and can be integrated more easily with other motion sensors. The increasing number of mobile phones with cameras represents a market opportunity of several hundred million devices per year, if we assume that currently 80% of the mobile phones also include a camera [75].

5.4.3.5 Pedestrian Navigation for Location-Based Services

More and more people resort to using personal navigation devices to navigate in unknown places, even in the city where they live. Thanks to the electromagnetic signals from several geo-stationary satellites and from the cellular network infrastructure, people can navigate reliably. Portable and vehicle-based navigation systems use GPS receivers to determine the position and provide route guidance; however, GPS signal reception is not always 100% reliable. In urban areas, where GPS signals are blocked due to tunnels, bridges and skyscrapers, accurate navigation becomes difficult. This is a big obstacle for telecommunications operators who want to sell location-based services such as zone-based advertising and augmented-reality experiences, and thus generate a new revenue stream [76]. In this context, microfabricated low-power and high-stability motion sensors can assist and substitute the GPS signal. In the event of signal loss, a dead-reckoning system continues to track movements when satellite signals are inaccessible or not sufficiently accurate. To implement dead reckoning, it is necessary to know the distance and the direction travelled. An ideal solution would be an integrated module for personal navigation, and many companies are trying to realize a complete sensor module capable of complementing the GPS signal. The ultimate goal is to realize a 10-axis module, i.e., a module comprising a 3× accelerometer plus a 3× magnetometer plus a 3× gyroscope and a pressure sensor, and everything coupled with the appropriate application software. The accelerometers in combination with pedometer software can tell the distance and the average speed of the user; the compass is able to pick the heading direction with respect to the magnetic north, the gyroscope is able to tell how much you change the direction of your path, and the pressure sensor can tell you the inclination of your path. This last feature may even become important when requesting emergency services. Thanks to GPS and pressure sensors in the phone from which the emergency call is made, the emergency responders will automatically know on which floor of the building the caller is. However, it will still be a few years until such 10-axis modules will appear on the market at an affordable price and in a small form factor. Six-axis modules, which combine a 3× accelerometer and a 3× axis magnetometer, are already available (see Fig. 5.18).

Several mobile phones already have an embedded pedometer feature, although it is still mainly being used for wellness and fitness, and not for personal navigation. Other personal navigation devices use a combination of compass and accelerometer to continuously readjust the map along the user's heading direction whenever the movements of the user are very slow so that they cannot be detected by the satellite. This, however, is only the beginning. In the years to come, pedestrian navigation will represent another big application area of microfabricated sensors in mobile devices.

5.4.4 The S&A Market: Quo Vadis?

Microfabricated sensors today are already used in numerous applications in the industrial and medical markets. Smart factory automation, optimization of water and energy consumption of household appliances, preventive machine diagnostics,

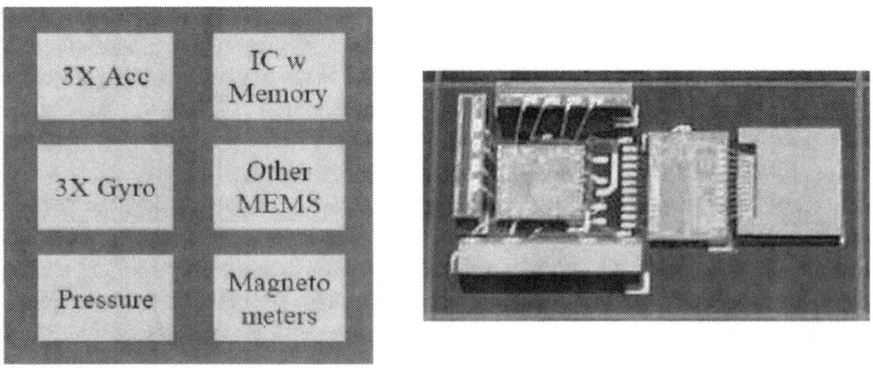

Fig. 5.18 On the left, schematic drawing of a 10-axis module for personal navigation. On right, first embodiment of a 6× module composed by 3× accelerometer (STMicroelectronics) and 3× Magnetometer (Microgate). The final dimensions of the 6× module are 4.4 × 7.5 × 0.9 mm^3

and continuous monitoring of patient activity are just a few examples of applications in these two markets. However, so far no single application has had the same market success as Nintendo's Wii™ or the Apple iPhone™. Moreover in the past few years, many other applications have been emerging in different market segments: for example, fall detection systems for elderly people that automatically activate an emergency call, context awareness of domestic robots to help with the housework, asset tracking and monitoring for more efficient logistic services, security modules for rescuers such as fire fighters, for people practicing different sports, and ambient assisted living applications [77].

All these new applications will likely require the deployment of wireless sensor networks, which today are still in their infancy [78]. A wireless sensor network results from the combination of different motes, i.e., wireless sensor modules consisting of some combination of a sensor, an integrated circuit controller, a wireless receiver, an antenna and a battery, or even better, an energy scavenger. It is difficult to predict the market success of these applications because a large-scale deployment still has to overcome some technological hurdles. However, the potential market for motes is limited only by our imagination. Once certain technical challenges related to energy harvesting have been overcome [79], motes will ultimately become a natural part of our lives. They could, for example, find many applications in consumer markets, with solutions ranging from security and biodetection to building and home automation, industrial control, pollution monitoring, and agriculture. Also, rising concerns for safety, convenience, entertainment, and efficiency factors, coupled with worldwide government regulations, could boost sensor usage to unprecedented levels, although not all of them are necessarily silicon-microfabricated.

In fact, motes will have to measure real-world variables such as pressure, temperature, heat, flow, force, vibration, acceleration, shock, torque, humidity, and

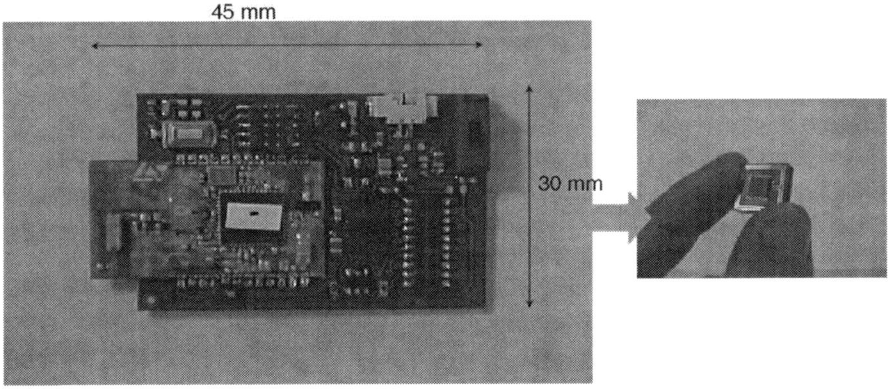

Fig. 5.19 Two examples of motes. Shown on the *left* is a mote in production and realized at board level by using existing discrete devices. On the *right*, in contrast, is a miniaturized mote from IMEC Research Center, Belgium, that is still under development

strain as well as record images [80, 81]. Some of them will use microfabricated solutions, whereas others will use conventional sensors that have been available for decades, although not yet optimized for this type of application. Suppliers must be ready to integrate different technologies in a modular format as well as to achieve high-volume manufacturability of motes. For this reason, companies are investing in state-of-the-art 8-in. fabs [82].

It is of course not clear which application will win in the long run, but it is clear that many applications will benefit from the tiny, low-power, microfabricated sensors (see Fig. 5.19) that are currently being developed for the consumer market.

5.5 CMOS Technology for S&As

Integration of S&As with integrated circuits is becoming increasingly important for performance and compactness reasons [83]. Although the fabrication technology of S&A and microelectronic devices share the same basic technology, optimization has necessitated the independent customization of their fabrication processes, often making them incompatible. In general, the fabrication of S&As uses specific processes referred to MEMS technology. Merging CMOS and MEMS technologies has led to a variety of technology platforms, ranging from customized CMOS technology to accommodate the S&A fabrication requirements, to specific add-on processes based on the CMOS process, to 3D integration by post-CMOS integration of S&As at chip or wafer level to form a so-called system on chip (SOC), and finally to a more traditional packaging technology for integrating the CMOS die and the S&A die at the package level (system on package, SOP).

Figure 5.20 illustrates the different levels of integration of S&As with CMOS. The majority of today's products rely on the two extreme integration cases: customized CMOS technology and SOP technology. Examples of these two integration technologies include imaging sensors (CCD) [84] and gyroscopes for consumer applications [33]. Both rely on proven technologies developed for pure electronics applications.

Customized CMOS-technology-based S&A fabrication typically addresses high-volume, cost-sensitive applications, whereas the SOP concept addresses small-volume applications in which the market requires modularity and short development times and for which cost pressure is less of an issue. However, many new market opportunities for S&A will rely on more advanced integration concepts that deviate from these two traditional models.

A wide range of technologies have emerged from customized CMOS technology with fabrication processes that diverge significantly from standard CMOS processes to optimize or, simply, enable a desired functionality [90]. From the SOP concept, technologies such as flip-chip or piggy-backing a S&A die to a CMOS die have been developed [91] to reduce packaging cost, shrink the system form factor, and improve performance by reducing interconnect length. Finally, wafer-level 3D integration technology is an attempt to merge the flexibility in materials and

Fig. 5.20 CMOS and S&A integration concept. (**a**) pure CMOS technology (Hall sensor, reprinted from [85], with permission from Elsevier.); (**b**) post-CMOS S&A process (thermal RMS converter [86]); (**c**) wafer-level stacking (cantilever array [87]); (**d**) flip-chip stacking (LETI's accelerometer [88]), and (**e**) SOP (Bosch's accelerometer [89])

processing available with flip-chip or SOP solutions with the high-performance, high-volume capability of CMOS-based technology [92].

The large variety of integration technologies highlights the difficulty of having a common technology platform for S&As on CMOS because, on the one hand, the extreme variety of physical effects used in S&As necessitates a unique set of processes and materials, and, on the other hand, there is the extreme fragmentation of the market with large differences in requirements regarding cost, time to market, form factor, operating environment, and performance. Hence, often a customized S&A technology and integration scheme is-developed for each specific application. Table 5.2 summarizes the characteristics of these different integration models. The S&As technology development trend is clearly toward an ever higher integration and performance level at lower cost. In this section, CMOS and S&As integration technologies and applications will be reviewed. SOP concepts will be discussed later in this chapter, together with packaging technology (Sect. 5.6).

5.5.1 Materials and Processes

Most MEMS material sets and processes are copied or derived from CMOS technology to take advantage of the huge investment made by the CMOS industry in tooling, process development and control, as well as characterization. There is a

Table 5.2 Characteristics of different S&As with respect to their CMOS integration technology

	Pure CMOS	CMOS with add-on S&A process	3D wafer-level integration	Flip-chip (chip-to-wafer or chip-to-chip) integration	Die integration at the package level
Maturity	++	+	--	+	++
Versatility/application field	-- Very limited set of materials and processes	- Some limitations in terms of materials and processes	++ Independent optimization of S&A and CMOS	++ Independent optimization of S&A and CMOS	++ Independent optimization of S&A and CMOS
Yield	++ CMOS maturity	+ Limited non-CMOS processes, yield multiplication	- Yield multiplication, lack of maturity	+ Limited non-*CMOS* processes	+ Technology maturity
Suitability for large-volume application	++	++	++	-- Serial process	-- Serial process
Suitability for low-volume application	-- Limited flexibility	-- Limited flexibility	-	++	++
S&A to CMOS interconnect density	++ CMOS wiring density	++ CMOS wiring density	+ Through wafer via, vertical interconnect studs	- Solder bump	-- Wire bond
S&A to CMOS interconnect parasitic	++	++	+	-	--
Form factor	++	++	++	-	--
Preservation of device orientation	+	+	+	-	+
Package simplicity	++ One-die solution	++ One-die solution	++ One-die solution	++ One-die solution	-

large body of literature describing CMOS fabrication and materia
Accordingly, we will focus in this section on specific materials anc
have been developed for MEMS. More detailed information on M
and processes can be found in [95, 96].

5.5.1.1 Specific Materials Used in S&A Applications

The most common materials in microfabricated S&As are materials used for CMOS chip fabrication because of the extensive know-how in material formation (deposition, growth, implantation, diffusion, polymerization), the structuring of such materials (dry or wet etching, lift-off, local oxidation of silicon or LOCOS, electro or electroless plating), and their characterization (electrical, mechanical, thermal). However, for specific S&A devices applications, many other materials are used that are not compatible with CMOS technology because of the risk of contamination or interdiffusion during processing that may effect the CMOS circuitry, or because thermal budget requirements for the material formation or processing exceed the maximum thermal budget of the CMOS process. Because of such incompatibilities, these materials are used when the integration of the S&A device with CMOS occurs after the CMOS processing and when the integration process will not alter the CMOS part. This is the case in hybrid integration technology or in post-CMOS S&A technology. Table 5.3 presents a summary of different materials used in S&A and the approach for integration with CMOS.

Table 5.3 Common S&A materials, their compatibility with CMOS technology and their applications

Materials	CMOS-compatible	Use in S&As	Type of devices
Monocrystalline silicon (bulk)	Yes, bulk CMOS wafer	Structural material	Spring, cantilever, membrane, proof mass, cap
		Electrical	Resistance, piezoresistance, electronic circuitry, electrode
		Thermal	Conductor/mass
		Optical	Waveguide, detector
		Magnetic	Sensor
Polysilicon	Yes, gate polySi	Structural material	Spring, cantilever, membrane, proof mass, sacrificial layer
		Electrical	Resistance, piezoresistance, electronic circuitry, electrode
		Thermal	Conductor, mass
Silicon oxide (film)	Yes (thermal oxide, BEOL oxide)	Structural material	Membrane, sacrificial layer
		Electrical	Isolation, dielectric layer
		Thermal	Isolation
		Optical	Waveguide, microlens, grating

(continued)

Table 5.3 (continued)

Materials	CMOS compatible	Use in S&As	Type of devices
Glass (bulk)	No	Structural material	Encapsulation (vacuum), cap
		Optical	Encapsulation with optical access
Metallization	Yes (BEOL metallization: Al, Cu, Ni, Au,...)	Structural material	Membrane, cantilever, sacrificial layer
		Electrical	Conductor, electrode
		Thermal	Conductor, mass
		Optical	Reflector
		Magnetic	Permanent magnet, electromagnet
Silicon nitride	Yes, bulk CMOS wafer	Structural material	Membrane, cantilever, hard coating
	Yes (BEOL passivation)	Electrical	Isolation, dielectric layer
		Thermal	Isolation
		Chemical	Passivation
Polymer	Yes (BEOL-compatible)	Structural material	Sacrificial layer, bonding layer
		Electrical	Isolation, dielectric layer
		Thermal	Isolation
		Chemical	Passivation
Solder	Yes (bumping technology)	Electrical	"Compliant" conductor, chip-to-chip interconnect
Ceramic (film)	Generally not (except AlN, ...)	Electrical	Dielectric layer, piezoelectric material (actuator and sensor)
		Chemical	Passivation
Ceramic (bulk)	No	Structural material	Encapsulation (vacuum)
Silicon carbide	No	Structural material	Spring, cantilever, membrane, proof mass, cap
III-V (GaAs, InP, ...)	No	Optical	Light emitter, detector
Diamond and diamond-like carbon	Yes (DLC)	Structural material	Sacrificial layer, hard coating

Lead zirconate titanate (PZT) piezoelectric material for actuators or filters is an example of a material that is not CMOS-compatible because of its high formation thermal budget [97]. Optoelectronic devices based on III/V semiconductors such as vertical-cavity surface-emitting lasers (VCSELs), made from (GaAs), and photodiodes (InP) are examples where the materials are very difficult to grow on silicon and thus are, to a large extent, not compatible with silicon technology [98–100]. For use in telecommunications or optical interconnect applications, these devices are integrated on CMOS using a hybrid technology. Figure 5.21 shows a photograph of an optoelectronic module used for board-level optical interconnect.

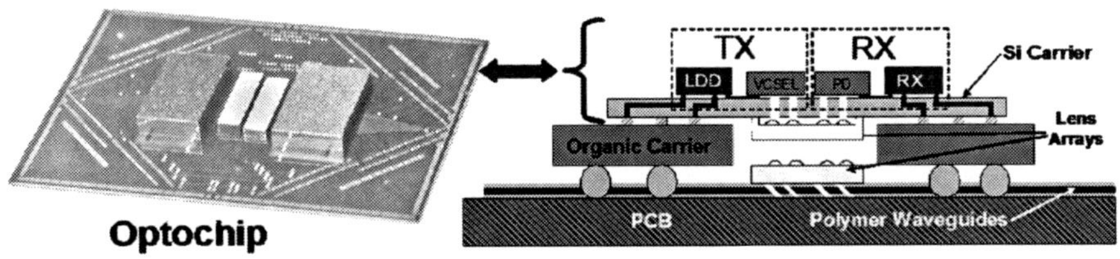

Fig. 5.21 Photograph of an assembled transceiver (IBM's Optochip) and conceptual side-view of the complete 850-nm transceiver package. The silicon carrier include wiring levels as well as electrical and optical trough wafer vias (TSVs) onto which CMOS laser diode driver (LDD), VCSEL and photodiode (PD) arrays, and CMOS transimpedance amplifier (RX) are mounted. The Optochip can be soldered to an organic carrier for subsequent attachment to a circuit board with integrated polymer optical waveguides [98], (c) 2008 IEEE Courtesy of J. Kash, IBM T. J. Watson Research Center

5.5.1.2 Specific MEMS Processes for S&A Applications

In general, MEMS fabrication methods can be classified into two groups: bulk and surface micromachining. The basic principle of these two approaches is illustrated in Fig. 5.22. In bulk micromachining, the microstructure is formed by etching the bulk of the silicon wafer, whereas surface micromachining refers to microstructures formed in thin films deposited on top of the substrate. MEMS processing technology historically started with silicon bulk wet anisotropic or isotropic etching that enabled the batch fabrication of complex 3D silicon microstructures.

Anisotropic wet etching uses solutions that have an etch rate that strongly depends on the silicon crystallographic plane as well as on the doping type and concentration. This allows the fabrication of all kinds of free-standing structures, such as membranes and cantilevers. Potassium hydroxide (KOH) is the most commonly used solution, and many microfabricated products are based on KOH etching [101].

More recently, also a reactive ion etching (RIE) based technology has been adapted to attain true bulk anisotropic etching. This technique, called deep reactive ion etching (DRIE), has become the standard etching technology for high-aspect-ratio structuring of silicon. The most widespread approach relies on a high-density plasma source and alternates between etching and passivation steps to control the verticality of the etching front in a process known as Bosch process [102]. Etch rates similar to those of wet etching (from a few of micrometers per minute to a few tens of micrometers per minute) can be achieved. Advantages of DRIE include the possibility to etch arbitrary patterns as well as trenches and holes with high aspect ratios (1:20 or more), which lead to surface area savings. Moreover, DRIE offers a much wider freedom in the choice of mask material, and simple resist masks can be used. Figure 5.23 shows the two phases of a Bosch process (etching and passivation) as well as the result of a complex structure etched with the DRIE technique.

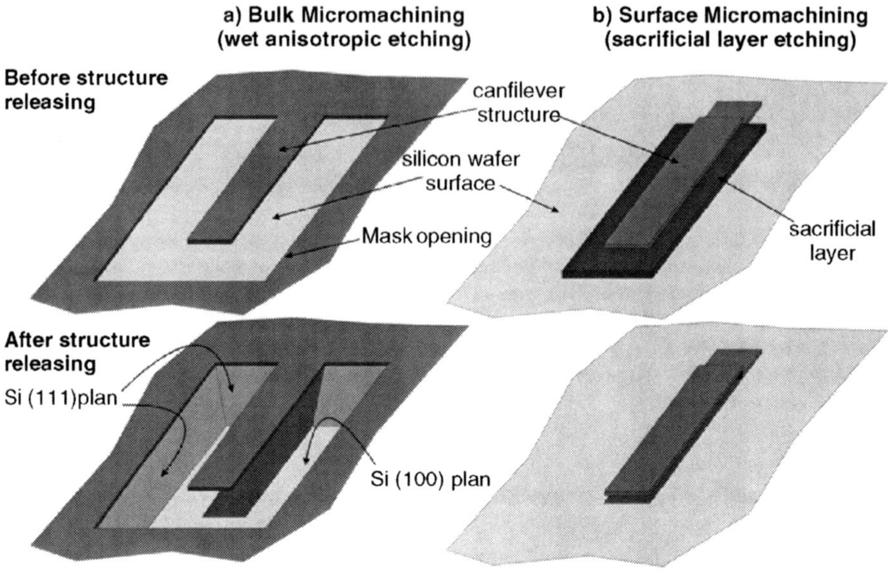

Fig. 5.22 Schematic illustrating the concepts of (**a**) bulk micromachining based on wet anisotropic etching and (**b**) surface micromachining based on sacrificial layer etching

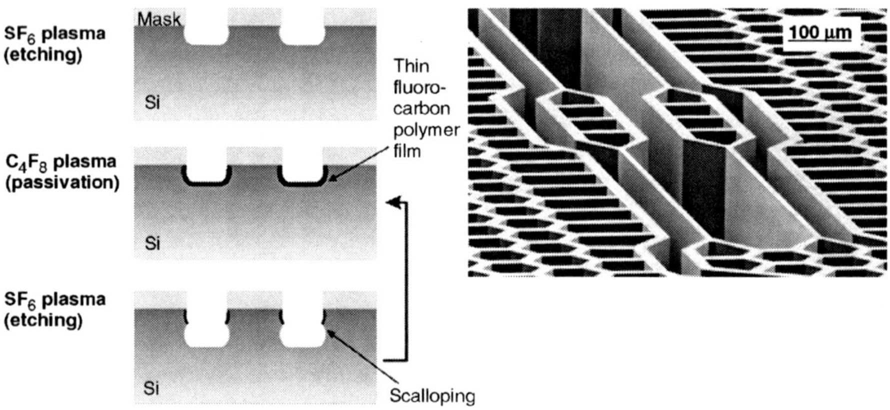

Fig. 5.23 DRIE Bosch process: Process principle (*left*) and an example of an etched complex structure (*right*)

Surface micromachining, in particular the use of polysilicon as a structural material, was a major paradigm shift for MEMS. Semiconductor engineers are very familiar with this material, processing cycle times are short, and smaller structures can be fabricated. The process consists of forming a mechanical structure on top of a sacrificial layer, which, after etching, forms free cantilevers or membrane-like structures. In the case of polysilicon, usually silicon dioxide is used as sacrificial material [103, 104]. However, also other material systems can be used, such as a sacrificial aluminum layer, to release dielectric structures (with an embedded metal layer) [105] or metallic structures released by etching a polymer sacrificial layer [106].

Another important technique used in S&A fabrication is through-mask electroplating, also known under the acronym LIGA (the German acronym for lithography electroplating molding, Lithographie Galvanoformung) [107]. This technique relies on a thick resist mask for electroplating metal microstructures. Historically, LIGA was based on X-ray lithography for exposing a thick (up to one centimeter) resist mask such as polymethylmethacrylate (PMMA) to fabricate metal microstructures with very high aspect ratios (more than 1:100). However, the high cost associated with synchrotron X-ray exposure is a serious check to its use for commercial applications. More recently, new UV-based photoresists such as SU-8 [108] have been developed, and currently most LIGA application development and products are based on UV LIGA, also called "poor man's" LIGA. Figure 5.24 illustrates the principle of the LIGA technique and shows a SEM view of a structure fabricated with X-ray LIGA.

Of course, also other specific processes for S&A fabrication have been developed. For example, hydrofluoric acid (HF) based electrochemical etching can be used to porosify silicon [109] to fabricate areas having a large surface-to-volume ratio for chemical sensing [110], to fabricate thick insulators [111] and high-aspect-ratio structures [112], or to use it as a sacrificial material [113]. Specific techniques have

Fig. 5.24 X-ray LIGA: Process principle for the fabrication of the metal template used for polymer injection. *Top row from left to right*: Exposure, structure after development, electroplating, and replication. The *bottom* picture shows an example of a fabricated structure used in a FTIR spectrometer. Images courtesy of the Institute for Microstructure Technology (IMT), Karlsruhe Institute of Technology

been developed for dry release of microstructures such as XeF_2 to etch silicon [114] or vapor HF to etch oxides [115]. The release of structures by means of dry etching is important in preventing free-standing structures from sticking to the surface during drying if wet processing is used.

5.5.2 Monolithic CMOS/S&A Technology

Several classes of microsystems can be completely formed using a pure CMOS or BiCMOS process, for example, magnetic, optical, and temperature sensors. However, an increasing number of applications rely on a combination of CMOS technology with CMOS-compatible MEMS technology. The fabrication steps can precede (pre-CMOS) or follow (post-CMOS) the regular CMOS process or can be performed in between the regular CMOS steps (intermediate-CMOS).

5.5.2.1 Pure CMOS S&A Technology

The key advantages of fabricating microsystems in pure CMOS technology are the availability of a large number of different foundries (fabless business model) and the possibility to cointegrate the sensor together with analog read-out electronics, signal-processing hardware, and communication interfaces. Modern CMOS processes use 30+ masks and an increasing number of different materials. Many processing steps in the CMOS processing sequence can be used to build sensors and microsystems without modifications of the CMOS process. Examples of recently added layers that can be used to build sensors include shallow and deep trenches in the front-end of the line (FEOL) processing of the silicon substrate, the addition of copper metallization in the back-end of the line (BEOL), various high-*k* and low-*k* dielectric layers, and many more. Classical examples of pure CMOS-based sensors are magnetic-field sensors based on Hall elements and temperature sensors. The first integrated Hall plates were fabricated in the late 1980s and early 1990s [116, 117]. Now they can be found in numerous products offered by, for example, Micronas [118], Infineon [119], Allegro [120], and Melexis [121].

There are currently two trends that can be observed for pure CMOS-based sensors. First, an increasing number of systems-on-chip (SOCs) designs require on-chip integrated sensors to compensate for the effects of temperature drift and to guarantee safe operation in the presence of large voltage and temperature variations. IBM's cell processor, for example, [122] includes 10 temperature sensors. Second, standard CMOS processes can initially be used to fabricate affordable products to benefit from a very short time to market, and later, when the market matures and volumes are sufficiently high, the standard CMOS process is tuned and modified to realize further cost reduction and performance optimization. This trend is best illustrated by the development of CMOS imagers. The first products were based on pure CMOS processes and were restricted to low-cost markets,

such as camera phones, because of their poor performance comp
imagers. The latest products use optimized CMOS processes and
the performance metric – outperform their CCD counterparts. In th
market is sufficiently large so that some foundries, e.g. IBM, o
wafer (MPW) services for imager processes through prototyping ser
MOSIS [84].

5.5.2.2 Pre-CMOS S&A Technology

In pre-CMOS technology, most MEMS processes occur before the CMOS process. This necessitates that the MEMS structures comply with the requirements of the CMOS process, such as stringent contamination criteria, surface flatness, and that they can withstand any thermal and chemical steps used in the CMOS process. To fulfill these requirements, the MEMS structure is typically sealed at the end of the MEMS process [91]. As an example of such pre-CMOS MEMS technology, Bosch has developed a technology called advanced porous silicon membrane (APSM) [123]. This technology makes use of porous silicon to form a cavity underneath the surface of the monocrystalline silicon, which helps to make the wafer manufacturing process fully CMOS-compatible and turns the traditional bulk-micromachining process into a surface technology. In Fig. 5.25, the

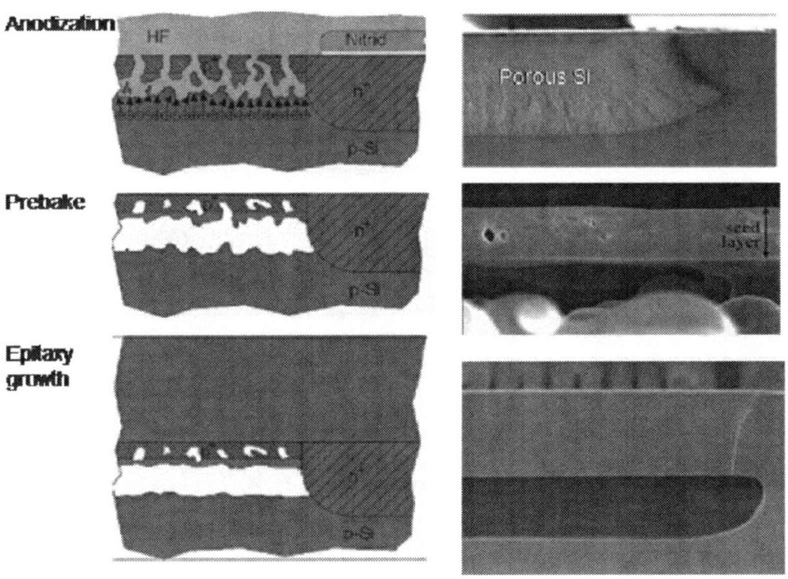

Fig. 5.25 Bosch's APMS process schematic (*left*) and the corresponding SEM cross section (*right*) [123]. Reproduced with permission of Robert Bosch GmbH

basic APMS process flow and a cross-sectional view of a fabricated membrane are shown.

Another pre-CMOS process uses embedded polysilicon microstructures, such as the technology platform developed by Sandia National Laboratory [104], where after the polysilicon structuring, the topography is filled with silicon dioxide followed by a chemical mechanical polishing (CMP) step to planarize the surface. Applications based on this technology include accelerometers [124] and other inertial sensors [125].

5.5.2.3 Intermediate-CMOS S&A Technology

Intermediate-CMOS S&A technology follows an approach in which the CMOS process sequence is interrupted for additional thin-film-deposition or micromachining steps. This interruption generally occurs either prior to the BEOL processes and is based on polysilicon microstructures or within BEOL to fabricate free-standing metal microstructures. Analog Devices used intermediate CMOS S&A technology based on polysilicon free-standing structures to fabricate gyroscopes and accelerometers, whereas Infineon and Freescale used a similar approach to fabricate pressure sensors.

An example of BEOL-type intermediate-CMOS S&A technology is the electrostatically actuated RF switch developed by IBM, see Fig. 5.26. Here copper is used as the mechanical material to fabricate a beam, and an organic material is used as sacrificial layer. The entire device is vacuum-sealed [106]. Figure 5.26 shows an optical image of the switch through the sealed layer and a schematic representing its cross section.

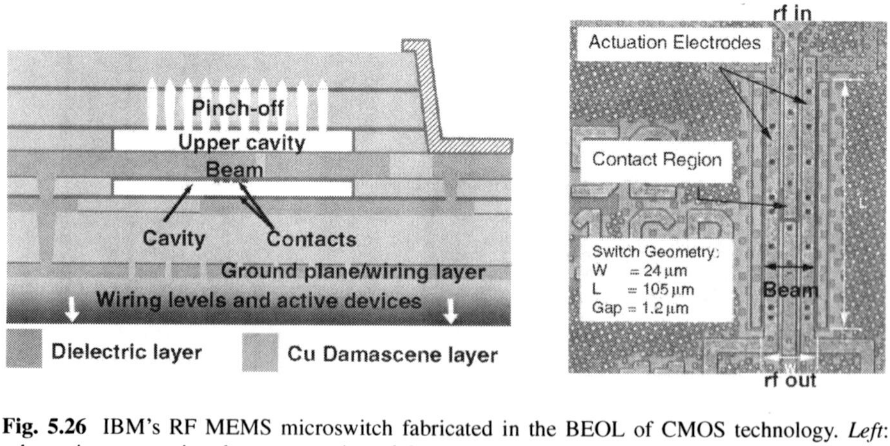

Fig. 5.26 IBM's RF MEMS microswitch fabricated in the BEOL of CMOS technology. *Left*: schematic representing the cross-section of the microswitch. *Right*: Top-view photograph of fabricated RF MEMS switch. Courtesy of J. Magerlein, IBM T.J. Watson Research Center

5.5.2.4 Post-CMOS S&A Technology

The best-known example in this category is the DLP by Texas Instruments™ [64]. The device integrates >1 million mirrors and has already been described earlier, in Sect. 5.4. Post-CMOS processing has two major advantages. First, it avoids the huge expense of a modern CMOS fab and enables start-up companies such as Sensirion [126, 127] to enter the market based on a fabless business models. Second, it decouples the slow development of the sensor technology from the fast-moving technology roadmap of mainstream CMOS technology. This allows S&A companies to take advantage of the features and cost reduction offered by the latest-generation CMOS technology without modifications in the sensor process or the CMOS process. The main disadvantage of S&As based on post-CMOS processes is the limited choice of processing steps that can be applied to a completed CMOS device. For example, the maximum temperature that a standard CMOS BEOL metal stack can withstand is typically around 400°C.

Besides the well-known DLP device, there are many examples of S&As based on post-CMOS processing. Moreover, an increasing number of devices use postprocessing to enhance CMOS-based sensors. The Hall-sensor-based device published by Sentron [128] uses electroplated magnetoconcentrators to achieve the magnetic-field sensitivities required for compass applications. A magnetic resolution and offset in the range of a few microtesla are beyond the specifications of standard integrated Hall elements. Other examples include devices such as the micro-lenses for CMOS imagers [129], fingerprint readers [130–132], and an increasing number of biotechnology applications (e.g., assays, DNA detection).

The main driving force for the monolithic integration of physical sensors is their price, and post-CMOS processing is probably the most cost-effective option. Monolithic integration uses only a single substrate, simplifies packaging, reduces the number of interconnects, and enables the use of advanced built-in self-test (BIST) features. Examples include accelerometers, gyroscopes, microphones, chemical sensors, and numerous other devices. Figure 5.27 shows a single-chip chemical sensor with mass-sensitive, capacitive, and calorimetric sensing elements. An interesting alternative was presented by Analog Devices [133]: their combination of pre-CMOS and post-CMOS processing avoids the bottleneck of the low-temperature restrictions of post-CMOS processing, but the challenge of introducing nonstandard wafers into a CMOS fab remains.

5.5.3 Hybrid CMOS/S&A Integration Technology

Hybrid CMOS/S&A integration technology or heterogeneous integration is used when a certain integration level is required for performance, form factor, or cost reasons, but a monolithical merging of S&A and CMOS is not possible or extremely complex.

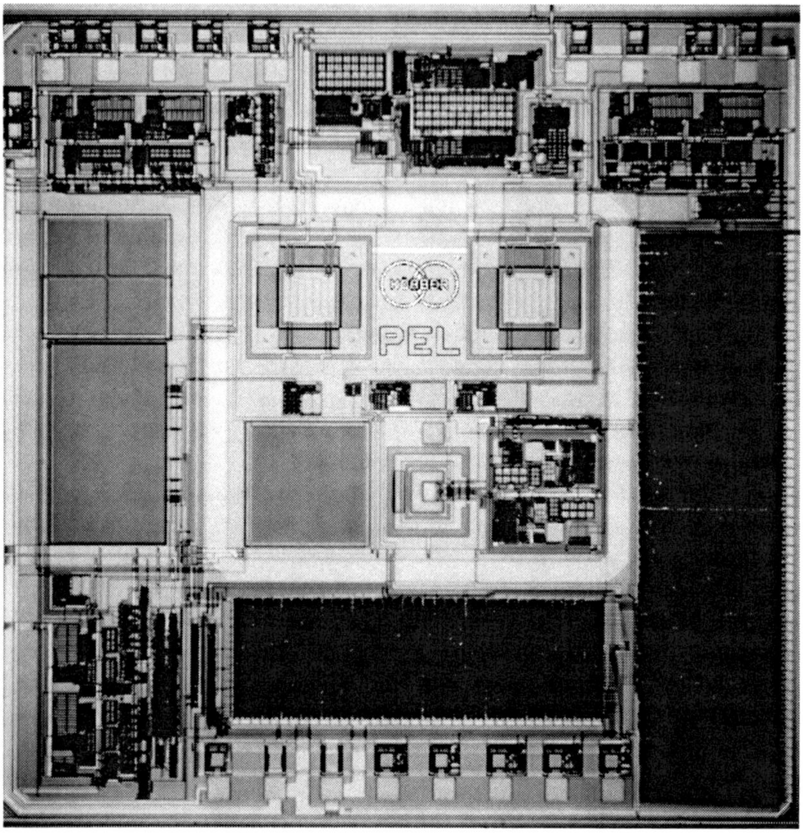

Fig. 5.27 Example of a monolithic S&A system with post-CMOS processing: Micrograph of a single-chip gas-sensing system including micromachined cantilevers for mass-sensitive detection, micromachined membranes for chalorimetric detection, and etched interdigitated metal fingers for capacitive detection. All sensors are functionalized by application of a sensitive polymer layer [134]

5.5.3.1 Wafer-Level Stacking Technology

Wafer-level integration has received a lot of interest from the research community, which resulted in the development of technologies to meet the high level of interconnect density needed. For both pure microelectronic and CMOS/S&A applications, the ability of integrating devices in a 3D fashion by stacking them offers several advantages in terms of overall system performance over traditional side-by-side chip placement (2D) [92].

Wafer-level with face-to-face joining technology is the most widely used wafer-level heterogeneous integration technology. Often it is also combined with a first-level packaging with or without hermetic sealing. An example is the accelerometer

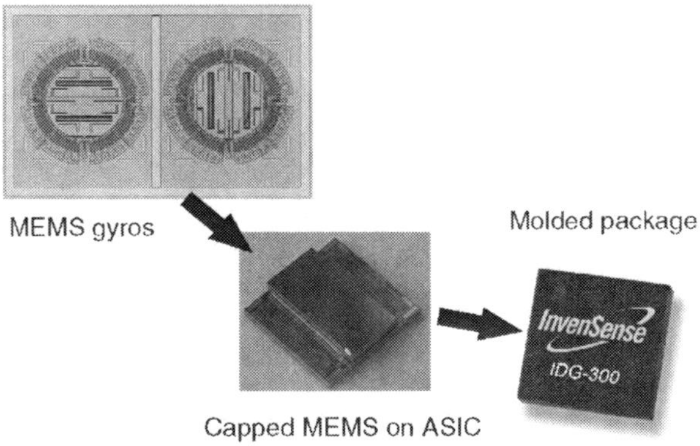

Fig. 5.28 InvenSense's Nasiri-Fabrication process to simultaneously seal and interconnect the Gyroscope MEMS structure to CMOS ASIC [136, 137]. Images courtesy of InvenSense Inc., reproduced with permission

from Austria Mikro Systeme (AMS) [135]. The CMOS part contains an ASIC with the actuation and sensing electrode. The MEMS part is a polysilicon beam released using KOH etching and bonded the CMOS wafer using eutectic metal bonding. InvenSense's gyroscope uses a similar sealed eutectic-based joining [136, 137], except that here the sensor part is fabricated in the silicon membrane of a SOI wafer. Figure 5.28 shows the different fabrication stages of InvenSense's gyroscope.

Face-to-face joining has the disadvantage of limiting the stacking to one level, which is sufficient for most S&A applications. However, the fact that the sensing or actuator device faces the CMOS surface limits the applications because it prevents access of the S&A to the environment. Hence 3D stacking with through-wafer vias or through-silicon vias (TSV) has been developed to preserve the device orientation and provide the S&A device with access to its environment. Silex Microsystems is developing a MEMS-compatible TSV technology for applications such as the micromirror arrays used in telecommunication networks as optical switches [138].

5.5.3.2 Wafer-Level Device-Transfer Technology

Although wafer-level stacking provides a solution for S&A and CMOS heterogeneous integration, it also raises a number of technological challenges. The compatibility of the process for fabricating a high density of TSVs and the technology used to fabricate devices is not obvious. Chip stacks with TSVs are much more sensitive to thermal cycling occurring during fabrication or due to changes in ambient temperature during device operation. The coefficient of thermal expansion (CTE) mismatch of the metal via and the chip material can cause nonnegligible local stress and serious reliability problems, with chip cracks occurring during thermal cycling.

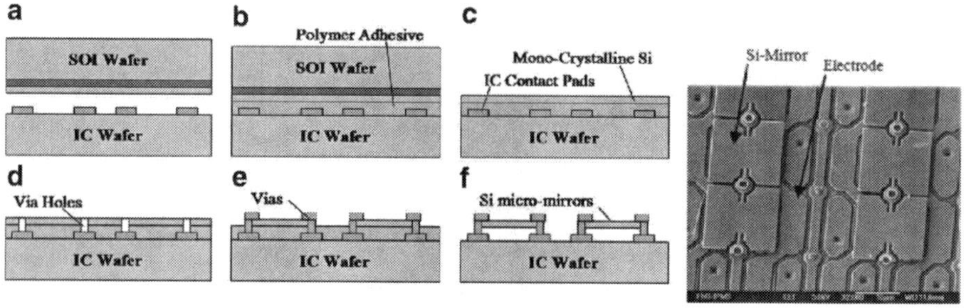

Fig. 5.29 Batch integration on CMOS wafer of thin monocrystalline silicon micromirrors for a pattern generator used in maskless lithography. Basic fabrication sequence (*left*) and SEM view of a section of the array (*right*) [140]

Another approach for wafer-level S&A/CMOS integration that can cope with these drawbacks is wafer-level device integration technology. This approach has the advantage of avoiding wafer stacking and the challenging fabrication of TSVs. The technology allows the heterogeneous integration of devices by transferring only the devices themselves, opening up new possibilities for 3D integration of SoCs. An example of this technology is its application for fabricating very-large-scale micromirror arrays integrated on CMOS electronics for use in deep ultraviolet (DUV) lithography and adaptive optics applications [139, 140]. Figure 5.29 shows the process sequence for fabricating the micromirror array and a SEM image of a section of the fabricated device. Such an array is used for the spatial modulation of an excimer laser beam at 248 nm wavelength and functions as the image-generating device in a system for writing photolithographic masks. A typical micromirror array comprises up to 512 × 2,048 individually addressable mirrors. Each mirror is suspended by two torsional springs and measures 16 × 16 μm^2.

A second example is the transfer of cantilever arrays used for the probe-based storage systems developed at IBM and LG Electronics [87, 141, 142]. A typical integration density achieved is 100 cantilevers/mm^2 and up to 1,000 electrical interconnects/mm^2. The arrays that IBM has fabricated consist of 64 × 64 (4,096) cantilevers, with each cantilever having three interconnects to the CMOS circuit. Each interconnect is made of a 15-μm-wide, 10-μm-high Cu/Sn-based solder stud. Figure 5.30 shows the basic process and an optical image of a section of such an array transferred on its CMOS driving electronics.

More recently, this technique has been extended for heterogeneous device integration and interconnects at the wafer-scale level, enabling the distribution of devices from one wafer to many wafers [140, 143]. Hence the cost of the devices transferred is distributed over the number of wafers populated. Depending on the distribution ratio and provided that the transfer itself can be made inexpensive, the additional cost associated with the devices transferred may become insignificant.

Other applications of wafer-level device transfer include the fabrication and integration of optical transducer arrays on standard CMOS wafers, i.e., uncooled infrared focal plane arrays (e.g., bolometer detectors, pyroelectric detectors) [144, 145], or optical transducer systems (e.g., spatial light modulators, tunable VCSELs) [146].

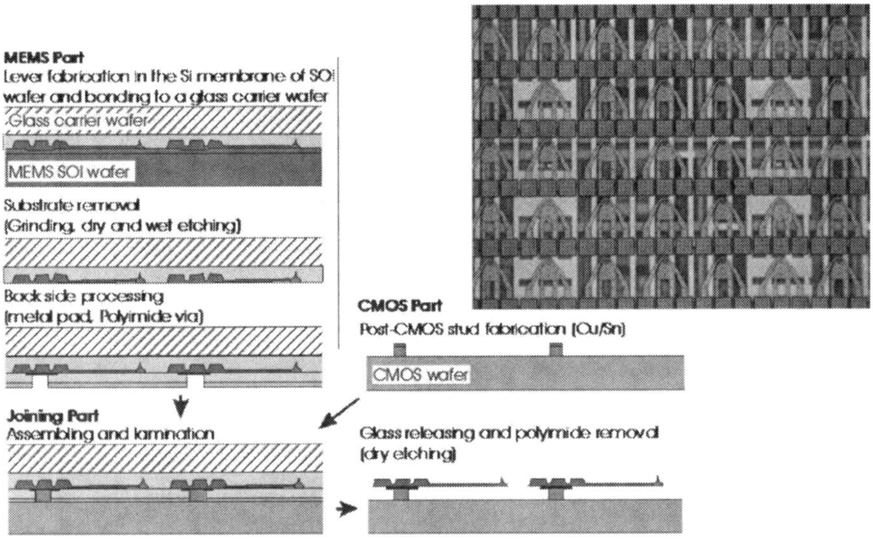

Fig. 5.30 Wafer-level cantilever transfer and interconnect for probe-based storage applications. Process sequence (*left*) and optical picture of transferred cantilever on CMOS wafer (*top right*)

5.5.3.3 Die-To-Die/Wafer Flip-Chip Integration

Despite the enormous interest in wafer-level 3D technology, more traditional chip-level integration methods such as flip-chip remain the technology of choice in the industry, thanks to their flexibility, and because for many applications, they are more cost-effective. Moreover, if the dice on both the source and the receiver wafer can be tested prior to the joining process, the functioning dies can be selected, thus avoiding the yield drop resulting from the multiplication of the yields of the waver-level process steps.

However, flip-chip integration suffers from a size limitation imposed by single-die handling. It is also a serial process, which, depending of the number of chips to be transferred per receiver wafer, can be time-consuming. Moreover, there is also a limitation in position accuracy, with a trade-off between placement speed and accuracy (~10 μm of accuracy at high speed) [92]. The minimum chip thickness might be also a disadvantage in form-factor-sensitive applications (e.g., mobile phones). Another disadvantage in many S&A applications is the fact that the chip is flipped over onto the receiver die and the devices transferred do not face the environment. In contrast, wafer-level integration using TSV stacking or wafer-level device transfer preserve the device orientation, which is important in applications such as DLP [64], chemical sensing, pressure sensors, CCD cameras, infrared cameras, and atomic force microscopy (AFM).

A recent example of hybrid CMOS/sensors application using flip-chip technology is the array of high-energy particle sensors interconnected to a CMOS pixel readout chip called Medipix [147]. Because of the flexibility of the hybrid

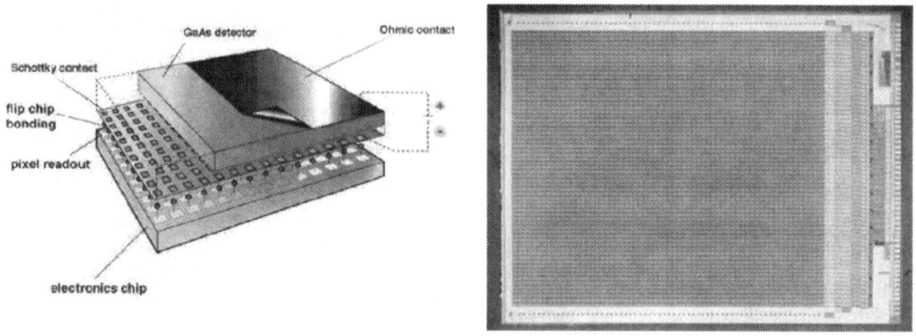

Fig. 5.31 Medipix chip [147] including a sensor array for high energy radiation: exploded view showing the different layers of the stacked device (*left*). Medipix2 schematic courtesy of the INFN, Napoli, Italy. *Right*: Optical image of the chip, reproduced by permission. Copyright © CERN

integration, the CMOS readout chip can be combined with different semiconductor sensors (Si, GaAs, CdZnTe) to convert the high-energy particles directly into detectable electrical signals to serve a wide range of applications in X-ray and gamma-ray imaging. The interconnect is made using a flip-chip bumping technology with a bump pitch of 55 μm. The sensor array has 256 × 256 pixels, representing an area of about 2 cm^2. Figure 5.31 shows the Medipix3 ASIC chip with the sensor array on top. Another example in which microstructures, such as resonators or rotary microactuators, are transferred to CMOS driving electronics uses cold welding (compression bonding) based on indium bumps [148]. In this case only the microstructure itself is transferred. The structure is held by tethers that break off in the chip separation step.

5.5.4 *Design Tools*

Designers working on CMOS circuitry and the MEMS community traditionally use a different set of electronic design automation (EDA) tools for their projects. Successful CMOS-based microsensors are only possible if a common design environment is defined at a very early stage of the project. Optimization of the overall system performance and functional verification are only possible if a unified simulation environment is used. Figure 5.32 shows an example of a combined design flow for CMOS-based sensors. Modern circuit design simulation suites (e.g., Spectre™) offer the possibility of simulating arbitrary combinations of high-level hardware-description languages (HDL, e.g., VerilogA, VHDL-AMS, C++ derivatives) together with digital circuitry. The MEMS structures can be incorporated by using equivalent circuit models [149], automatically reducing the complex finite-element models (FEM) to a multi-domain behavioral model using model-order reduction (MOR) [150].

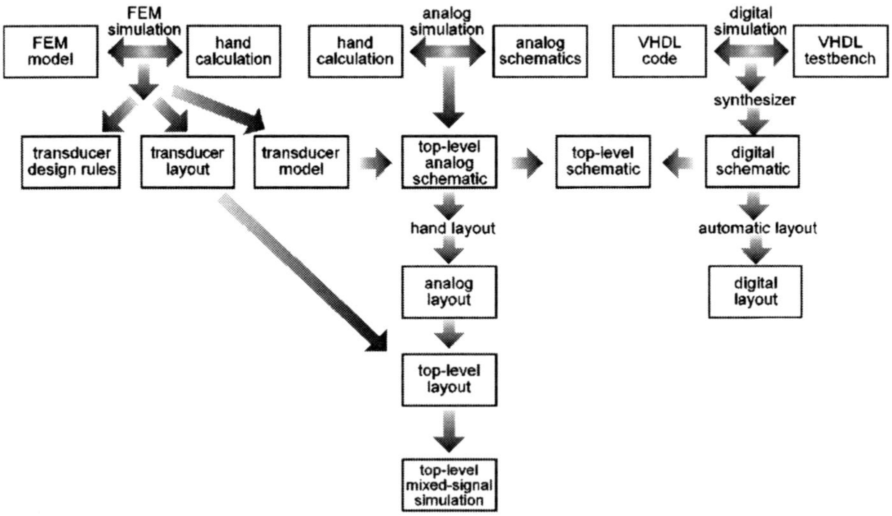

Fig. 5.32 Design flow for CMOS-based sensors

A second consideration for CMOS-MEMS concerns design verification. CMOS circuit designs are usually verified by a design-rule check DRC and layout-vs.-schematic (LVS) verification that are provided by the CMOS foundry. Although some EDA providers for MEMS design software are starting to offer interfaces to the standard software packages used for CMOS verification (e.g., Coventor [151], Tanner [152]), in most cases custom solutions for verification must be developed.

5.6 Packaging, Assembly, and Testing of S&As for Large-Volume Production

Packaging is the process of enclosing a semiconductor chip or device inside a package. The role of the package is to offer a proper interface between the microsystem and the external application board as well as to provide mechanical and environmental protection, thus guaranteeing reliable operation of the device during its lifetime. Assembly refers to the collection of industrial processes that are used in a specific sequence to create a package housing the microsystem [153]. Testing is the sum of all electrical and nonelectrical tests performed to verify product functionality and to calibrate the microsystem itself. The packaging, assembly, and testing of S&As depend to a large extent on the application. Therefore, this section will focus on recent developments and trends for the few large-volume MEMS-based S&A products that are built on CMOS platforms. For a more detailed discussion of S&A packaging, the reader is referred to [153–157].

MEMS packaging techniques are mainly derived from those developed for microelectronics. In contrast to integrated circuits (IC), where the package protects the electronic device from the outside world, the MEMS package sometimes may have to provide direct physical interaction with the external world. At the same time, the MEMS package needs to protect the microsystem from damage or failures in the assembly line and when S&As are deployed in the field. A first classification can be made according to the microsystem packages, where we distinguish hermetic and open.

Hermetic packages are intended to protect chips from the external environment, such as moisture corrosion and particle contamination, as these agents can prevent the devices from working properly. Inertial sensors, such as accelerometers and gyroscopes, are examples of devices that require hermetic packages. The hermetic package family is well consolidated. First examples based on ceramic package solutions appeared in the market in the early 1990s, mainly for automotive applications. Later on, with the onset of "the MEMS consumerization wave," more cost-effective solutions based on plastic molded packages started to be developed by microsystem manufacturers.

Testing equipment for this class of products is designed to provide proper physical stimulation to the packaged microsystems. Linear accelerometers must be calibrated under the effect of linear accelerations, usually measured in units of g, the Earth's gravity acceleration. Gyroscopes, in contrast, must be calibrated by applying an external angular rate, ranging from 100°/s for image stabilization to 1,000°/s for user interfaces. Depending on the application, acceleration levels have to be tuned from few g's for mobile phone applications to hundreds of g's for side-impact airbags in cars.

Open MEMS packages provide an interface between the device and the external world through one or more ports. Typical examples of open-package MEMS sensors are pressure sensors and microphones. More generally speaking, open packages are required whenever the external stimulus requires physical contact with the microsystem to be detected and measured. Apart from accelerometers, gyroscopes, magnetic sensors, temperature, and infrared radiation sensors, almost all sensors require an open package to work properly. Moreover, each sensor requires a customized open package to fit the specific application it is designed for. This class of packages involves special manufacturing designs and equipments. Several suppliers offer services that also cover this class of packages, but the manufacturing industry still is far from the standardization wave that happened for hermetic MEMS packages in the past decade.

5.6.1 Packaging: History, Current Technologies, and Trends

Apart from the functional classification between hermetic and open packages, another fundamental difference exists in the material employed for the package: ceramic and plastic [65]. In the plastic package family, a further difference can be made between full molded and premolded packages. Figure 5.33 shows the adoption history of different hermetic packages for inertial sensors from ceramic to plastic packages.

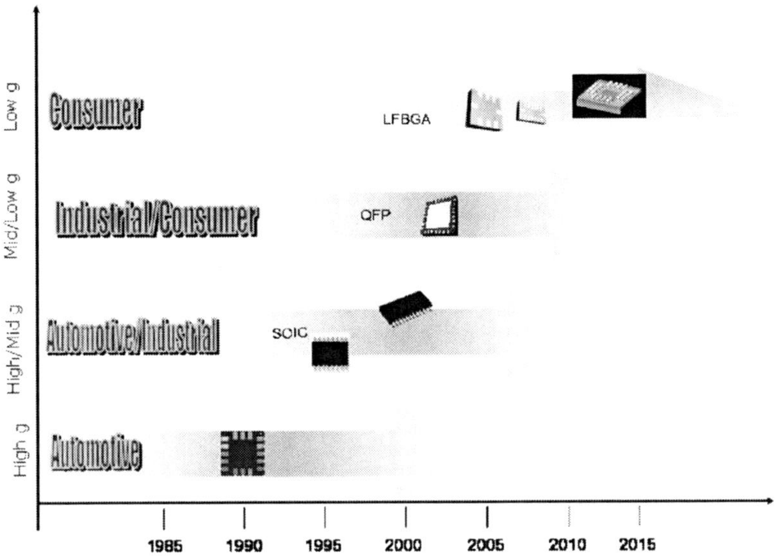

Fig. 5.33 Adoption history of hermetic packages across different market segments: From ceramic and SOIC to QFPN and LGA

Today, the manufacturing of high-volume MEMS products is linked to the use of molded packages. In the field of consumer-related products, the only cost-effective and volume-scalable solutions for MEMS products are full molded plastic packages. In the late 1990s, leadframe solution packages based on SOIC (small outline integrated circuit) packaging was the first approach for mass production of a full molded package for accelerometers. Several technical challenges [158] were successfully overcome to improve the performance in terms of thermal stability over the required temperature range and time stability over the device lifetime, which was limited because of changes in humidity and mechanical stress coming from the application board. For example, gel overcoating of the silicon dice is used to stress-decouple the microsystem from the enclosing epoxy filling.

The omnipresent need for miniaturization led MEMS suppliers to look for packages that were still based on leadframe technology, but were smaller from a volumetric point of view, such as quadrate flat packaging (QFPN). QFPN packages were first adopted by STMicroelectronics for the small analog low-g 3-axis (3×) accelerometer LIS3L02AQ [159], although there was a general consensus that QFPN packages are not applicable to MEMS products. Nowadays QFPN packages are also successfully employed by Analog Device, Freescale and Kionix, and the packages meet different levels of the JEDEC Solid State Technology Association standards specifications, ranging from consumer to automotive grade (Moisture Sensitive Level 1).

Leadframe-based packages are not easily customizable. As the pin-out of the leadframe is given, the layout of the silicon dice must be changed for any customization at this level. This is an important disadvantage in the fragmented MEMS market, because it increases not only the time to market but also, and significantly so, the cost at front-end level. At STMicroelectronics, this bottleneck on the way to large-volume manufacturing was solved by the introduction of the L/BGA matrix platform (i.e., LGA and BGA, Land/Ball Grid Array) in 2004. In the meantime, the LGA package has become the de-facto standard for inertial sensors in the consumer market. For the L/BGA packages, the metallic leadframe material was replaced by a polymer substrate material built by an organic core made from bismaleimide tiazine (BT) and several metal layers. These metal layers can easily be rerouted according to design requests, and thus guarantee a higher and faster level of customization. Therefore, the development cost for a L/BGA package is significantly lower than that for a leadframe-based solution. Finally, the L/BGA platform allows the use of the same manufacturing steps and tools both for hermetic and open packages.

Figure 5.34 shows some examples of LGA packages developed by STMicroelectronics both for hermetic and open microsystems. For hermetic packages, a standard L/BGA platform is used. The micromachined silicon is embedded into the molding compound and, therefore, protected from environmental stress, as shown in the picture on the left. For open packages, the communication port can be placed either on top of package (Holed LGA or HLGA™) or on the substrate itself for Rear Hole LGA (RHLGA™), as shown in the center and at the right in Fig. 5.34.

A microsystem is the combination of electrical and nonelectrical functions. Some suppliers integrate the two functions on the same silicon chip, whereas others use two separate silicon dice. These dice may be placed in different ways inside the package itself: stacked or side by side, see Fig. 5.35. The side-by-side approach is the preferred solution for slim packages, whereas stacked dice are preferred for high-performance, high-stability products, and smaller footprints. The future trend is to continue to move MEMS packaging out from the "jewelry" space of custom solutions to standard production rules and controls. The goal is to minimize manual interventions through the use of automatic equipment. This step is necessary to increase both the level of global production and the level of quality. As conventional

Fig. 5.34 Three possible solutions for MEMS package. LGA for hermetic products. HLGA™ and RHLGA™ for non-hermetic ones

Fig. 5.35 LGA packaged products: In the *left picture* the two chips are stacked, whereas on the *right* they are side by side

3D packaging technology, based on BGA with stack dice, Package on Package (PoP), and System in Package (SiP), is nearing full maturity, future MEMS products will use wafer-level packaging and direct bare die bumping on the application board [160]. Enabling technologies, such as TSVs and metallic bonding, will drive the next miniaturization wave, boosting market adoption of MEMS in the consumer market and enabling wireless sensor node applications.

5.6.2 Packaging of the LIS3L02AL Micromachined Accelerometer: A Case Study

Accelerometers for automotive applications have been on the market for more than two decades, and their assembly and packaging technology has seen making constant progress and change. Sensor hermeticity was the first – and the main – topic addressed to achieve high-volume manufacturability.

MEMS sensors can be designed to be capped one by one in the packaging line or directly in the cleanroom with the wafer-to-wafer bonding (W2W) technology process [161]. Low-melting-point glass (glass frit) is used to seal the microstructure at the wafer level. This is the most commonly used technique by several suppliers today. In the 1980s and 1990s, the sensors were assembled and sealed in the BEOL by ceramic hermetic lid soldering. This approach did not scale well in mass production because of cost and quality issues. Sealing at the package level implies the use ceramic/premolded packages. These packages are not able to meet all the requirements of mass production. Moreover, outgassing of solvents contained in organic materials used during assembly processes resulted in significant quality issues. The outgassing solvent could cause stiction of the acclerometer due to temperature cycling during transport or regular use. This was solved by resorting to a reliable wafer-bonding process for MEMS encapsulation, which is now being used by all large MEMS silicon manufacturers [69].

The flexibility in the design of the polymer substrate [162] of the L/BGA matrix platform (see Sect. 5.6.1) is important to avoid degradation of the accelerometer performance due to the package. The metals layers and vias of the substrate must be properly designed and positioned to reduce stresses on the sensing element. Thermal and mechanical stress can be induced during the manufacturing processes (i.e., molding and soldering of the devices) as well as during life in the field (i.e., thermal ageing, drops, shock damage). Dedicated simulations during the design phase and customized experiments at the laboratory level are necessary to tailor and optimize the design of the metal layers as well as the positions of the balls and pads on the silicon dice.

Although the substrate technology provides an easily customizable package base for 3D MEMS packaging, it is not sufficient to guarantee robust and stable products. The use of low-stress glue to attach the silicon chip to the substrate or to another die (die-attach operation) and the selection of a proper mold material are additional key elements to obtain good product performance. Both glue and mold materials are intimately close to the MEMS dice. Furthermore, the mismatch of the thermal expansion coefficients of silicon and organic materials is a relevant source of stress that must be handled at the design and production level. To address this issue, MEMS manufacturers have been working with material suppliers to design special dedicated low-stress materials [162].

Figure 5.36 shows an example of a FEM of an LGA package $5 \times 5 \times 1.5$ mm^3 in size (LIS3L02AL [162]), which is now in production at STMicroelectronics for all left controllers of the Nintendo Wii™. The sensor die is first bonded to the BT substrate with adhesive tape. The package footprint is minimized by stacking the logic die on top of the sensor die. After wire bonding (see Fig. 5.37), the devices are encapsulated by means of molding as shown in Fig. 5.36b.

3D FEM simulation tools were used to simulate the package warpage because of material thermal mismatch. Typical material properties used in the simulations are listed in Table 5.4. The thickness of sensor, cap, and electronics dice also affect the stress. Figure 5.36 shows that the compound molding warpage will result in

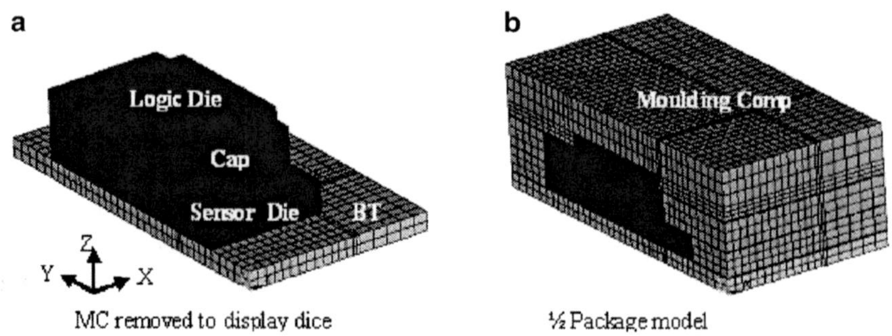

Fig. 5.36 FEM model for MEMS package

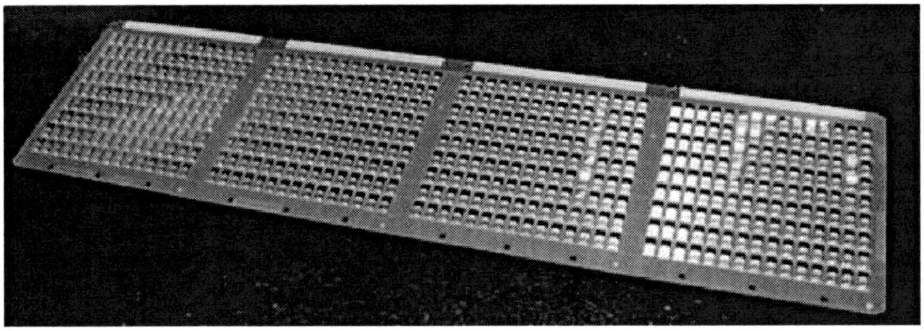

Fig. 5.37 Examples of L/BGA strip after the die-attach and wire-bonding processes. The strip displayed is ready for molding

Table 5.4 Material properties used in FEM analysis

Materials	E (GPa)	T_g (°C)	α_1 (ppm/°C)	α_2 (ppm/°C)
Silicon	131	–	2.8	–
BT Core	*xy*:26 *z*:11	–	*xy*:15.2 *z*:40	–
Solder mask	2.4	101	50	160
Copper	117	–	17.7	–
Die attach	1.5	128	80	170
Mold compound	23.5	125	8	34
Glass frit	70	–	7	–

sensor warpage, and thus ultimately influence the performance and stability of the final product. The warpage of the sensor substrate affects the signal performance of the accelerometer. Typically, the warpage effect after molding and postmold curing is compensated by signal calibration prior to shipping products to customers. The temperature variations in field applications also induce thermal mismatch and thus deformation of the sensor microstructures. This will cause a drift of the sensor signal. Therefore, it is not sufficient to calibrate the device at a single temperature, but the dependence of the warpage on the temperature and its influence on the signal must be calibrated.

MEMS testing is typically performed on the final device after completion of all assembly-related process steps. A first screening of device functionality is already done at the wafer level by means of electrical stimuli (electrical wafer sorting or EWS). Despite the moving structures present on the wafer, there is no difference between the EWS of the microsystems and that of a standard IC. MEMS testing at the packaging level aims at calibration of the devices and/or verification of the correct performance. Calibration and verification are performed either at ambient temperature (for most consumer market applications) or at different temperatures (for almost all automotive applications) according to customer needs. Stability over the device lifetime as well as drift vs. temperature are key parameters that must be guaranteed, not only just by design, but also through testing.

5.6.3 Production Testing

Innovative technologies typically require dedicated novel testing equipment. This rule is only partially true for MEMS products. Dedicated assembly equipment is rarely required because a large part of the machinery typically used for standard L/BGA products can be used for all MEMS products in L/BGA package. Software upgrades, adjustments of existing equipment, and customization of the tooling for proper handling and placement of the stress-sensitive MEMS dice are sufficient. Therefore, the investment necessary to convert a standard manufacturing line into a MEMS-dedicated assembly is moderate if the MEMS product is well designed.

The specificity of S&A testing arises from the need to stimulate the products by means of a nonelectrical sensor signal. This nonelectrical stimulus can range from acceleration, vibration or pressure to sound waves and magnetic fields. For accelerometers, key parameters, such as sensitivity or zero-*g* level (output with no applied acceleration), must be guaranteed for each component through spatial movements. Vibration, rotational, and linear accelerations are used to investigate and validate correct operation of the inertial sensor.

Therefore, a MEMS handler differs from a traditional handler. The device-under-test (DUT) board is not fixed, but can move or shake in space along the *X*, *Y*, and *Z* directions. The DUT board typically is decoupled mechanically from the other machine components. Furthermore, decoupling elements between the testing equipment and the testing floor area are used to reduce environmental vibrations and disturbances during product calibration (Fig. 5.38).

The handlers and sockets are also customized to minimize mechanical deformation and stress on the sensor element. DUT insertion and its placement in the socket are carefully monitored because nonplanar insertion or incorrect fixing of the package in the socket could result in a calibration error. Recently, testing equipment suppliers in cooperation with MEMS manufacturers are starting to introduce a novel testing strategy based on the strip format for the L/BGA package family.

Fig. 5.38 MEMS tester and handler

Strip testers present the following advantages:

- They reduce handling and stress issues related to the testing of single small packages, like the 3 × 3 × 0.9 mm^3 package used for the LIS331AL and LIS331DL products [159].
- They improve the quality and speed up the manufacturing process, thanks to increased parallelism by going from 4/20 devices on classical robot handlers to 24/160 devices. This represents a factor of almost ten in productivity increase.
- They offer the possibility to switch easily between different classes of products. By means of a simple modification of the stimulus chamber, the same equipment can be used to test inertial MEMS (e.g., accelerometers, gyroscopes) and differential MEMS (e.g., pressure sensors, microphones).
- They enable the testing of sensor modules requiring the application of several physical stimuli.

5.6.4 Next-Generation Challenges

The introduction of the customizable matrix L/BGA packaging technology represented a breakthrough for the volume manufacturing of MEMS products at STMicroelectronics. In both assembly and testing, the possibility to handle these products as a matrix allowed a dramatic reduction of the manufacturing cost and, at the same time, an increase in process yield and product quality. Miniaturization and sensor clustering are the driving forces in MEMS development. Wafer-level packaging complemented by a bumping process for direct soldering on the board might be the ultimate answer to the miniaturization need of the market. The second big challenge is the ability to integrate and test different MEMS products and standard ICs inside the same L/BGA package.

5.7 Nanostructures for S&As

The scaling of microfabricated S&As to the nanometer scale is an exciting research area that promises significant performance improvements and new device possibilities, but at the same time poses new challenges.

There are two basic approaches to making nanoscale devices: top-down and bottom-up. Top-down manufacturing utilizes the lithography and etching techniques used in CMOS fabrication. To fabricate nanostructures using these techniques, state-of-the-art lithographic techniques such as e-beam are required. A drawback of this approach is the inherent side-wall roughness, which plays an increasingly important role at the nanoscale, as will be discussed later. In the bottom-up approach, chemical techniques are used to synthesize well-defined nanoscale structures, such as nanodots, nanowires, nanobelts, or nanotubes. Such structures are typically single-crystal and can have very smooth side walls. However, as there typically is also a spread in the dimensions and other properties of the structures, reducing these variations and developing techniques for postsynthesis sorting are active areas of research. The list of materials that can be synthesized into nanostructures is large and growing, offering the capability to integrate high-quality single-crystal structures made from novel material sets into nanoscale devices. It is also possible to modulate the composition of a material during growth or even to create core shell structures. The main challenge in bottom-up manufacturing is to assemble the synthesized structures into a device or a set of interconnected devices. This is another active area of research, but a detailed discussion is beyond the scope of this chapter.

Nanoscale actuators are inherently mechanical devices, whereas in the area of sensing there are two basic device classes: mechanical and FET-based. The advantages of scaling electromechanical systems such as S&As to the nanoscale include an increase in the fundamental resonance frequencies of the devices into the gigahertz regime and beyond, improved sensitivity, and reduced power consumption [163]. High-frequency resonators with high Q-factors show promise for applications in filtering. Much of the current research in this area is driven by its potential application to mobile phones. In the area of sensing, scaling devices to the nanoscale can result in orders of magnitude improvement in sensitivity. For example, Mamin et al. [164] demonstrated force sensing with a sensitivity of less than 1 aN (10^{-18} N). Yang et al. [165] used a nanoscale mechanical resonator to demonstrate mass sensing with a sensitivity of 7 zg (i.e., 7×10^{-21} g), and predict that the ultimate resolution limit of the technique is as low as 1.7 zg (i.e., 1.7×10^{-24} g). Finally, in the area of calorimetry, Fon et al. [166] have demonstrated a resolution below 1 aJ/K using a nanomechanical sensor.

In addition to the difficulties in the fabrication of nanoscale mechanical devices, sensing the motion of such devices is also challenging. Many of the techniques used for sensing mechanical motion on the microscale do not scale well to the nanoscale. For example, optical techniques are limited by the diffraction limit of the wavelength of the light used. Capacitive techniques also suffer because of the reduction in capacitance as the size of the device is reduced such that stray capacitance tends to dominate the measured signal. To take advantage of the potential response-time improvements of nanoscale mechanical devices, the transduction technique employed to go from the mechanical to the electrical domain should also have a high bandwidth. Promising solutions for sensing in nanoscale structures include floating-gate FET-based detection, magneto-motive detection, and superconducting quantum interference devices (SQUIDs) [163]. Another challenge in nanoscale S&As arises from the dramatic increase in the surface-to-volume ratio that occurs when structures are scaled to very small dimensions. This increase can result in an increase in noise related to surface defects or to an increase in mechanical damping. As such, the importance of controlling the surface quality and state of such structures increases as the size is reduced. Roughness due to lithography or other fabrication techniques can limit the performance of such devices, and therefore, improving and controlling surface quality is an active research area.

The importance of surface quality in nanoscale devices is one of the factors motivating research into bottom-up approaches for fabricating nanoscale S&As. Here, the most common approach is to fabricate a floating-gate FET structure from a nanowire or nanotube using e-beam lithography. In this approach, the quality and state of the surface can be controlled by the chemical synthesis process used to fabricate the nanowire or nanotube, rather than by the limits of lithography. These devices function by sensing a change in charge that occurs when a molecule binds to the floating gate, which in turn results in a change in conductance of the transistor. In such devices, the gate region is typically functionalized with specific receptor groups such that the device is only sensitive to a specific molecule or class of molecules. Nanoscale FET-based sensors have similar advantages as nanomechanical systems, including improved response time and sensitivity and reduced power consumption. Example applications include protein detection [167], virus detection at the single virus level [168], and DNA detection with femtomolar sensitivity [169, 170]. The small size of such devices gives rise to the possibility of fabricating large arrays of sensors, each functionalized to be sensitive to a different molecule, with potential applications in high-throughput screening of DNA, proteins, or viruses [171]. Nanowires and nanotubes have also been used in the area of nanomechanical sensors. For example, Grow et al. used the piezoresistance effect in a carbon nanotube (CNT) to fabricate pressure sensors with gauge factors of up to 850, approximately 4× higher than conventional silicon piezoresistors. Stampfer et al. reported similar CNT-based pressure sensors [172] as well as a more general force sensor [173] based on the piezoresistance in single-wall CNTs, which is shown Fig. 5.39. In addition to the higher gauge factor, CNTs have a much lower thermal coefficient of resistance than silicon does [174], making them potentially less sensitive to temperature effects. Other applications include resonators for filtering applications, relays, and memory elements.

To date, most nanoscale S&As are standalone devices without integrated electronics. Research is focused on demonstrating proof of principle by using individual, custom-fabricated devices and exploring the fundamental physical limits of these

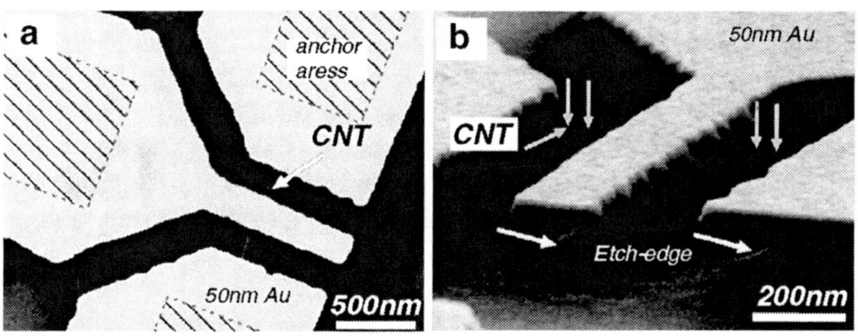

Fig. 5.39 Scanning electron microscope (SEM) images of a single-wall carbon nanotube (CNT) force sensor. (**a**) Top view of the structure: The CNT is indicated by the *arrow*. (**b**) SEM image under an angle of 15°. Reprinted with permission from [173] (c) IEEE 2006

approaches. The state of the field is similar to the early days of microscale sensor and actuator research, with many exciting demonstrations of devices and much potential for future advances and commercial applications as the field moves to batch fabrication and the direct integration of electronics.

5.8 Outlook and Summary

Although S&A functionalities are already quite well explored, the integration and interconnection to a CMOS platform are technologically of great importance, because they strongly affect performance, yield, and cost. Small companies, such as Sensirion [126], have successfully established an excellent market position for various sensor types based on monolithically integrated sensors on CMOS platforms. As an example, a CMOS-based humidity sensor is about 40 times smaller in size, performs up to 10 times better, and costs about 10 times less than its PCB-based counterpart. This illustrates the great potential of CMOS-based S&As. In most of these cases, post-CMOS sensor fabrication has been successfully used to integrate sensors onto a CMOS platform. For the mid-term future, we will see an even stronger trend toward sensor integration on CMOS platforms. As Sensirion has successfully demonstrated, even small companies can be successful in this market by making use of CMOS foundry services for the CMOS platform with subsequent dedicated in-house post-CMOS sensor fabrication. Hence, large investments are not necessarily required for new MEMS fabrication lines as there is a potential for good synergy with existing lines. However, when very large volumes or very complex fabrication techniques are required, specialized MEMS lines that are either CMOS-line extensions or separate operations will be needed.

The integration aspects (monolithic and hybrid) of S&As onto CMOS (ASIC) platforms remain an important challenge and focus for the years to come. High-yield and hence low-cost integration and interconnect techniques will open up great opportunities. This also includes nonsilicon-based S&As offering novel functionalities or lower cost through, for example, the use of III/V or plastic materials, as well as arrays of S&As having the same or even different functionalities. In addition, new sensor types such as nanowire and CNT-based devices have already demonstrated their potential for improved sensitivity and better performance. In the future, further improvements and new functionalities can be expected. The fabrication processes for these new nanodevices are still far from being compatible with large-scale-production. Processes have to be developed to integrate such new sensing elements onto a CMOS platform and into devices, systems, and applications.

The VLSI of arrays of thousands or even millions of S&As on a CMOS platform has been successfully demonstrated by the Texas Instruments Digital Micromirror Device product [64] and by IBM's AFM-probe-based data-storage research prototype [175]. Complex individual control for each S&A is possible as each element is interconnected to its own dedicated electronics. New system functionalities and highly parallel operation of very large arrays can be achieved by such systems, giving rise to novel applications. However, large volumes and markets are necessary to support the development costs of these rather complex technologies and systems. On the basis of these concepts and techniques, it can be envisioned that future nanoscale fabrication tool boxes will be possible that enable local atomic or molecular modifications at large scale and with high throughput.

In summary, S&As on CMOS platforms are a reality and in several cases even product success stories. This success is based on the great maturity and rapid advancement of CMOS technology, which greatly inspired and supported silicon-based S&A development. Modified microelectronics fabrication techniques led to the development of wet chemical and DRIE-based micromachining techniques, which enable the fabrication of complex micro/nanomechanical structures to be used in S&As. S&As on CMOS platform are good examples of the potential and opportunities for "more than Moore" that will be greatly supported by the continuation of "more Moore."

Acknowledgment With great pleasure and appreciation, the authors acknowledge the encouragement of Paul Seidler and Werner Bux, the Department Managers of Science & Technology and Systems, respectively, and the editorial support of Charlotte Bolliger from the Publications Group of the IBM Zurich Research Laboratory. In addition, numerous interesting and stimulating discussions and collaborations with the many colleagues from the teams at IBM and STMicroelectronics were important cornerstones for the content of this chapter. Thank you all very much.

References

1. Lion KS (1969) *Transducers: problems and prospects*. IEEE Transactions on Industrial Electronics and Control Instrumentation IECI–16:2–5.
2. Middelhoek S, Noorlag D (1981) *Three-dimensional representation of input and output transducers*. Sensors and Actuators 2:29–41.
3. Waser R (2003) *Nanoelectronics and information technology*. Wiley-VCH, Weinheim.
4. Gopel W, Hesse J, Zemel J (1989) *Sensors a comprehensive survey*. VCH, Weinheim.
5. Sze SM (1994) *Semiconductor sensors*. Wiley, New York.
6. Atkinson GM, Pearson RE, Ounaies Z, et al. (2003) *Novel piezoelectric polyimide MEMS*. Proceedings of the IEEE International Solid-State Sensors and Actuators Conference, Boston, MA, USA, vol. 1, 782–785.
7. Wang LP, Deng K, Zou L, et al. (2002) *Microelectromechanical systems (MEMS) accelerometers using lead zirconate titanate thick films*. IEEE Electron Device Letters 23:182–184.
8. Shibata T, Unno K, Makino E (2003) *Diamond AFM probe with piezoelectric sensor and actuator*. Proceedings of the IEEE International Solid-State Sensors and Actuators Conference, Boston, MA, USA, vol. 1, 500–503.
9. Okuyama M, Yamashita K, Noda M (2003) *Infrared and ultrasonic sensors using ferroelectric thin films*. Proceedings of the IEEE International Solid-State Sensors and Actuators Conference, Boston, MA, USA, vol. 1, 226–229.
10. Kanda Y (1991) *Piezoresistance effect of silicon*. Sensors and Actuators A (Physical) A28:83–91.
11. Smith CS (1954) *Piezoresistance effect in germanium and silicon*. Physical Review 94:42.
12. Huang S, Li X, Wang Y, et al. (2003) *A piezoresistive accelerometer with axially stressed tiny beams for both much increased sensitivity and much broadened frequency bandwidth*.

Proceedings of the IEEE International Solid-State Sensors and Actuators Conference, Boston, MA, USA, vol. 1, 91–94.
13. Dao DV, Toriyama T, Sugiyama S, et al. (2003) *A MEMS-based microsensor to measure all six components of force and moment on a near-wall particle in turbulent flow*. Proceedings of the IEEE International Solid-State Sensors and Actuators Conference, Boston, MA, USA, vol. 1, 504–507.
14. Charlot B, Parrain F, Galy N, et al. (2004) *A sweeping mode integrated fingerprint sensor with 256 tactile microbeams*. Journal of Microelectromechanical Systems 13:636–644.
15. Charlot B, Parrain F, Galy N, et al. (2003) *A sweeping mode integrated tactile fingerprint sensor*. Proceedings of the IEEE International Solid-State Sensors and Actuators Conference, Boston, MA, USA, vol. 2, 1031–1034.
16. Seo C-T, Gong S-S, Sim J-H, et al. (1999) *Novel differential pressure sensor with silicon beams embedded in a silicone rubber membrane*. Proceedings of the IEEE International Solid-State Sensors and Actuators Conference, Sendai, Japan, 354–357.
17. Seo C-T, Kim Y-J, Shin J-K, et al. (2004) *A novel comb-type differential pressure sensor with silicon beams embedded in a silicone rubber membrane*. Japanese Journal of Applied Physics 43:2046–2049.
18. Svedin N, Kalvesten E, Stemme G (1999) *A new edge-detected lift force flow sensor*. Proceedings of the IEEE International Solid-State Sensors and Actuators Conference, Sendai, Japan, 1078–1081.
19. Svedin N, Kalvesten E, Stemme G (2003) *A new edge-detected lift force flow sensor*. Journal of Microelectromechanical Systems 12:344–354.
20. Baxter LK (1997) *Capacitive sensors design and applications*. IEEE Press, New York.
21. Tang WC-K, *Electrostatic comb drive for resonant sensor and actuator applications*, Ph.D. dissertation, University of California, Berkeley, CA, USA, 1990.
22. Puers R (1993) *Capacitive sensors: when and how to use them*. Sensors and Actuators A (Physical) A37–A38:93–105.
23. Dai C-L (2007) *A capacitive humidity sensor integrated with micro heater and ring oscillator circuit fabricated by CMOS-MEMS technique*. Sensors and Actuators B (Chemical) 122:375–380.
24. Malcovati P, Haberli A, Mayer F, et al. (1995) *Combined air humidity and flow CMOS microsensor with on-chip 15 bit sigma-delta A/D interface*. Proceedings of the Symposium on VLSI Circuits, Kyoto, Japan, 45–46.
25. Yotter RA, Baxter RR, Ohno S, et al. (2003) *On a micromachined fluidic inclinometer*. Proceedings of the IEEE International Solid-State Sensors and Actuators Conference, Boston, MA, USA, vol. 2, 1279–1282.
26. Dauderstadt UA, de Vries PHS, Hiratsuka R, et al. (1995) *Silicon accelerometer based on thermopiles*. Sensors and Actuators A (Physical) A46:201–204.
27. Binnig G, Despont M, Drechsler U, et al. (1999) *Ultrahigh-density atomic force microscopy data storage with erase capability*. Applied Physics Letters 74:1329–1331.
28. Lantz MA, Binnig GK, Despont M, et al. (2005) *A micromechanical thermal displacement sensor with nanometre resolution*. Nanotechnology 16:1089–1094.
29. Duerig U (2005) *Fundamentals of micromechanical thermoelectric sensors*. Journal of Applied Physics 98:1–14.
30. Trimmer WSN, Gabriel KJ (1987) *Design considerations for a practical electrostatic micromotor*. Sensors and Actuators 11:198–206.
31. Hoen S, Bai Q, Harley JA, et al. (2003) *A high-performance dipole surface drive for large travel and force*. Proceedings of the IEEE International Solid-State Sensors and Actuators Conference, Boston, MA, USA, vol. 1, 344–347.
32. Legtenberg R, Groeneveld AW, Elwenspoek M (1996) *Comb-drive actuators for large displacements*. Journal of Micromechanics and Microengineering 6:320–329.
33. Oboe R, Antonello R, Lasalandra E, et al. (2005) *Control of a z-axis MEMS vibrational gyroscope*. IEEE/ASME Transactions on Mechatronics 10:364–370.

34. Kim CJ, Pisano AP, Muller RS, et al. (1990) *Polysilicon microgripper*. Proceedings of the IEEE Solid-State Sensor and Actuator Workshop, Hilton Head Island, SC, USA, 48–51.
35. Del Sarto M, Sassolini S, Baldo L, et al. (2003) *Batch fabricated silicon electrostatic micropositioner for dual stage actuation*. Proceedings of the SPIE 4979:155–164.
36. Li ZH, Zhang DC, Li T, et al. (2000) *Bulk micromachined relay with lateral contact*. Journal of Micromechanics and Microengineering 10:329–333.
37. Xu Y, MacDonald NC, Miller SA (1995) *Integrated micro-scanning tunneling microscope*. Applied Physics Letters 67:2305–2307.
38. Lu Y, Pang CK, Chen J, et al. (2005) *Design, fabrication and control of a micro X-Y stage with large ultra-thin film recoding media platform*. Proceedings of the IEEE/ASME International Conference on Advanced Intelligent Mechatronics, Monterey, CA, USA, 19–24.
39. Hoen S, Merchant P, Koke G, et al. (1997) *Electrostatic surface drives: theoretical considerations and fabrication*. Proceedings of the International Solid State Sensors and Actuators Conference, Chicago, IL, USA, 41–44.
40. Klaasse G, Puers B, Tilmans HAC (2004) *Piezoelectric actuation for application in RF-MEMS switches*. Proceedings of the SPIE 5455:174–180.
41. Park JH, Lee HC, Park YH, et al. (2006) *A fully wafer-level packaged RF MEMS switch with low actuation voltage using a piezoelectric actuator*. Journal of Micromechanics and Microengineering 16:2281–2286.
42. Park JY, Yee YJ, Nam HJ, et al. (2001) *Micromachined RF MEMS tunable capacitors using piezoelectric actuators*. Proceedings of the IEEE MTT-S International Microwave Symposium, Phoenix, AZ, USA, vol. 3, 2111–2114.
43. Maillefer D, van Lintel H, Rey-Mermet G, et al. (1999) *A high-performance silicon micropump for an implantable drug delivery system*. Proceedings of the IEEE International Workshop on Micro Electro Mechanical Systems - MEMS, Orlando, FL, USA, 541–546.
44. Kobayashi T, Itoh T, Sawada R, et al. (2007) *Tunable optical microscanner driven by piezoelectric actuator*. Japanese Journal of Applied Physics Part 1-Regular Papers Brief Communications & Review Papers 46:6429–6432.
45. Koganezawa S, Uematsu Y, Yamada T, et al. (1999) *Dual-stage actuator system for magnetic disk drives using a shear mode piezoelectric microactuator*. IEEE Transactions on Magnetics 35:988–992.
46. Guckel H, Klein J, Christenson T, et al. (1992) *Thermo-magnetic metal flexure actuators*. Proceedings of the IEEE Solid-State Sensor and Actuator Workshop, Hilton Head Island, SC, USA, 73–75.
47. Lakdawala H, Fedder G (1999) *Analysis of temperature-dependent residual stress gradients in CMOS micromachined structures*. Proceedings of the IEEE International Solid-State Sensors and Actuators Conference, Sendai, Japan, 526–529.
48. Zheng LS, Lu MSC (2007) *A large-displacement CMOS micromachined thermal actuator with comb electrodes for capacitive sensing*. Sensors and Actuators A (Physical) 136:697–703.
49. Nguyen NT, Ho SS, Low CLN (2004) *A polymeric microgripper with integrated thermal actuators*. Journal of Micromechanics and Microengineering 14:969–974.
50. Noworolski JM, Klaassen EH, Logan JR, et al. (1996) *Process for in-plane and out-of-plane single-crystal-silicon thermal microactuators*. Sensors and Actuators A (Physical) A55:65–69.
51. Field LA, Burriesci DL, Robrish PR, et al. (1996) *Micromachined 1*2 optical-fiber switch*. Sensors and Actuators A (Physical) A53:311–316.
52. Comtois JH, Bright VM (1997) *Applications for surface-micromachined polysilicon thermal actuators and arrays*. Sensors and Actuators A (Physical) A58:19–25.
53. Trimmer WSN (1989) *Microrobots and micromechanical systems*. Sensors and Actuators 19:267–287.

54. Zhang YH, Ding GF, Shun XF, et al. (2007) *Preparing of a high speed bistable electromagnetic RF MEMS switch*. Sensors and Actuators A (Physical) 134:532–537.
55. Ahn CH, Kim YJ, Allen MG (1993) *A planar variable reluctance magnetic micromotor with fully integrated stator and coils*. Journal of Microelectromechanical Systems 2:165–173.
56. Choi JJ, Park H, Kim KY, et al. (2001) *Electromagnetic micro x-y stage with very thick Cu coil for probe-based mass data storage device*. Proceedings of the SPIE 4334:363–371.
57. Hsieh HT, Su GD (2007) *Reliability of a MEMS actuator improved by spring corner designs and reshaped driving waveforms*. Sensors 7:1720–1730.
58. Rothuizen H, Despont M, Drechsler U, et al. (2002) *Compact copper/epoxy-based electromagnetic scanner for scanning probe applications*. Proceedings of the IEEE International Conference on Micro Electro Mechanical Systems – MEMS, Las Vegas, NV, USA, 582–585.
59. Rothuizen H, Drechsler U, Genolet G, et al. (2000) *Fabrication of a micromachined magnetic X/Y/Z scanner for parallel scanning probe applications*. Microelectronic Engineering 53:509–512.
60. Lantz MA, Rothuizen HE, Drechsler U, et al. (2007) *A vibration resistant nanopositioner for mobile parallel-probe storage applications*. Journal of Microelectromechanical Systems 16:130–139.
61. Judy JW, Muller RS, Zappe HH (1995) *Magnetic microactuation of polysilicon flexure structures*. Journal of Microelectromechanical Systems 4:162–169.
62. Judy JW, Muller RS (1996) *Magnetic microactuation of torsional polysilicon structures*. Sensors and Actuators A (Physical) A53:392–397.
63. Vigna B (2005) *More than Moore: micro-machined products enable new applications and open new markets*. Proceedings of the International Electron Devices Meeting, Washington, DC, USA, 8 pp.
64. DLP by Texas Instruments, Dallas, TX, USA. *http://www.dlp.com*. Accessed 04 Jan 08.
65. Nagel DJ, Zaghloul ME (2001) *MEMS: micro technology, mega impact*. IEEE Circuits and Devices Magazine 17:14–25.
66. Vigna B (2007) *Physical sensors drive MEMS consumerization wave*. Proceedings of the International Micromachine/Nanotech Symposium, Tokyo, Japan.
67. Bourne M (2007) *A consumer's guide to MEMS and nanotechnology*. Bourne Research LLC.
68. Murari B (2003) *Bridging the gap between the digital and real worlds: the expanding role of analog interface technologies*. Proceedings of the IEEE International Solid-State Circuits Conference, San Francisco, CA, USA, 30–35.
69. Gola A, Pasolini F, Chiesa E, et al. (2003) *A 2.5-rad/s(2) resolution digital output MEMS-based rotational accelerometer for HDD applications*. IEEE Transactions on Magnetics 39:915–919.
70. Johnson R (2007) There's more to MEMS than meets the iPhone. EETimes, *http://www.eetimes.com/showArticle.jhtml?articleID=200900669*.
71. Toedosio M (2007) Making MEMS the new 'mouse'. EETimes Asia, *http://www.eetasia.com/ART_8800460968_499495_NT_734b413b.HTM*.
72. Marriott M (2006) At the heart of the Wii, micron-size machines. New York Times, *http://www.nytimes.com/2006/12/21/technology/21howw.html*. Accessed 04 Jan 08.
73. Yoshida M (2006) For Nintendo, MEMS the word. EETimes, *http://www.eetimes.com/showArticle.jhtml?articleID=187201578*.
74. Nuttal C (2006) Video games take the next virtual leap. Financial Times, *http://search.ft.com/ftArticle?queryText=%22Shigeru+Miyamoto%22&id=060509009782&ct=0&nclick_check=1*. Accessed 04 Jan 08.
75. Sachs D, Nasiri S, Goeh D (2007) Image stabilization technology overview. invensense.com, *http://www.invensense.com/shared/pdf/ImageStabilizationWhitepaper_051606.pdf*.
76. Marriott M (2005) Getting around, made easier. New York Times, *http://travel.nytimes.com/2005/12/15/technology/circuits/15basics.html*. Accessed 04 Jan 08.

77. Makimoto T, Doi TT (2002) *Chip technologies for entertainment robots - present and future*. Proceedings of the IEEE International Electron Devices Meeting, San Francisco, CA, USA, 9–16.
78. Chalard L, Helal D, Verbaere L, et al. (2007) *Wireless sensor networks devices: overview, issues, state of the art and promising technologies*. ST Journal of Research 4:4–18.
79. Wang ZL (2008) *Self-powered nanotech*. Scientific American 50:66–71.
80. Allan R (2005) *Wireless sensors land anywhere and everywhere*. Electronic Design 53:65–70.
81. Allan R (2004) *The future of sensors*. Electronic Design 52:51–55.
82. Cheyney T (2007) As MEMS "consumerization wave" surges, developers address production challenges. Small Times, *http://www.smalltimes.com/articles/article_display.cfm?article_id=309262*.
83. 2007 Industry report: Focus on Integration. MEMS Industry Group, *http://www.memsindustrygroup.org*.
84. The Mosis Service, CA, USA. *http://www.mosis.com/*.
85. Schurig E, Demierre M, Schott C, et al. (2002) *A vertical Hall device in CMOS high-voltage technology*. Sensors and Actuators A (Physical) 97(8):47–53.
86. Klaassen EH, Reay RJ, Kovacs GT (1996) *Diode-based thermal RMS converter with on-chip circuitry fabricated using CMOS technology*. Sensors and Actuators A (Physical) 52:33–40.
87. Kim YS, Jang S, Lee CS, et al. (2007) *Thermo-piezoelectric Si_3N_4 cantilever array on CMOS circuit for high density probe-based data storage*. Sensors and Actuators A (Physical) 135:67–72.
88. Hilbert C, Curtis H (2000) MEMS/Microsystems Device and Process Technologies. LETI, France, *http://www.wtec.org/loyola/mcc/mems_eu/Pages/Chapter-5.html*. Accessed 04/Jan/08.
89. Marek J, Illing M (2002) *Micromachined sensors for automotive applications*. Proc. IEEE Sensors, Orlando, FL, USA, vol. 2, 1561–1564.
90. Brand O (2005) *Fabrication technology*. In: Brand O, Fedder GKs (eds.) *CMOS-MEMS: advanced micro & nanosystems*, Wiley-VCH, Weinheim.
91. Lutz M, Partridge A, Gupta P, et al. (2007) *MEMS oscillators for high volume commercial applications*. Proceedings of the International Solid-State Sensors, Actuators and Microsystems Conference, Lyon, France, 49–52.
92. Arkalgud S (2007) 3-D integration options. Semiconductor International, *http://www.semiconductor.net/webcastsDetail/2140065776.html*.
93. Sze SM (1988) *VLSI technology*. McGraw-Hill, New York.
94. Ghandhi SK (1994) *VLSI fabrication principles silicon and gallium arsenide*. Wiley, New York.
95. Madou MJ (1997) *Fundamentals of microfabrication*. CRC Press, Boca Raton.
96. Gad-el-Hak M (2002) *The MEMS handbook*. CRC Press, Boca Raton.
97. Lee H-J, Park J-Y, Bu J (2005) *Piezoelectrically actuated RF MEMS DC contact switches with low voltage operation*. IEEE Microwave and Wireless Components Letters 15:202–204.
98. Schow CL, Doany FE, Tsang CK, et al. (2008) *300-Gb/s, 24-channel full-duplex, 850-nm, CMOS-based optical transceivers*. Proceedings of the Optical Fiber Communications Conference, San Diego, CA, USA, 1–3.
99. Schares L, Kash JA, Doany FE, et al. (2006) *Terabus: terabit/second-class card-level optical interconnect technologies*. IEEE Journal of Selected Topics in Quantum Electronics 12:1032–1044.
100. Goossen KW, Walker JA, D'Asaro LA, et al. (1995) *GaAs MQW modulators integrated with silicon CMOS*. IEEE Photonics Technology Letters 7:360–362.
101. Petersen KE (1982) *Silicon as a mechanical material*. Proceedings of the IEEE 70:420–57.
102. Laermer F, Schilp A, Funk K, et al. (1999) *Bosch deep silicon etching: improving uniformity and etch rate for advanced MEMS applications*. Proceedings of the International Workshop on Micro Electro Mechanical Systems - MEMS, Orlando, FL, USA, 211–216.

103. Core TA, Tsang WK, Sherman SJ (1993) *Fabrication technology for an integrated surface-micromachined sensor*. Solid State Technology 36:39-ff.
104. Smith JH, Montague S, Sniegowski JJ (1995) *Material and processing issues for the monolithic integration of microelectronics with surface-micromachined polysilicon sensors and actuators*. Proceedings of the SPIE 2639:64–73.
105. Westberg D, Paul O, Andersson GI, et al. (1996) *Surface micromachining by sacrificial aluminium etching*. Journal of Micromechanics and Microengineering 6:376–384.
106. Jahnes CV, Cotte J, Lund JL, et al. (2004) *Simultaneous fabrication of RF MEMS switches and resonators using copper-based CMOS interconnect manufacturing methods*. Proceedings of the IEEE International Conference on Micro Electro Mechanical Systems – MEMS, Maastricht, The Netherlands, 789–792.
107. Spiller E, Feder R, Topalian J, et al. (1976) *X-ray lithography for bubble devices*. Solid State Technology 19:62–67.
108. Despont M, Lorenz H, Fahrni N, et al. (1997) *High-aspect-ratio, ultrathick, negative-tone near-uv photoresist for MEMS applications*. Proceedings of the IEEE International Workshop on Micro Electro Mechanical Systems - MEMS, Nagoya, Japan, 518–522.
109. Lang W, Steiner P, Sandmaier H (1995) *Porous silicon: a novel material for microsystems*. Sensors and Actuators A (Physical) 51:31–36.
110. Archer M, Christophersen M, Fauchet PM (2005) *Electrical porous silicon chemical sensor for detection of organic solvents*. Sensors and Actuators B (Chemical) 106:347–357.
111. Nam CM, Kwon YS (1997) *High-performance planar inductor on thick oxidized porous silicon (OPS) substrate*. IEEE Microwave and Guided Wave Letters 7:236–238.
112. Lehmann V, Honlein W, Reisinger H, et al. (1996) *A novel capacitor technology based on porous silicon*. Thin Solid Films 276:138–142.
113. Lammel G, Renaud P (2000) *Free-standing, mobile 3D porous silicon microstructures*. Sensors and Actuators A (Physical) 85:356–360.
114. Brazzle JD, Dokmeci MR, Mastrangelo CH (2004) *Modeling and characterization of sacrificial polysilicon etching using vapor-phase xenon difluoride*. Proceedings of the IEEE International Conference on Micro Electro Mechanical Systems – MEMS, Maastricht, The Netherlands, 737–740.
115. Lee YI, Park KH, Lee J, et al. (1997) *Dry release for surface micromachining with HF vapor-phase etching*. Journal of Microelectromechanical Systems 6:226–233.
116. Baltes HP, Baltes HP, Popovic RS (1986) *Integrated semiconductor magnetic field sensors*. Proceedings of the IEEE 74:1107–1132.
117. Popovic RS, Baltes HP (1983) *A CMOS magnetic field sensor*. IEEE Journal of Solid-State Circuits 18:426–428.
118. Micronas, Zurich, Switzerland. *http://www.micronas.com/products/index.html*. Accessed 04/Jan/08.
119. Infineon Technologies, Munich, Germany. *http://www.infineon.com/sensors/*.
120. Allegro Microsystems, Worcester, MA, USA. *http://www.allegromicro.com/hall/*.
121. Melexis Microelectronic Systems, Belgium. *http://www.melexis.com/prod_hall.asp*.
122. Pham D, Aipperspach T, Boerstler D, et al. (2006) *Overview of the architecture, circuit design, and physical implementation of a first-generation cell processor*. IEEE Journal of Solid-State Circuits 41:179–196.
123. Lammel G, Armbruster S, Schelling C, et al. (2005) *Next generation pressure sensors in surface micromachining technology*. Proc. International Conference on Solid-State Sensors, Actuators and Microsystems, Seoul, South Korea, 35–36.
124. Seshia AA, Palaniapan M, Roessig TA, et al. (2002) *A vacuum packaged surface micromachined resonant accelerometer*. Journal of Microelectromechanical Systems 11:784–793.
125. Bhave SA, Seeger JI, Xuesong J, et al. (2003) *An integrated, vertical-drive, in-plane-sense microgyroscope*. Proceedings of the IEEE International Solid-State Sensors and Actuators Conference, Boston, MA, USA, vol. 1, 171–174.
126. Sensirion Inc., Staefa, Switzerland. *http://www.sensirion.com*. Accessed 04/Jan/08.

127. Mayer F (2008) *CMOS-based sensors: From a sample to real products*. Proceedings of the 21st IEEE International Conference on Micro Electro Mechanical Systems – MEMS, Tucson, AZ, USA, 1–5.
128. Schott C, Schott C, Racz R, et al. (2007) *CMOS single-chip electronic compass with microcontroller*. IEEE Journal of Solid-State Circuits 42:2923–2933.
129. Toshikiyo K, Yogo T, Ishii M, et al. (2007) *A MOS image sensor with microlenses built by sub-wavelength patternings*. Proceedings of the IEEE International Solid-State Circuits Conference, San Francisco, CA, USA, 514–515.
130. Jung S, Thewes R, Scheiter T, et al. (1999) *CMOS fingerprint sensor with automatic local contrast adjustment and pixel-parallel encoding logic*. Proceedings of the Symposium on VLSI Circuits, 161–164.
131. Manaresi N, Rambaldi R, Tartagni M, et al. (1999) *A CMOS-only micro touch pointer*. IEEE Journal of Solid-State Circuits 34:1860–1868.
132. Seung-Min J, Jin-Moon N, Dong-Hoon Y, et al. (2005) *A CMOS integrated capacitive fingerprint sensor with 32-bit RISC microcontroller*. IEEE Journal of Solid-State Circuits 40:1745–1750.
133. Lewis S, Lewis S, Alie S, et al. (2003) *Integrated sensor and electronics processing for >10^8 "iMEMS" inertial measurement unit components*. Proceedings of the IEEE International Electron Devices Meeting, 39.1.1–39.1.4.
134. Hagleitner C, Hierlemann A, Lange D, et al. (2001) *Smart single-chip gas sensor microsystem*. Nature 414(6861):293–296.
135. Brandl M, Kempe V (2001) *High performance accelerometer based on CMOS technologies with low cost add-ons*. Proceedings of the IEEE International Conference on Micro Electro Mechanical Systems - MEMS, Interlaken, Switzerland, 6–9.
136. Nasiri S (2007) A critical review of MEMS gyroscopes technology and commercialization status. invensense.com. *http://www.invensense.com/shared/pdf/MEMSGyroComp.pdf*
137. Invensense, CA, USA. *http://www.invensense.com/products/index.html.*
138. Bauer T (2007) High density through wafer via technology. Silex. *http://www.silex microsystems.com/docs/Silex_High_Density_Through_Wafer_Via_Technology.pdf*
139. Bakke T, Volker B, Friedrichs M, et al. (2006) *Micromirror array of monocrystalline silicon*. Proceedings of the IEEE/LEOS International Conference on Optical MEMS and Their Applications, Big Sky, MT, USA, 128–129.
140. van der Wijngaart W, Guerre R, Despont M, et al. (2008) Recent advances in wafer scale transfer bonding technologies. MST News, *http://www.mstnews.de/*. Accessed 04/Jan/08.
141. Despont M, Drechsler U, Yu R, et al. (2004) *Wafer-scale microdevice transfer/interconnect: its application in an AFM-based data-storage system*. Journal of Microelectromechanical Systems 13:895–901.
142. Kim YS, Nam HJ, Jang S, et al. (2006) *Wafer-level transfer of thermo-piezoelectric Si_3N_4 cantilever array on a CMOS circuit for high density probe-based data storage*. Proceedings of the IEEE International Conference on Micro Electro Mechanical Systems - MEMS, Istanbul, Turkey, 922–925.
143. Guerre R, Drechsler U, Jubin D, et al. (2007) *CMOS-compatible wafer-level microdevice-distribution technology*. Proceedings of the International Solid-State Sensors, Actuators and Microsystems Conference, Lyon, France, 2087–2090.
144. Niklaus F, Kalvesten E, Stemme G (2001) *Wafer-level membrane transfer bonding of polycrystalline silicon bolometers for use in infrared focal plane arrays*. Journal of Micromechanics and Microengineering 11:509–513.
145. Niklaus F, Stemme G, Lu JQ, et al. (2006) *Adhesive wafer bonding*. Journal of Applied Physics 99(3):031101.
146. Haasl S, Niklaus F, Stemme G (2003) *Arrays of monocrystalline silicon micromirrors fabricated using CMOS compatible transfer bonding*. Proceedings of the IEEE International Conference on Micro Electro Mechanical Systems – MEMS, Kyoto, Japan, 271–274.

147. Ballabriga R, Campbell M, Heijne EH, et al. (2007) *The Medipix3 prototype, a pixel readout chip working in single photon counting mode with improved spectrometric performance*. IEEE Transactions on Nuclear Science 54:1824–1829.
148. Singh A, Horsley DA, Cohn MB, et al. (1999) *Batch transfer of microstructures using flip-chip solder bonding*. Journal of Microelectromechanical Systems 8:27–33.
149. Senturia SD (1998) *CAD challenges for microsensors, microactuators, and microsystems*. Proceedings of the IEEE 86:1611–1626.
150. Hagleitner C, Bonaccio T, Rothuizen H, et al. (2007) *Modeling, design, and verification for the analog front-end of a MEMS-based parallel scanning-probe storage device*. IEEE Journal of Solid-State Circuits 42:1779–1789.
151. Coventor, MA, USA. *http://www.coventor.com/*.
152. Tanner Research, CA, USA. *http://www.tanner.com/*.
153. Tummala RR (2001) *Fundamentals of microsystems packaging*. McGraw-Hill, New York.
154. Brand O, Baltes H (2002) *Microsensor packaging*. Microsystem Technologies 7:205–208.
155. Bright VM, Stoldt CR, Monk DJ, et al. (2004) *Packaging of advanced micro- and nanosystems*. In: Baltes H, Brand O, et al. (eds.) *Enabling technology for MEMS and nanodevices*, Wiley-VCH, Weinheim.
156. Gilleo K (2002) *MEMS packaging and assembly challenges. Area array packaging handbook*, McGraw-Hill, New York.
157. Reichl H, Grosser V (2001) *Overview and development trends in the field of MEMS packaging*. Proceedings of the IEEE International Conference on Micro Electro Mechanical Systems – MEMS, Interlaken, Switzerland, 1–5.
158. Yazdi N, Ayazi F, Najafi K (1998) *Micromachined inertial sensors*. Proceedings of the IEEE 86:1640–1659.
159. STMicroelectronics, Geneva, Switzerland. *www.st.com/mems*.
160. Johnson R (2007) 'Chip-on-MEMS' enables wafer-level calibration. EEtimes Asia, *http://www.eetasia.com/ART_8800493826_480500_NT_b48f765c.HTM*.
161. Knechtel R (2005) *Glass frit bonding: an universal technology for wafer level encapsulation and packaging*. Microsystem Technologies 12:63–68.
162. Zhang X, Tee TY, Luan J (2005) *Comprehensive warpage analysis of stacked die MEMS package in accelerometer application*. Proceedings of the International Conference on Electronics Packaging Technology, Shenzhen, China, 581–586.
163. Roukes ML (2000) *Nanoelectromechanical systems*. Proceedings of the IEEE Solid-State Sensor and Actuator Workshop, Hilton Head Island, SC, USA, 367–376.
164. Mamin HJ, Rugar D (2001) *Sub-attonewton force detection at millikelvin temperatures*. Applied Physics Letters 79:3358–3360.
165. Yang YT, Callegari C, Feng XL, et al. (2006) *Zeptogram-scale nanomechanical mass sensing*. Nano Letters 6:583–586.
166. Fon WC, Schwab KC, Worlock JM, et al. (2005) *Nanoscale, phonon-coupled calorimetry with sub-attojoule/Kelvin resolution*. Nano Letters 5:1968–1971.
167. Cui Y, Wei QQ, Park HK, et al. (2001) *Nanowire nanosensors for highly sensitive and selective detection of biological and chemical species*. Science 293:1289–1292.
168. Patolsky F, Zheng GF, Hayden O, et al. (2004) *Electrical detection of single viruses*. Proceedings of the National Academy of Sciences of the United States of America 101:14017–14022.
169. Patolsky F, Zheng G, Lieber CM (2006) *Nanowire sensors for medicine and the life sciences*. Nanomedicine 1:51–65.
170. Heer F, Keller M, Yu G, et al. (2008) *CMOS electro-chemical DNA detection array with on-chip ADC*. Proceedings of the IEEE International Solid-State Circuits Conference, San Francisco, CA, USA.
171. Patolsky F, Timko BP, Zheng GF, et al. (2007) *Nanowire-based nanoelectronic devices in the life sciences*. MRS Bulletin 32:142–149.

172. Stampfer C, Jungen A, Linderman R, et al. (2006) *Nano-electromechanical displacement sensing based on single-walled carbon nanotubes*. Nano Letters 6:233–237.
173. Stampfer C, Jungen A, Hierold C (2006) *Fabrication of discrete nanoscaled force sensors based on single-walled carbon nanotubes*. IEEE Sensors Journal 6:613–617.
174. Grow RJ, Wang Q, Cao J, et al. (2005) *Piezoresistance of carbon nanotubes on deformable thin-film membranes*. Applied Physics Letters 86:093104.
175. Vettiger P, Albrecht T, Despont M, et al. (2003) *Thousands of microcantilevers for highly parallel and ultra-dense data storage*. Proceedings of the IEEE International Electron Devices Meeting, Washington, DC, USA, 32.1.1–32.1.4.

Chapter 6
Biochips

C. Vauchier and P. Puget

Abstract This chapter gives an introduction to the basics of general architecture of microsystem for biological and chemical applications. A typical microfluidics system may contain many "technological blocks" relating to various stages of the sample preparation and analysis. Commercial success of microfluidic systems requires development of key technologies for fabrication, fast prototyping, and system integration. The development of technologies for integrated biochips and medical microsystems, therefore, represent an exciting challenge for the near future. Applications of such integrated systems include in-vitro diagnostics, environment monitoring, homeland security, and life sciences.

Keywords Microfluidics • Integration • Lab-on-chip • DNA • Molecular diagnostic • Microarray • PCR • Sample preparation • Electrowetting • EWOD • Chemical analysis

6.1 Introduction

Reliable detection and fast measurement of the concentration of biological or chemical species in a raw sample is a key issue for many fields such as environmental analysis, medical diagnosis, homeland security or agro food industry. The answer to this problem is done classically according to two approaches. The first suggested by the biological (biosensors) and chemical sensors aims at carrying out a measurement 'in situ' in the sample. Constraints of detection limits, specificity, and stability are entirely deferred on the sensor. The second approach, suggested by the Total Analysis Systems, μTAS, or lab-on-a-chip, consists in treating the raw sample according to a complete protocol based on several elementary steps such as sampling, extraction, purification, concentration, and at last detection. The latter is done

C. Vauchier (✉) and P. Puget
CEA-LETI-Minatec, 17 rue des Martyrs, 38054 Grenoble Cedex 9, France
e-mail: claude.vauchier@cea.fr

G.Q. Zhang and A.J. van Roosmalen (eds.), *More than Moore: Creating High Value Micro/Nanoelectronics Systems*,
DOI 10.1007/978-0-387-75593-9_6,

under conditions more favourable than previously and makes it possible to obtain a better measurement, which can justify the additional complexity of the treatment. It is this approach which is generally implemented for pathogen detection from biological (case of in-vitro diagnosis (IVD) in particular) or environmental samples. Diverse applications areas are targeted from clinical and point of care diagnostic devices to the biotechnology market with devices for high-throughput screening (HTS), gene expression analysis, phramacogenomics, and portable environmental detection instrumentation, showing one or more following characteristics:

- On-site analysis, i.e. instruments with small size, low power consumption, low cost
- Easy and friendly to use, as possible by a non-specialized personnel
- If required, to be able to treat a great number of samples in a limited time (typically for HTS applications)
- Small size samples

Thus ensures a high level of performances in terms, in particular, of (1) detection limit, which must be lowest possible, (2) specificity (capacity to detect a particular molecule in a complex sample), and (3) speed.

Regarding these needs, since about 30 years, the development of the micro and nanotechnologies allowed the extraordinary evolution of electronics (increase in the performances and lower costs). Microtechnologies are characterized by the fact that the objects based on characteristic size about micron are manufactured in a collective way on a silicon substrate (wafer), thus making it possible to divide by several orders of magnitude the cost of several fabrication steps. Beyond strict microelectronics and integrated circuits, these manufacturing processes were implemented for other applications, to carry out in particular miniaturized devices (microsystems), which represent credible technological solutions for new analysis instruments presented earlier. The contribution of microtechnologies for the biological and chemical analysis is large and varied. We can give a non-exhaustive list of examples:

- Technologies of micromachining and collectives assembling allowing to manufacture objects including elementary size structures of about the micron or lower
- Principles of mechanical, magnetic or electromechanical actuation (e.g. electrostatic or piezoelectric)
- Optoelectronics functions, as magnetic detectors or sources (interesting for all detections based on fluorescence or chemiluminescence)
- Other methods of transduction, for example, magnetic (with giant magnetoresistance-GMR, able to detect in a very significant way the presence of magnetic micro or nanoparticles, classically used in biological and medical analysis) or electrochemical
- Possibility of fulfilling analogical or numerical electronic functions for acquisition step or data treatment

It is in this context, advances in microtechnologies and needs in new tools for biological analysis, that the concept of lab-on-a-chip or micro total analysis systems (μTAS) for the first time was proposed by Andreas Manz in 1990 [1]. Since this date, lab-on-a-chip is defined thus as microsystems performing one or more steps of a biological or chemical protocol analysis. Lab-on-a-chip domain was largely developed in the nineties, in terms of number of actors, scientific publications, patents in the nineties, companies, and commercial products. Several articles of review exist in the literature [2–5], as well as specialized works [6].

Application domains, which pushed for ten years developments relating to lab-on-a-chip, are mainly in the field of life sciences. In the field of IVD, in particular, analyses (especially in Europe) are prescribed by a doctor and are carried out in a laboratory. Eventually, sampling and analysis are not done at the same place. This organization induced (1) a significant time between the test prescription and obtaining the result, and (2) a complex management of samples and results (source of medical errors). So, there is a need for fast tests, making it possible to the doctor to make quasi immediate decisions during a traditional consultation, or in emergency medicine or operating room. For this kind of tool, we talk about "point-of-care" device [7]. These requirements in medical diagnosis can be transposed rather directly to the field of the veterinary medicine where there exists also a requirement in fast analysis systems usable easily in a farm, for example. Within the framework of agro-food control, detection systems are not only directed towards the search for pathogenic contaminants [8], but also, because of regulation on food quality control, towards the detection of genetically modified organisms (GMO) [9].

In the field of the environment control, for air [10] or water [11] analysis, we not only seek portable systems of high efficiencies, for the search of pollutants of industrial origin, but also within the framework of biological or chemical protection in the field of homeland security or against bioterrorism [12, 13]. Space research, in particular, within the framework of search of extraterrestrial life, finances the development of automated system for trace detection of amino acids in the ground of the planet Mars [14].

Lab-on-a-chip, finally, is interesting for HTS in pharmaceutical companies to provide new tools for analysis where a chemical or biological test is repeated a great number of times while varying the inlet reagents. In this case, constraints of portability are left here to privilege:

- Capacity to carry out the tests in parallel to minimize the duration of a test campaign
- Cost of the elementary test by decreasing the quantity of reagents used

Devices, contrary to the preceding applications, are not of single use but are used in general for a great number of tests.

6.2 General Architecture of Microsystem for Biology and Chemistry

As we saw earlier, a lab-on-a-chip carries out one or more operations of a protocol analysis. In the case of biological analysis, a complete protocol generally can be divided into three great operations (Fig. 6.1):

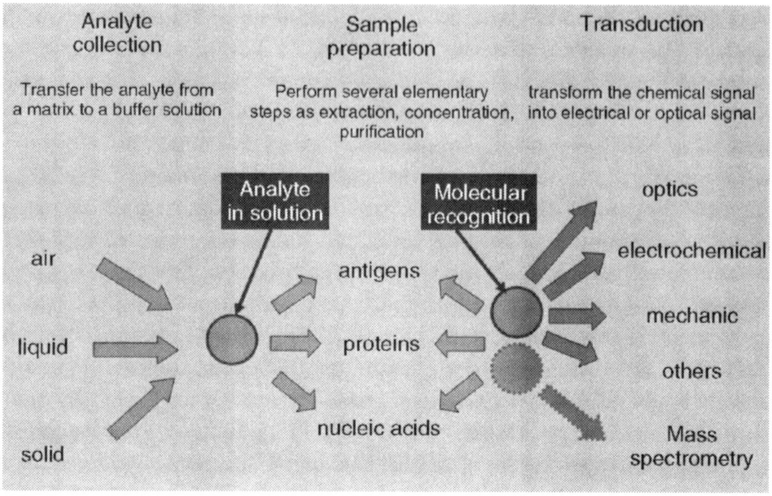

Fig. 6.1 Main steps of biological analysis protocol

- *Analyte collection*: Objective of this step is to collect analytes from a raw sample in air, liquid (blood, water, swab, etc.) or solid (biopsy, food, etc.) state. At the end of this step, analyte is transferred to a buffer solution.
- *Sample preparation*: The purpose of sample preparation is to bring the analyte under the optimal physicochemical conditions for the molecular recognition: extraction, concentration, purification, pH, temperature, addition of others chemical reagents, etc. In the particular case where the analyte is a nucleic acid, the preparation of sample can comprise a stage of molecular amplification, generally by polymerase chain reaction (PCR) [15].
- *Transduction*: More critical step concerns the molecular recognition of analyte, which is called the "target". Generally, the target is a protein or a fragment of nucleic acid, and it reacts with a known molecule, present in the system, the "probe", which is able to react specifically with the required target. This probe can be an antibody (in case of the immuno-assays), another protein, or a sequence of DNA, complementary to the sequence of nucleic acid required.

In the general case, beyond the biological analysis, the established protocols can differ from the preceding diagram but the principles remain similar and the operations of bases, detailed later, are common.

The diagram given in Fig. 6.1 applies to the molecular analyses. There are also others lab-on-a-chip developed especially for cell analysis (cellchip) [16, 17].

6.3 Microfluidics

This diversity of applications has given rise to a large number of technologies based on the manipulation of small volumes of liquids. As such, microfluidics has become a key research theme involving multi-disciplinary competencies, from mechanics, physical chemistry, biology, to microtechnology. Thus, we have witnessed the emergence of a huge variety of fluidic methods for the manipulation and control of fluids at a small scale. These various microfluidic approaches have all advantages and inconvenients, and it is rather difficult to predict which solutions for which application will prevail.

A large number of chemical and biological reactions have already been performed on microfluidic chips, albeit in research laboratory settings. Only a few commercialized microfluidic systems, such as the Agilent 2100 Bioanalyzer [18], have been widely adopted. The reason behind this successful integration can be explained by the simplicity of the fluidic protocols, the ease of use, and the robustness of the method. Thus a remaining challenge is the successful integration of a complex protocol within a single component. Indeed, the system (instrumentation, microcomponent, and protocol) needs to be autonomous, portable, and reliable while handling numerous fluidic functions such as calibrated volume dispensing, sub-volume fragmentation, coalescence, mixing, reagent storage.

Liu et al. [19] illustrates this integration concept, where a credit-card-sized chip performs a dozen elementary functions in series. However, just one sample is processed on a fairly complex protocol. One may furthermore want to perform several analyses in parallel to study the presence of several pathogens. If this is the case, chip design and associated instrument using pump and valves may become quickly difficult to integrate.

From a fluidic standpoint, it is not only a matter of moving a liquid, but also of implementing a set of basic fluidic functions leading to a fully integrated system. The integration complexity of the system must be looked not only at the microcomponent level but also at the instrumentation level. As we can see the choice of a microfluidic method displaying a strong promise of integration will be key for lab on chip dissemination.

Today it is possible to have a choice in a microfluidics toolbox as which solution is the most adapted for a specific application for fluid handling control. There are three main microfluidics concepts useful for lab-on-chip development. They can be described in Fig. 6.2.

6.3.1 *Continuous Flow and Monophasic Microfluidics*

A monophasic microfluidic device is composed of one or more etched channels with at least one dimension less than 1 mm. The volume of fluids within these channels is very small, usually several nanoliters, and the amount of reagents and analytes used is quite small. This is especially significant for expensive reagents. Microtechnologies used to construct silicon or glass microfluidic devices are relatively inexpensive and are very amenable both to highly elaborate, multiplexed devices and also to mass production.

The flow of a fluid through a microfluidic channel can be characterized by the Reynolds number, defined as

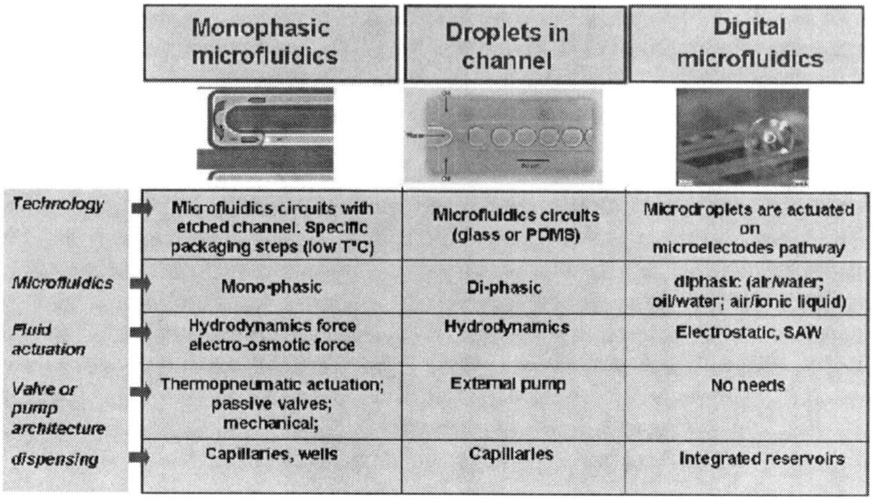

Fig. 6.2 Main microfluidics concepts for lab-on-chip device

$$\mathrm{Re} = \frac{\rho v L}{\mu},$$

where L is the characteristic length, μ is the dynamic fluid viscosity, ρ is the density of the fluid, and ρ is the mean fluid velocity of the flow. Because of the small dimensions of microchannels, the Re is usually much less than 100, often less than 1.0. In this Reynolds number regime, flow is completely laminar and no turbulence occurs. Laminar flow provides a means by which molecules can be transported in a relatively predictable manner through microchannels.

There are two common methods by which fluid handling through microchannels can be achieved.

In pressure-driven flow, the fluid is pumped through the device via positive displacement pumps, such as syringe pumps. One of the basic laws of fluid mechanics for pressure-driven laminar flow, the so-called no-slip boundary condition, states that the fluid velocity at the walls must be zero. This produces a parabolic velocity profile within the channel. Pressure-driven flow can be a relatively inexpensive and quite reproducible approach to pump fluids through microdevices. With the increasing efforts at developing functional micropumps, pressure-driven flow is also possible to miniaturize.

Another way for pumping fluids is that of electro-osmotic pumping. If the walls of a microchannel have an electric charge, an electric double layer of counter ions is formed at the walls. When an electric field is applied across the channel (between inlet and outlet of the channel), the ions in the double layer move towards the electrode of opposite polarity. This creates motion of the fluids near the walls and transfers viscous forces into convective motion of the bulk fluid. The velocity profile is uniform across the entire width of the channel.

6.3.2 Digital Microfluidics

The concepts and applications of ElectroWetting On Dielectric (EWOD) based droplet manipulation were first demonstrated at UCLA [20], Duke University [21], and CEA LETI [22]. Many articles in this domain have confirmed that the approach is promising and the technology is maturing.

Indeed, electrowetting can modify the contact angle of a droplet on a surface, thanks to electrostatic effects controlled by a set of electrodes placed within the component (Fig. 6.3). This wettability modification is reversible and acts locally at each reactive volume. Intuitively, one can understand that electrostatic forces are a good fit for objects of small dimensions, as capillarity forces resulting from Laplace law become stronger when the scale decreases. Energy needed to modify the shape of a droplet increases as the diameter of a droplet decreases, unless it directly affects the capillarity and, in particular, the wetting properties. Thus EWOD actuation operates independently and locally on every droplet. EWOD is, therefore, a better fit than a global actuation mode when implementing complex protocols.

EWOD has an enormous advantage from an integration point of view. Actuation is exclusively electric and does not require elements either prone to deformation (gates or valves) or in movement (pumps, syringes, etc.). Indeed, EWOD fluidic actuation necessitates only electrical connections, which are much more simple and standardized than fluidic connections for pressure-driven flow or pneumatic interface as used for controlling PDMS valves. Likewise, the control system requires only a PC and electrical switches easily miniaturized and integrated. A portable EWOD device has been described by S.-K. Fan [23], where droplets are controlled via a wireless PDA.

The ease with which individual droplets/reactions are manipulated enables the implementation of so-called self-organizing protocols consisting of several feed back loops. Thus given intermediate results, it will be possible to envision several alternatives: redo an analysis to confirm results, continue or stop a protocol, etc.

To summarize, EWOD allows for local actuation for individual droplet manipulations that is more adapted to complex protocol implementation than a global approach associated with droplet-based channel microfluidics. EWOD requires only electrical switches for actuation with no moving parts. This technology holds great promise for integration and miniaturization of numerous biological applications (Fig. 6.4).

Fig. 6.3 EWOD principle and pictures of a droplet without and with EWOD actuation

6.3.3 Diphasic Microfluidics in Microchannels

Digital microfluidics often consists in using two immiscible phases, where reactions are formed and moved in an immiscible fluid by pumps or pressure-driven flow. This method is called droplet-based channel microfluidics and is characterised by plug or slug flow. For certain flow rates, the two phases may be continuously injected with droplet formations ensured through hydrodynamic effects. Indeed, droplet-based channel microfluidics is an easy method that appears very efficient for high-speed droplet formation with well-controlled characteristics. Although it is possible to divide or sort droplets, it appears difficult to independently manipulate every droplet in terms of droplet-specific reagent addition or when droplet-specific protocols necessitating feed back loops are used. Indeed, individual droplet management for droplet-based channel microfluidics implies complex fluidic injection systems capable of switching reagents. Thus, droplet-based channel microfluidics is well adapted when droplets can be managed collectively in a batch mode all having the same protocol and reagents. This limitation of droplet-based channel microfluidics is mainly due to the global actuation mode, which is governed externally by pressure conditions existing at the channel entries or exits (Fig. 6.5).

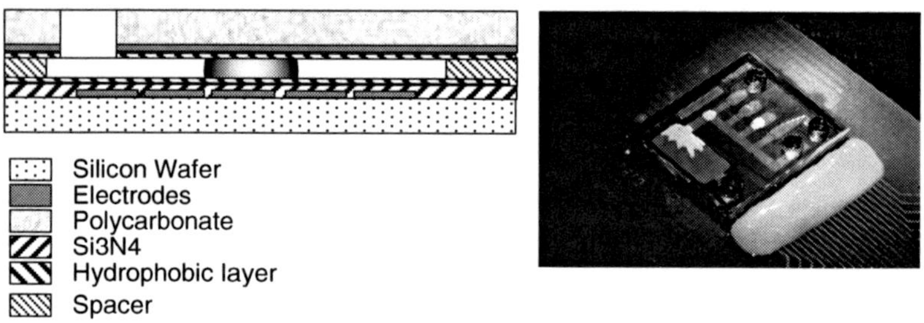

Fig. 6.4 Diagram and photo of closed EWOD chip (cea leti)

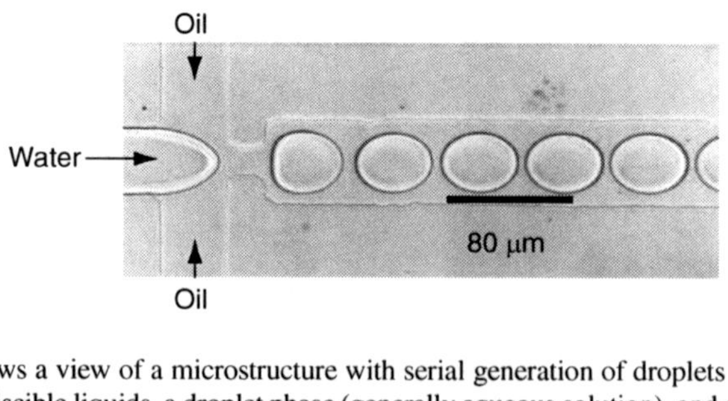

Fig. 6.5 This shows a view of a microstructure with serial generation of droplets in a microchannel in which two immiscible liquids, a droplet phase (generally aqueous solution), and a continuous phase (generally oil), are supplied to the central and side channels of the device, respectively. The liquids are forced through a narrow orifice in which a thread of liquid breaks up and releases droplets [24]

6.4 Embedded Functions

6.4.1 Sample Preparation

Basic functions relating to extraction of analytes and sample preparation are extrapolated from well-known methods in analytical chemistry and are miniaturized, thanks to microtechnologies. Well-documented publications are focused on these operations. In particular [25], most general and systematic, gives a lot of examples of realization, whereas [26] sticks more to the physicochemical principles (thermodynamics, kinetics). The general scheme of pre-analytical steps for a complete biological or chemical analysis is given in Fig. 6.6. A generic architecture is not possible to define, because of the large variety of samples (air, water, blood, biopsy, etc.). Moreover, the sample size needs to be statistically significant. It is not always possible to reduce sample size even with highly sensitive detection methods. Concentration value is an important data and very often samples are rather big at the microelectronic scale. That means, for a large volume sample, it is necessary to have a macro-sample preparation module (above 0.1 mL) connected to a microsample preparation module (from few microlitres to 100 μL). Concentration step increases assay detection sensitivity, and purification reduces background to give less interference. In this paragraph, we will see how the principal functions related to the sample preparation can be implemented on a chip.

6.4.2 Mechanical Filtering

The purpose of filtering is to eliminate the solid particles (or for cell sorting) present in a solution. Filters can be produced within a lab-on-a-chip in several

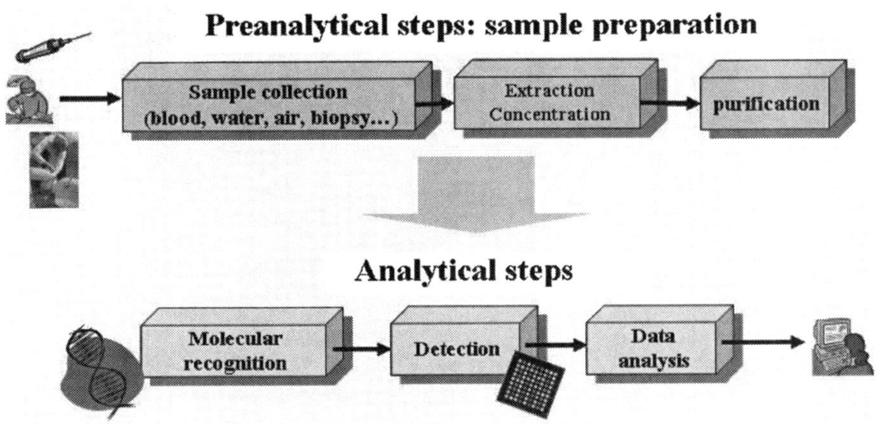

Fig. 6.6 Standard biological analysis protocol

ways. Simplest from the point of view of the principle consists in engraving structures in the form of pillars spaced of a distance lesser than the diameter of the particles to be retained. An example of such a realization is described in [27]. Figure 6.2 shows the details of such a filter intended to retain particles. According to a different principle, and with a more complex technology, the filter can be presented as a membrane produced in the plan of the substrate [28].

Limitation usually mentioned for these two approaches comes from the fact that the performances of the filter are narrowly dependent on the space resolution with which the structures are carried out. They are the characteristics of the fabrication process, including at the same time possible steps of lithography and micromachining (dry etching, oxidation, moulding, etc.), which will determine the final space resolution obtained. The current processes make it possible to reach a space resolution of about a micron, which fixes the limit of the minimal size of the particles that one can retain with these filters (about 1 μm). Standard fabrication process makes it possible to reach space resolutions controlled much better according to the thickness of the layers deposited, i.e. according to the normal direction in the plan of the substrate. The realization of sacrificial layers thus makes it possible to lower the size of the particles being able to be retained by the filters. Such a realization is proposed in [29]. Such filter makes it possible to obtain the pores of approximately 25 nm with a very good reproducibility (Fig. 6.7).

6.4.3 Active Filter

Although the preceding approach is founded on a mechanical principle, it can be possible to also separate molecules or nanoparticules from larger entities by exploiting other phenomena that do not imply any contact of the microfluidic

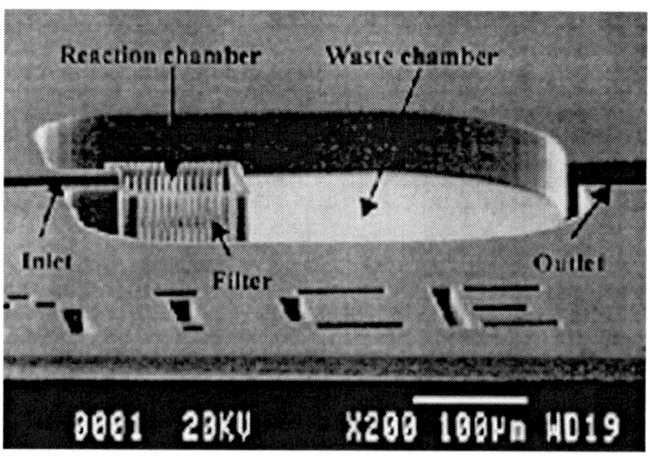

Fig. 6.7 Structured micropillars used to beads filtration (from [27])

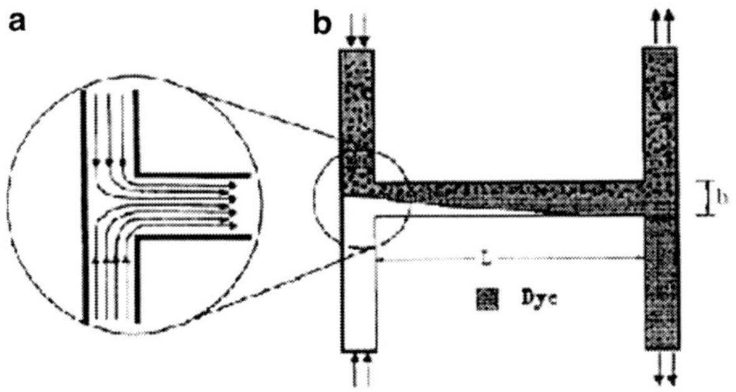

Fig. 6.8

circuit with the solid fractions of the sample, thus avoiding, for example, the risks of clogging of the filter.

A first principle used is discrimination by the diffusion in a laminar flow [30]. Indeed, flows in submillimeter capillaries at speeds lower than centimetre per second are such that flows are always laminar and completely free from turbulence. In microchannels, structure (H shape) such as that of Fig. 6.8, the two flows entering the fluidic circuit mix only by diffusion in the centre section. Consequently, a particle introduced by the inlet 1 will have a higher probability to go out by outlet 2 if its diffusion constant is more important. More precisely, the average times of two objects (molecules or particles) to traverse the same length *L* are inversely proportional to the ratio of their respective diffusion constants. For a protein in pure water at room temperature, the diffusion constant is about 5×10^{-11} m^2/s, for a DNA single bin of 20 bases, it is of 15×10^{-11} m^2/s, whereas it is of 10^{-13} m^2/s for a sphere of 1 μm in diameter, which is 500 and 150 times weaker than for the preceding molecules, respectively. It is possible thus to create conditions of flow such as at the outlet of the device of Fig. 6.8, the concentration in light fractions is practically uniform in the totality of the capillary, whereas the heaviest particles practically did not diffuse in the "low" part of the capillary (outlet 2), thus making it possible to collect part of the sample exempting the largest particles. This principle was also exploited for the realization of an immunological test [31]. In this work, the authors exploit the difference in diffusion constants between an antigen (light) and an antibody, free or grafted on a particle. They show that the gradient profile of concentration between the flows of solutions of antibody and antigens is not only detectable but also depends on the antigen concentration, which thus makes it possible to carry out an immunological test.

An "active" alternative of this principle consists in combining the flow with an electrostatic attraction on the species to be separated [32]. The electric field is perpendicular to the direction of the flow. Thus particles are deviated of their rectilinear trajectory of a more or less large angle according to the electrophoretic mobility of the molecules. This principle of separation, invented initially in "macroscopic" geometries, was validated in microfluidic devices.

6.4.4 Example of Microfluidic System for Extracting Nucleic Acids for DNA and RNA Analysis

Hui et al. [33] illustrates this integration of sample preparation steps on a silicon chips. They develop a microfluidic chip for extracting genomic deoxyribonucleic acid (DNA) and viral ribonucleic acid (RNA) from blood sample. For viral RNA extraction, it is necessary to remove both white blood cells (WBC) and red blood cells (RBC) and also to avoid diluting viruses in the plasma. Hence, a submicron filter was designed for handling a large volume of blood to obtain sufficient copies of viruses. The submicron filter is constituted by oxidised trench gap (0.8 μm) obtained by deep reactive ion etching (DRIE). To extract RNA from viruses (or DNA from WBC; 1 μg/μL of blood), the system have to lyse the viruses (chemical lysis). Then, a binder was used to capture RNA or DNA from the lysed cells. After a cleaning step, DNA or RNA is eluted as purified DNA or RNA. To perform a full automation protocol, the fluidic circuit of the chip includes an integrated mixer, two paraffin valves, the filter, and the binder (Fig. 6.9).

6.4.5 Liquid-Liquid Extraction

The liquid–liquid extraction is a current process in chemical engineering. Two immiscible liquid phases (e.g. an aqueous solution and an organic solvent) are put in contact in a structure (H shape) as seen in Fig. 6.8. Molecules present in the first phase and having a higher affinity for the second will pass from the first to the

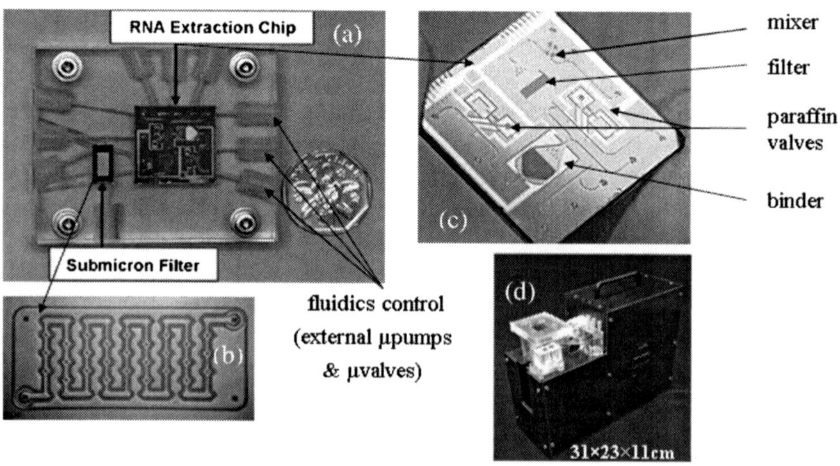

Fig. 6.9 The viral RNA extraction test block (**a**) with the submicron filter (**b**) and the micro-DNA extraction chip (**c**). The portable system for viral RNA extraction (**d**)

second phase and thus could be extracted. In this case, still the output of the extraction will depend on the geometrical factors of the zone of contact and the diffusion constants of the molecules of interest in the liquids concerned. In particular, the reduction in the sizes, by decreasing times of diffusion and by increasing the surface to volume ratio, allows a very important increase in the output of extraction [34].

In a recent work, Tran et al. [35] describe a system with a primary microchannel containing a flowing aqueous carrier liquid and a secondary organic storage fluid circulating in an adjacent channel. The interfaces between the two immiscible fluids are stabilized by vertical micropillars. The system encompasses three functions: (i) extraction of the target molecules from the carrier fluid through the pillar-stabilized interfaces, (ii) concentration of the targets in the secondary organic solvent due to its very low - or zero - velocity, (iii) on-line detection via optical spectrometry. Two constraints determine the physical behaviour of the system. The first constraint is that the interfaces must be stable and attached to the pillars at all times. The second constraint is that the interfacial area must be sufficient to provide an efficient mass transfer. The first constraint is associated with a maximum length L_{st} of the extraction channel. It is impossible to have interface stability if the length of the extraction channel is larger than L_{st}. Next, the second constraint can also be translated in a characteristic length L_e. It is clear that the longer the extraction channel, the larger the efficiency of extraction. They note L_e as the length sufficient to obtain the capture of 70% of the flowing targets. The other parameters for sensitivity refer to chemistry, i.e. a good association constant between targets and ligands diffusing in the solvent (Fig. 6.10).

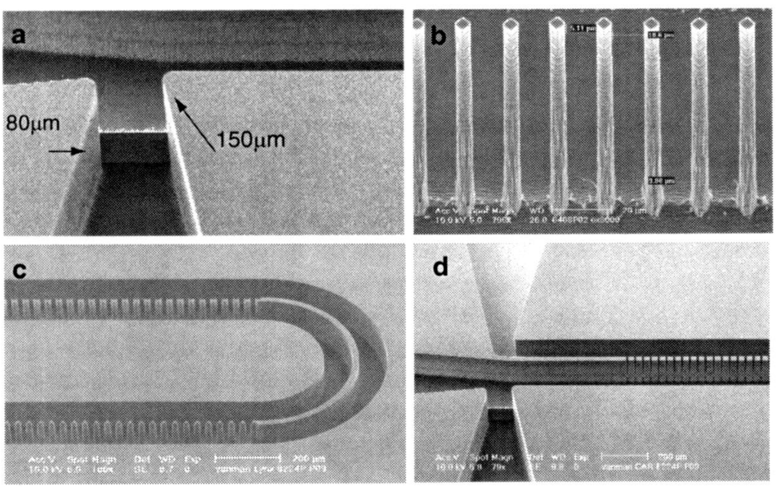

Fig. 6.10 SEM images of micro-extractor showing the two levels of deep etching (**a**), view of the micropillars (**b**), a continuous wall can be fabricated so that the channels can make turns (**c**), view of the channels and pillars: the carrier fluid channel is at the *top*, the solvent channel at the *bottom*, and the space for the optical fiber in the *middle* (**d**) (cea-Leti)

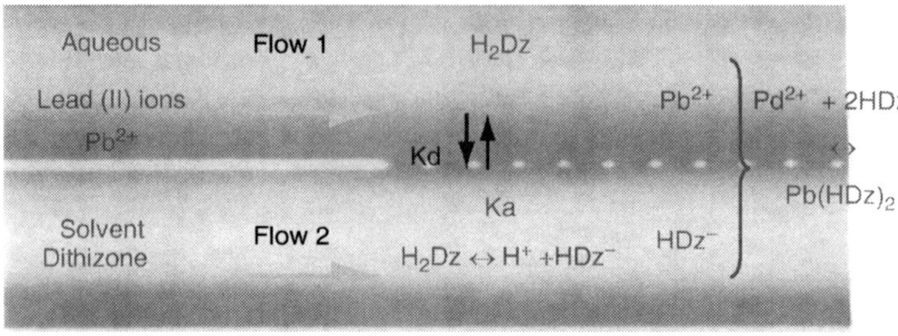

Fig. 6.11 Lead extraction process in micro-extractor (cea leti)

This concept was applied to demonstrate the extraction, concentration, and detection of lead ions (Pb^{2+}) from a water flow. The spectrophotometric determination of lead ions uses dithizone (diphenylthiocarbazone) as an organic reagent. In the primary channel, the water is flowing with a maximum flow rate of 2–3 μL/min, whereas the solvent (chloroform) is at rest. Dithizone (H_2Dz) molecules are mixed to the solvent to extract the lead ions on the interface. The maximum absorbance of H_2Dz is located at the wavelength 605–606 nm. After reaction with Pb^{2+}, a pink colour appears due to the formation of lead dithizonate complex, which absorbs at wavelength of 520–525 nm.

It has been shown that the microchip achieves the extraction of lead ions at a concentration of 1–5 ppm but detection in the range 10–100 ppb is expected in the future (Fig. 6.11).

6.4.6 Separation on Solid Phase

The principle of affinity chromatography is to retain on the surface of a solid phase the molecules of interest present in a solution, while undesirable compounds or solvents are eluted. In the second time, the fluidic channel is eluted by a low volume of a buffer solution in which analytes are in a purified state and in general concentrated. In the case of the high pressure liquid chromatography, the solid phase interacts with the analytes present in the sample so that their time of passage in the column differs from the one with the other, thus making it possible to separate them. These two principles were established successfully in lab-on-a-chip. Several different approaches were proposed. Jointly, they have the objective to maximize the interactions between the molecules present in the solution and the solid phase constituting the device. The various stationary phases and materials are compared in [36].

The first approach consists of coating the solid phase, generally by etching of silicon substrate. The most typical structure is a microchannel filled with a forest of micropillars to increase the surface to volume ratio. The sample is injected at high pressure (about 20 bars) in the chip, thanks to an external nano LC pump.

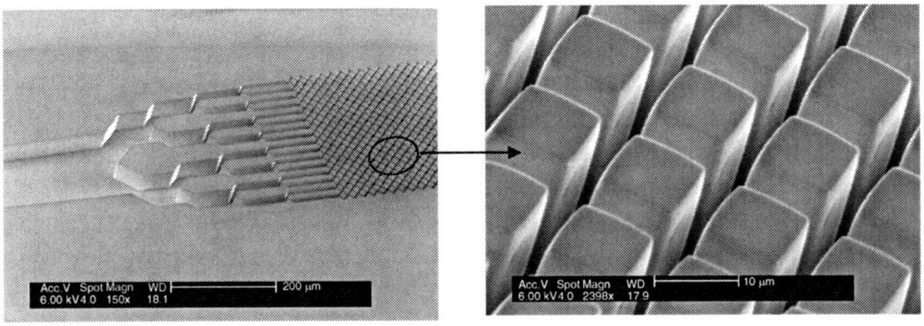

Fig. 6.12 SEM views of a structured microcolumn for high pressure liquid chromatography (cea leti)

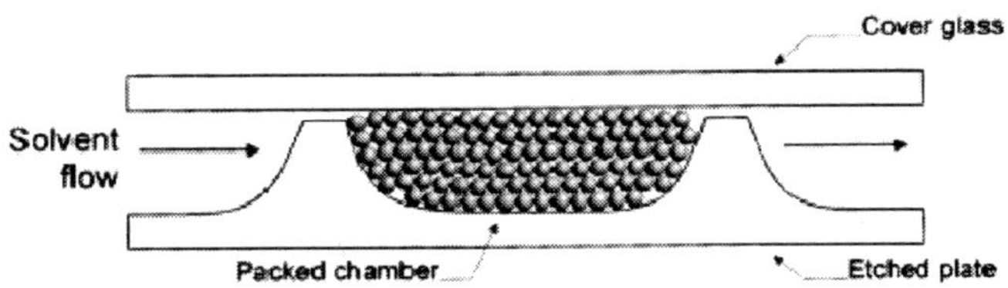

Fig. 6.13 Microreactor with embedded microbeads

The component of Fig. 6.12 shows such a structure, functionalized with silane molecule as C18 to carry out a liquid chromatography microcolumn. [37]

The second approach consists to fill a reaction chamber with prefunctionalized microbeads. In the case of [38], the authors micromachined a reactor of a volume of 330 pL (200 µm in width and 10 µm in depth), limited by restriction "walls" at the oulet and the inlet, leaving an aperture of 1 µm only between the fluidic substrate and the cover, to block microbeads (Fig. 6.13).

The third approaches finally, proposed since 2000 by the team of Jean Fréchet at the University of Berkeley, consists in including a porous polymer in the component, which plays the role of solid phase. After channel micromachining, they are filled of a solution of monomers then polymerized by UV insulation. The choice of the monomers allows the porous polymer design having well-controlled characteristics of hydrophobicity, surface charge, various functionalizations, etc. [39, 40].

Between these three approaches, expected surface to volume ratio, are according to [41], lower than $10^6\,m^{-1}$ for the functionalized structures, about $10^7\,m^{-1}$ for the immobilized functionalized beads, and between 10^6 and $10^7\,m^{-1}$ for porous polymer.

6.4.7 *Magnetic Particles*

The preceding approaches of filtering and chromatography were based on the use of an immobilized solid phase. It is also possible to use a fractioned solid phase as particles, of size about a few microns. These particles are in general functionalized.

Compared with an immobilized solid phase, they have the advantage of remaining in suspension. Chemical reactions between analytes and the surface of the particles are thus faster, especially if the liquid is agitated. However, if we want to rinse these particles and elute analytes (like in the case of affinity chromatography), it is necessary to fix these particles. This is why magnetic particles are used. With the sizes considered, they are "super paramagnetic", which enables them to form a deposit under the effect of a magnetic field and thus to disperse again in the absence of field. Actuation of these microbeads can be done by external permanent magnets [42] or integrated microcoils [43].

6.4.8 Polymerase Chain Reaction

For the nucleic acids, amplification, in particular by PCR, is the key step of the biological analysis. Indeed, amplification simplifies detection by decreasing constraints in term of sensitivity limits. Characteristic of the PCR, in terms of its implementation, is the need for performing to the sample, a succession of thermal cycles at 2 or 3 different temperatures, and this to about 30 times. Integration of PCR in a lab-on-a-chip, in addition to the usual aspects of storage of reagents, their mixing, and biocompatibility, also represents a difficult problem of thermics. Several principles exist to perform thermal cycles to the sample. The first is naturally to heat and cool either the device in its totality or only area where the sample is. Another solution is to create within the lab-on-a-chip several zones at different temperatures, corresponding to the temperatures of the various steps. Then, the sample circulates sequentially on the different thermal areas [44].

There exist also several means of heating and cooling. The contact with a thermostated plate (Peltier effect) is the most current solution, at the same time for the heating and cooling. Joule effect heating is obviously usable and in general easily integrable. Other approaches are more original: illumination by infra-red radiation [45] or use of endo or exothermic chemical reactions [46]. Information additional could be found in [47].

At last, silicon is an ideal candidate to perform PCR reaction. Silicon presents a very good thermal conductivity, and heaters can be integrated on chip. The work of ST Microelectronics to integrate PCR and DNA chip on a single silicon chip is a very promising approach for IVD application ("in-Check" platform) [48]. As for the work of Hui et al., In-Check is an example of integration of microelectronics and micromachining technology, with microfluidic and optical features, biochemical surface functionalization and molecular biology. It comprises a core lab-on-chip device, control and reading instrumentation, a complete suite of software modules, and application protocols.

The lab-on-chip core device embeds a combination of nucleic acid amplification, by PCR and target identification and typing by DNA microarray (Fig. 6.14). The cross section view of the lab-on-chip is shown in Fig. 6.15.

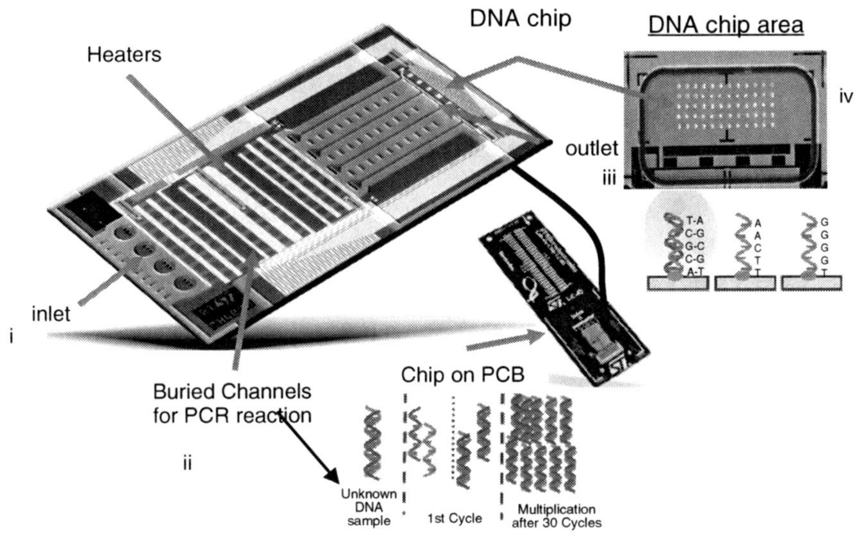

Fig. 6.14 Architecture of the Lab-on-Chip: (i) Input microfluidics ports for sample loading, (ii) PCR DNA amplification region, (iii) microfluidics interconnection, (iv) DNA microarray region (ST Microelectronics)

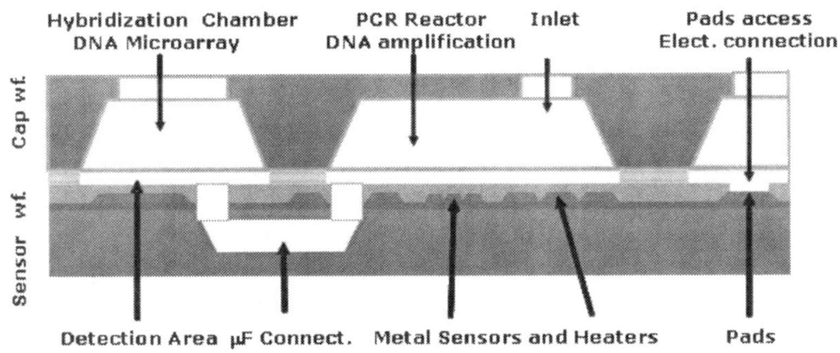

Fig. 6.15 Cross-sectional view of the Lab-on-Chip device (ST Microelectronics)

The device has multiple microfluidic input ports from 1 to 4 and can load through different biological samples or PCR mixes. In the region (ii) the DNA amplification takes place. In this section, the biological sample undergoes the thermal cycling typical of PCR amplification protocol. With proper enzyme and other reagents design an efficient reaction that takes place with exponential DNA concentration amplification over the cycle numbers. The reactor volume can be designed according to the application specification (4–25 μL). Such flexibility is very important to address different biological applications, and larger volumes are necessary when there is a low concentration of input specimen, i.e. infection disease, to get the maximum sensitivity out of the PCR protocols. The (iv) area integrates the

DNA microarray, where an array of DNA probes is grafted on the surface and can hybridize with their complementary strands. The microarray chamber is designed with an overall hybridization volume (20–30 μL). The microarray has mid-range complexity, and the number of probes that can be grafted can vary from 126 to 400 (with typical dimensions of 120 μm in diameter and 250 μm in pitch). Such array complexity makes the device suitable for many diagnostic assays. In between the PCR reactors and microarray, there is a microfluidic interconnection layer (iii), in the form of a microchannel, that provides a capillary stop for the PCR load to naturally flow into the microarray, and that can be actuated by a microfluidic PDMS valve to provide sealing and prevent evaporation during the high temperature PCR cycles.

6.4.9 Transduction

The problem of transduction in a lab-on-a-chip is the generation of a measurable signal (optical or electrical), which is able to translate whether a specific chemical reaction takes place or not. The modes of transduction being able to be implemented [49] are:

- *Optics:* Where the reaction can induce the generation of photons (chimiluminescence), the colouring of the reactional solution, the accumulation or the activation of fluorescent markers, and the modification of the optical characteristics of an optical interface
- *Mechanics:* Where a reaction of hybridization on the surface of a material disturbs the characteristics of this one. It can be the damping factor or the proper frequency of a quartz oscillator (Quartz microbalance QMB), a surface acoustic wave (SAW), or the deflection of a micro-cantilever [50]
- *Electric:* Where the chemical reaction can modify an electrochemical potential on an electrode, an impedance between electrodes or creates a field effect on the grid of a transistor [51, 52]
- *Magnetic:* Where the target molecule is labelled by a submicrometric magnetic particle, which can be detected by a magnetic sensor integrated in the microsystem [53].

The choice of a detection system is done by combining several criteria of performances, cost, or portability, depending on the applications. To compare the advantages and respective disadvantages of all these methods, it can be interesting to plot them on a two-dimensional graph. The first dimension is the complexity imposed by detection and the other is the protocol of sample preparation. The most unfavourable case is that where the principle of detection requires a chemical labelling step of the molecule of interest. This additional step results in an increased complexity of the protocol embedded and consequently of the device to be realized. The other axis translates the potential portability of the reading system: at first, with a reading system, which can be completely integrated on the microsystem, and at

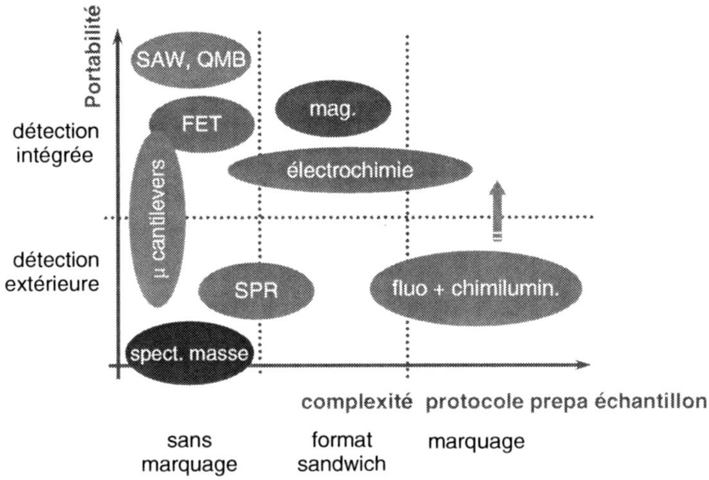

Fig. 6.16 Comparison of the various methods of detection. Optical, electric, mechanical, and magnetic methods are in green, blue, pink, and purple, respectively

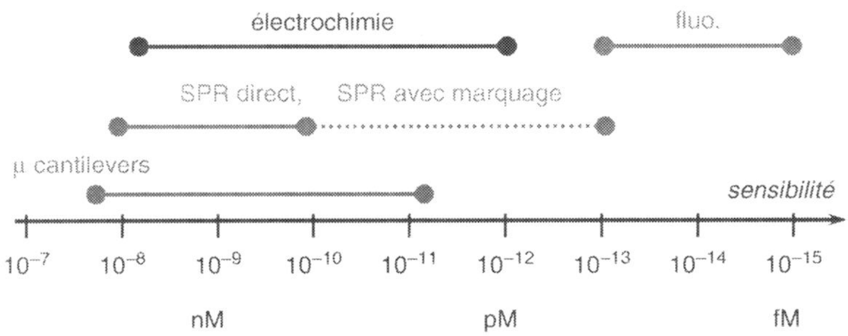

Fig. 6.17 Compared sensitivity of the various methods of detection. The values were taken in the literature (comparison are not made on a single biological model)

last a heavy reading system, which can be used only on a bench. From the graph in Fig. 6.15, we can see that fluorescent methods, which are most widespread, are also more penalizing from a system point of view, in terms of sample preparation and portability. Conversely, the mechanical methods seem to be more interesting on the two points of view (Fig. 6.16).

However, when we consider performances of these methods, in particular, in terms of sensitivity (Fig. 6.17), we can see that fluorescence is definitely more efficient than the concurrent methods. For this reason today, the competition between methods is very open. Other methods implementing functionalized nanostructures such as nanofils [54, 55] or biomimetic membranes functionalized by membrane proteins [56, 57], allowing free labelling detection of molecule, seem to be promising methods.

A particular case of detection is the mass spectrometry, which presents high level of performances in terms of sensitivity and discrimination capacity, and is used for nucleic acids [58] or proteins detection [59]. More and more, miniaturization of mass spectrometer appeared to be a challenge for the future.

6.4.10 Technological Aspect

After having considered the lab-on-a-chip under a functional aspect, it is also useful to consider them under a technological aspect. We will present in this section various technologies which are associated there and which make it possible to microstructure them, to assemble them, and to functionalize their surfaces [60, 61].

More and more lab-on-a-chip technology needs to use different kind of materials. Very often silicon chip is the heart of the system, but the final chip is a hybrid assembling of different materials (plastic, glass, silicon, etc.). The choice of materials used and the development processes result from a multi-criteria optimization, taking in to account the functions and the expected performances of the device, its cost, and volume of production. The following must be considered in particular:

- Chemical properties (biocompatibility and chemical functionalization of surfaces), thermal properties, and possibly optics and electric properties (for detection mainly) of materials
- Microfabrication processes, assembling, packaging, chemical functionalization of surfaces, which must be compatible between them and compatible with the biological material used.
- Full costs of production of the microsystem, taking in to account at the same time the cost of rough material and the cost of the different processes. The production costs of a unit device can rather strongly depend on volume of production.

The different materials most usually used with silicon are: (1) polymers in various forms [62] and (2) glass, quartz, silicon [63]. The polymers have as an advantage their cost and their simplicity of implementation, in particular, on the level of their working. Plastic are generally used for package silicon chip, to obtain a cartridge format. Silicon has as an advantage the capacity to integrate electric or electronic functions (e.g. for detection), its very good thermal conduction, and the possibility of microstructuration of size about the micron with a perfect process controlled [64, 65].

A complete technological process for fabrication of lab-on-a-chip requires necessarily several steps:

- Design and fabrication of the microstuctured silicon chip, using standard microtechnologies facilities. At this step, the microcomponent has no biological function implemented.
- Surface treatment (e.g. silanization step on silicon oxide) and functionalization to graft molecules (biological probes) to ensure a molecular recognition reaction (e.g. such as the case of the DNA chips), grafting can be made at a wafer level.

- Assembling and packaging make it possible to assemble the various parts, which constitute the complete lab-on-a-chip. A traditional operation is the assembly of a cover, including outlet and inlet, on a fluidic circuit wafer. When the fluidic circuit wafer includes embedded biological reagents or probes, specific low temperature bonding processes are to be used, such as screen printing technology.

6.5 Conclusion

In terms of the R&D effort, the field of lab-on-a-chip is extremely dynamic. In about 15 years, many elementary functions were studied and established on miniaturized devices. These functions relate to the molecular recognition and transduction, and also processes relating to sample preparation. For each elementary function, several different approaches are proposed.

The integration of a complete protocol of analysis remains a major challenge, and few achievements are validated, even on the simple level of proof of concept. We can give [66–67] as the most remarkable. The sequence of different steps of the protocol passes by a complete control of the microfluidique, which remains a key point.

From an industrial point of view finally, a certain number of products were born during the last years. Most popular are the "Bioanalyser" of Agilent. This company also marketed more recently a liquid chromatography microcolumn. These products fulfil only one particular function and not still a complex protocol. Other companies, like "start-up or spin-off", offer more complex products specialized for a particular application.

Integration of several biological functions on silicon chip and packaging with hybrid materials are a key challenge to address a future mass market, which could be IVD, agro-food control, or environment controls domains.

References

1. Manz, A., N. Graber, and H.M. Widmer, Miniaturized total chemical analysis systems: A novel concept for chemical sensing. Sensors and Actuators B: Chemical, 1990, 1(1–6): pp. 244–248.
2. Bange, A., H.B. Halsall, and W.R. Heineman., Microfluidic immunosensor systems. Biosensors and Bioelectronics, 2005, 20: pp. 2488–2503.
3. Vilkner, T., D. Janasek, and A. Manz, Micro total analysis systems. Recent developments. Analytical Chemistry, 2004, 76(12): p. 3373.
4. Reyes, D.R., et al., Micro total analysis systems. 1. Introduction, theory, and technology. Analytical Chemistry, 2002, 74(12): pp. 2623–2636.
5. Auroux, P.-A., et al., Micro total analysis systems. 2. Analytical standard operations and applications. Analytical Chemistry, 2002, 74(12): pp. 2637–2652.
6. Oosterbroek, R.E. and A. van den Berg, Lab-on-a-chip, miniaturized systems for (bio)chemical analysis and synthesis, R.E. Oosterbroek and A. van den Berg, eds. 2003, Elsevier, Amsterdam, p. 394.

7. Tüdõs, A.J., G.A.J. Besselink, and R.B.M. Schasfoort, Trends in miniaturized total analysis systems for point-of-care testing in clinical chemistry. Lab on a Chip, 2001, 1(2): pp. 83–95.
8. Hobson, N.S., I. Tothill, and A.P.F. Turner, Microbial detection. Biosensors & Bioelectronics, 1996, 11(5): pp. 455–477.
9. Ahmed, F.E., Detection of genetically modified organisms in foods. Trends in biotechnology, 2002, 20(5): pp. 215–223.
10. Koester, C.J., S.L. Simonich, and B.K. Esser, Environmental analysis. Analytical Chemistry, 2003, 75: pp. 2813–2829.
11. Richardson, S.D., Water analysis: Emerging contaminants and current issues. Analytical Chemistry, 2003, 75(12): pp. 2831–2857.
12. Iqbal, S.S., et al., A review of molecular recognition technologies for detection of biological threat agents. Biosensors & Bioelectronics, 2000, 15(11–12): pp. 549–578.
13. Wang, J., Microchip devices for detecting terrorist weapons. Analytica Chimica Acta, 2004, 507(1): pp. 3–10.
14. Skelley, A.M., et al., Development and evaluation of a microdevice for amino acid biomarker detection and analysis on Mars. Proceedings of the National Academy of Sciences of the United States of America, 2005, 102(4): pp. 1041–1046.
15. White, T.J., The future of PCR technology: Diversification of technologies and applications. Trends in Biotechnology, 1996, 14(12): pp. 478–483.
16. Le Pioufle, B., M. Frenea, and A. Tixier, Biopuces pour le traitement de cellules vivantes: micromanipulation des cellules par voie electrique ou microfluidique. Comptes Rendus Physique, 2004, 5(5): pp. 589–596.
17. El-Ali, J., P.K. Sorger, and K.F. Jensen, Cells on chips. Nature, 2006, 442(7101): p. 403.
18. A. R. Kopf-Sill, Successes and challenges of lab-on-a-chip, Lab on a Chip, 2002, 2: pp. 42n–47n.
19. R. H. Liu, J. N. Yang, R. Lenigk, J. Bonanno, and P. Grodzinski, Self-contained, fully integrated biochip for sample preparation, polymerase chain reaction amplification, and DNA microarray detection, Analytical Chemistry, 2004, 76: pp. 1824–1831.
20. J. Lee and C. J. Kim, Surface-tension-driven microactuation based on continuous electrowetting, Journal of Microelectromechanical Systems, 2000, 9: pp. 171–180.
21. M. G. Pollack, R. B. Fair, and A. D. Shenderov, Electrowetting-based actuation of liquid droplets for microfluidic applications, Applied Physics Letters, 2000, 77: pp. 1725–1726.
22. Fouillet Y., et al., "EWOD digital microdluidic for lab-on-chip", Proceedings of ASME ICNMM2006 4th International Conference on Nanochannels, Microchannels and Minichannels, June 19–21, 2006, Limerick, Ireland.
23. S.-K. Fan, C. Hashi, and C.-J. Kim, "Manipulation of multiple droplets on N/spl times/M grid by cross-reference EWOD driving scheme and pressure-contact packaging," presented at Micro Electro Mechanical Systems, 2003. MEMS-03 Kyoto. IEEE The Sixteenth Annual International Conference on, 2003.
24. M. Joanicot and A. Ajdari, Droplet control for microfluidics, Science, 2005, 309.
25. Lichtenberg, J., N.F. de Rooij, and E. Verpoorte, Sample pretreatment on microfabricated devices, Talanta, 2002, 56(2): pp. 233–266.
26. Pawliszyn, J., Sample preparation: Quo Vadis? Analytical Chemistry, 2003, 75(11): pp. 2543–2558.
27. Andersson, H., et al., Micromachined flow-through filter-chamber for chemical reactions on beads. Sensors and Actuators B-Chemical, 2000, 67(1–2): pp. 203–208.
28. Xing, X., et al., Micromachined membrane particle filters. Sensors and Actuators a-Physical, 1999, 73(1–2): pp. 184–191.
29. Desai, T.A., et al., Nanoporous anti-fouling silicon membranes for biosensor applications. Biosensors and Bioelectronics, 2000, 15(9–10): pp. 453–462.
30. Brody, J.P. and P. Yager, Diffusion-based extraction in a microfabricated device. Sensors and Actuators A: Physical, 1997, 58(1): pp. 13–18.
31. Hatch, A., et al., A rapid diffusion immunoassay in a T-sensor. Nature Biotechnology, 2001, 19(5): pp. 461–465.

32. Raymond, D.E., A. Manz, and H.M. Widmer, Continuous sample pretreatment using a free-flow electrophoresis device integrated onto a silicon chip. Analytical Chemistry, 1994, 66(18): pp. 2858–2865.
33. Hui W. C., et al., Microfluidic systems for extracting nucleic acids for DNA and RNA analysis, Sensors and Actuators A, 2007, 133: pp. 335–339.
34. Tokeshi, M., et al., Continuous-flow chemical processing on a microchip by combining microunit operations and a multiphase flow network. Analytical Chemistry, 2002, 74(7): pp. 1565–1571.
35. Tran V.M, et al, "Micro-extractor for liquid-liquid extraction, concentration and in-situ detection of lead", 10th International Conference on Microreaction Technology (IMRET – 2008), 2008, New Orleans, USA.
36. Stachowiak, T.B., F. Svec, and J.M.J. Frechet, Chip electrochromatography. Journal of Chromatography A, 2004, 1044(1–2): pp. 97–111.
37. Sarrut, N., et al. "Enzymatic digestion and liquid chromatography in micro-pillar reactors, hydrodynamic versus electroosmotic driven flow", Photonics West, Microfluidics, BioMEMS and Medical Microsystems III, 2005, San Jose, CA. SPIE-Int. Soc. Opt. Eng.
38. Oleschuk, R.D., et al., Trapping of bead-based reagents within microfluidic systems: On-chip solid-phase extraction and electrochromatography. Analytical Chemistry, 2000, 72(3): pp. 585–590.
39. Yu, C., et al., Monolithic porous polymer for on-chip solid-phase extraction and preconcentration prepared by photoinitiated in situ polymerization within a microfluidic device. Analytical Chemistry, 2001, 73(21): pp. 5088–5096.
40. Stachowiak, T.B., et al., Fabrication of porous polymer monoliths covalently attached to the walls of channels in plastic microdevices. Electrophoresis, 2003, 24(21): pp. 3689–3693.
41. Lion, N., et al., Microfluidic systems in proteomics. Electrophoresis, 2003, 24(21): pp. 3533–3562.
42. Fan, Z.H., et al., Dynamic DNA hybridization on a chip using paramagnetic beads. Analytical Chemistry, 1999, 71(21): pp. 4851–4859.
43. Choi, J.-W., et al., Development and Characterization of Microfluidic Devices and systems for magnetic bead-based biochemical detection. Biomedical Devices, 2001, 3(3): pp. 191–200.
44. Hashimoto, M., et al., Rapid PCR in a continuous flow device. Lab on a Chip, 2004, 4(6): pp. 638–645.
45. Giordano, B.C., et al., Polymerase chain reaction in polymeric microchips: DNA amplification in less than 240 seconds. Analytical Biochemistry, 2001, 291(1): pp. 124–132.
46. Guijt, R.M., et al., Chemical and physical processes for integrated temperature control in microfluidic devices. Lab on a Chip, 2003, 3(1): pp. 1–4.
47. Roper, M.G., C.J. Easley, and J.P. Landers, Advances in polymerase chain reaction on microfluidic chips. Analytical Chemistry, 2005, 77(12): pp. 3887–3893.
48. Palmieri, M., et al, Develops the "In-Check" platform for diagnostic applications, Proceedings of SPIE, 2008, 6886, p. 688602.
49. Schwarz, M.A. and P.C. Hauser, Recent developments in detection methods for microfabricated analytical devices. Lab on a Chip, 2001, 1(1): pp. 1–6.
50. Fritz, J., et al., Translating biomolecular recognition into nanomechanics. Science, 2000, 288(5464): pp. 316–318.
51. Drummond, G., M.G. Hill, and J.K. Barton, Electrochemical DNA sensors. Nature biotechnology, 2003, 21(10): pp. 1192–1199.
52. Bakker, E. and Y. Qin, Electrochemical sensors. Analytical Chemistry, 2006, 78(12): pp. 3965–3984.
53. Miller, M.M., et al., Detection of a micron-sized magnetic sphere using a ring-shaped anisotropic magnetoresistance-based sensor: A model for a magnetoresistance-based biosensor. Applied Physics Letters, 2002, 81(12): pp. 2211–2213.
54. Cui, Y., et al., Nanowire nanosensors for highly sensitive and selective detection of biological and chemical species. Science, 2001, 293: pp. 1289–1292.

55. Patolsky, F., G. Zheng, and C.M. Lieber, Nanowire-based biosensors. Analytical Chemistry, 2006, 78(13): p. 4261.
56. Gu, L.-Q., S. Cheley, and H. Bayley, Capture of a single molecule in a nanocavity. Science, 2001, 291: pp. 636–640.
57. Saleh, O.A. and L.L. Sohn, An artificial nanopore for molecular sensing. Nano Letters, 2003, 3(1): pp. 37–38.
58. Gut, I.G., Automation in genotyping of single nucleotide polymorphisms. Human Mutation, 2001, 17: pp. 475–492.
59. Aebersold, R. and M. Mann, Mass spectrometry-based proteomics. Nature, 2003, 422: pp. 198–207.
60. Hierlemann, A., et al., Microfabrication Techniques for Chemical/Biosensors. Proceedings of the IEEE, 2003, 91(6): pp. 839–863.
61. Verpoorte, E.M.J. and N.F. de Rooij, Microfluidics Meets MEMS. Proceedings of the IEEE, 2003, 91(6): pp. 930–953.
62. Becker, H. and L.E. Locascio, Polymer microfluidic devices. Talanta, 2002, 56(2): pp. 221–378.
63. Spearing, S.M., Materials issues in microelectromechanical systems (MEMS). Acta Materialia, 2000, 48(1): pp. 179–196.
64. Kovacs, G.T.A., N.I. Maluf, and K.E. Petersen, Bulk micromachining of silicon. Proceedings of the IEEE, 1998, 86(8): pp. 1536–1551.
65. Bustillo, J.M., R.T. Howe, and R.S. Muller, Surface micromachining for microelectromechanical systems. Proceedings of the IEEE, 1998, 86(8): pp. 1552–1574.
66. Liu, R.H., et al., Self-contained, fully integrated biochip for sample preparation, polymerase chain reaction amplification, and DNA microarray detection. Analytical Chemistry, 2004, 76(7): pp. 1824–1831.
67. Easley, C.J., et al., A fully integrated microfluidic genetic analysis system with sample-in-answer-out capability. Proceedings of the National Academy of Sciences, 2006, 103(51): pp. 19272–19277.

Chapter 7
Optoelectronics

Jaroslav Kováč , František Uherek, Daniel Donoval, Jaroslav Kováč jr., and Alexander Šatka

Abstract Optoelectronics covers the design, manufacture, and characterization of hardware devices that convert electrical signals into photon signals and vice-versa. The interaction between light and matter lies at the heart of optoelectronics, and progress in this area impacts all fields of application, including light sources, detectors, and optical communication systems. A brief review of the performance and development of up-to-date light-emitting diodes (LEDs) and photodetectors applicable in information technology is presented. After a short introduction into the physics and technology of semiconductor optoelectronic devices, their potential importance for future optoelectronics and photonics applications are discussed. Examples, where III–V compound semiconductors are used in LEDs, resonant cavity enhanced and avalanche photodetector structures, as well as new materials and structures for efficient visible light emitting diodes based on organic semiconductors are discussed. Key problems that are still to be addressed with regard to the need for interdisciplinary integration consistent with More than Moore domain in ENIAC strategy research agenda are also identified.

Keywords III–V semiconductors • Organic semiconductors • Optoelectronic devices • LEDs • OLEDs • Photodetectors • RCE p-i-n • APD • Optical communication systems • Silicon photonics

7.1 Introduction

If you use electric lights, remote-controlled televisions, compact disc players, computers, or the fiber optic networks, which link modern telephones, optoelectronics has become integral part of today's life. The concept of using optical processes to

---

J. Kováč (✉), F. Uherek, D. Donoval, J. Kováč jr., and A. Šatka
Faculty of Electrical Engineering and Information Technology STU,
Ilkovičova 3, 81219 Bratislava, Slovakia
e-mail: jaroslav.kovac@stuba.sk

G.Q. Zhang and A.J. van Roosmalen (eds.), *More than Moore: Creating High Value Micro/Nanoelectronics Systems*,
DOI 10.1007/978-0-387-75593-9_7, © Springer Science+Business Media, LLC 2009

perform useful functions is known as optoelectronics, electro-optics, or photonics, and it is a new area for future. Photonics harnessing the potential of light across a broad spectrum of applications including communications and information processing and optoelectronics is the alliance of optics and electronics. Optoelectronics covers the design and manufacture of hardware devices that convert electrical signals into photon signals and vice-versa. In the large frequency band from infrared to ultraviolet, it deals with problems of light generation, detection, transmission, technology of advanced materials needed to manufacture optoelectronic components, and their utilizations not only in various technologies including fiber optic communications, laser systems, cameras, remote sensing systems, data storage, and optical information systems but also with systems for medicine, environment protection, industrial and military applications. Photonics has been identified as a key technology with far reaching influences in communications, transportation, medicine, manufacturing, construction, computing, and defense. It is generally expected that optoelectronics and light (photons) will replace electrons as the principal information carrier in the near future [1].

Optoelectronics as applied to telecommunications, data communications, CATV, data storage displays and imaging is characterized by a prodigious global growth and spectacular technological innovations. Two main driving forces lie behind the development of the optoelectronic devices and integrated optoelectronic components and structures. First, our information age brings a daily increasing demand on data transfer, storage, and imaging. This pushes forward – among others – optical communication technologies that are capable of transmitting huge amounts of information over distances ranging from millimeters to thousands of kilometers. A concept for increasing transfer rates of the existing and future optical links is the wavelength division multiplex (WDM), where data are transmitted by light of different wavelengths simultaneously via the same optical fiber [2]. Besides this motivation, the second driving force for the development of the entire range of modern optoelectronic devices is the rapid development of modern semiconductor technologies. This allows continuous research and development of lasers, detectors and fiber-optical components, powerful transmitter and receiver modules for medium and long-range communication. Additionally, the development of new materials and structures for efficient visible light-emitting diodes (LEDs) based on wide band gap III-nitride semiconductors and organic semiconductors such as organic light-emitting diodes (OLEDs) is of growing interest.

Novel approaches are explored daily to improve the current optoelectronics state of-the-art. Such improvements will extend the usage and the efficiency of new light sources including solid-state-lighting, support the rising information technology age for displays and high density optical data storage, and enhance the environmental awareness capabilities of humans by utilization of ultraviolet and visible photon detectors and sensors [3].

In addition, a new field of nanotechnology enables all components of a transmitter or receiver to be integrated on a single chip and very likely this integration process will lead to increased performance and reduced production costs, as was the case in integrated micro and optoelectronics.

The application of nanotechnology in optoelectronics is not limited to active devices. Modern epitaxial techniques such as MBE and metalorganic chemical vapour deposition (MO CVD) allow the growth of precisely defined thin semiconductor layers with perfectly flat interfaces. They provide technological basis for novel optoelectronic devices with enhanced optical properties. The application of quantum wells (QWs) instead of bulk materials as an active layer of LEDs, photodetectors, and semiconductor lasers have significantly improved their performance in both static and dynamic modes of operation. A further improvement of semiconductor devices performance is expected with the introduction of multidimensional quantum confinement to wire-like quantum wires (QWr) and box-like quantum dots (QD) of sufficient small lateral dimensions less than a few nanometers [4]. In particular, microcavity devices such as vertical cavity surface emitting lasers (VCSELs), resonant cavity enhanced (RCE) LEDs, photodetectors, and modulators have recently received increased interest in optical interconnect technology [5]. They benefit from enclosing a conventional semiconductor structure into an optical resonant cavity grown usually in the same epitaxial process as the device itself.

Recently, development of new materials and structures for efficient visible LEDs is of growing interest. The world at the end of the twentieth century has witnessed a blue rush toward the development of violet-blue-green LEDs and laser diodes (LDs) based on GaN wide band gap III-nitride semiconductors [6]. First, the hard work has culminated with the demonstration of commercial high brightness blue and green LEDs and of commercial violet LDs.

New types of LEDs, lasers, detectors, and other photonic devices will become possible by the use of inorganic and recently of new organic semiconductors. Organic LEDs and organic FETs have underwent dramatic improvements in performance in the last 5 years [7]. The existing achievements in device area have greatly stimulated the studies in the related aspects of research in semiconductor organic materials. A bright future is forecasted for organic LEDs. Various applications ranging from organic displays, organic solid-state-lighting, to integrated sensor systems are expected to produce a huge impact on our future daily life.

7.2 Optoelectronics Devices

7.2.1 III–V semiconductor Materials for Optoelectronic Devices

One of the most important aspects of III-V compound semiconductors is the availability of bulk crystalline substrates of GaAs and InP, which can be obtained from commercial sources. The most commonly used III–V compounds are shown in Fig. 7.1, which plots their band gap as a function of the lattice constant, where the solid and dash lines represent direct and indirect band gaps, respectively.

An important aspect of a direct band gap semiconductor is that the minimum energy in the conduction band Brillouin zone occurs at the same k value as the

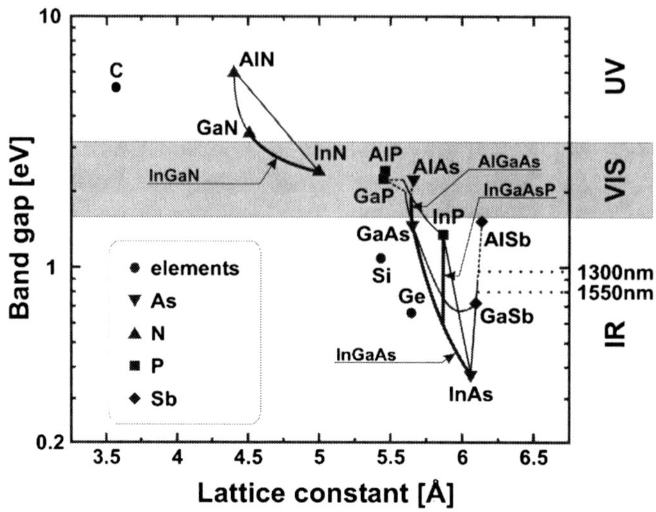

Fig. 7.1 Plot of the energy gap vs. lattice constant to some important semiconductors. *Lines* connecting points on the graph show the band gap and lattice constant variation

maximum energy of the valence band. This means that electrons can make transitions between the two bands without having to lose or gain momentum in the process. Therefore, in direct band gap semiconductor, the electron-hole pairs can recombine directly and emit photons. The majority of light emitting devices use direct band gap semiconductors such as GaAs, InP, GaN. In the case of indirect band gap semiconductor, such as Si or GaP, the minimum energy in the conduction band is not directly above the maximum energy of valence band but it is displaced on the *k* axis [8]. Any necessary loss or gain of the momentum always involves a third entity, a phonon for example, and this makes the transition much less probable. Hence, a direct band gap material is much more efficient for light sources than an indirect band gap material [9]. Different ranges of band gap energy can be obtained by mixtures of binary compounds. For example, the ternary compound $Al_xGa_{1-x}As$ can be made to have a band gap between 1.4 and 2.2 eV. By varying the ratio of Al to Ga, it becomes an indirect gap at a high Al content. The most important semiconductors for optical communications are mixed crystals based on GaAs and InP [8]. The quaternary $In_{1-x}Ga_xAs_yP_{1-y}$ can be grown on InP by lattice matching and its band gap can be tuned over the range 1.35–0.75 eV (0.95–1.67 μm), as shown in Fig. 7.1. Thus band gap engineering will generally choose the substrate, and then a suitable alloy, which has the required band gap. Band gap engineering is now a sophisticated and extremely valuable technology providing materials for optoelectronic devices.

Binary, ternary, and quaternary III–V alloy layers can be grown mostly using the epitaxial deposition techniques, such as molecular beam epitaxy (MBE) and MO CVD. In the MBE technique, the substrate is exposed to a thermal beam of atoms and molecules that compose the semiconductor in ultrahigh vacuum typically 5×10^{-11} mbar or better. The advantages of MBE grown structures include highly abrupt junctions between different materials, good control over the thickness of layers at about 1 monolayer per second, and reasonable reproducibility [10]. In the MOCVD technique, the heated substrates are exposed to a hot stream of gaseous compounds. The composition of the gases can be varied rapidly to control the composition of the material grown. MOCVD has a reputation of producing better optoelectronic devices than MBE. It is also faster and has been successfully used for commercial production, growing on a large number of wafers simultaneously [11].

7.2.2 Quantum Confined Effect in Low-Dimensional Semiconductor Structures

In the last decades, much effort has been devoted to the study of two-dimensional (2D) semiconductor QWs. Most of the effort has concerned with understanding the basic quantum mechanical effects in QWs and their potential applications for optoelectronic devices [12]. If a semiconductor with a smaller band gap (GaAs) is sandwiched between layers of a larger band gap semiconductor (AlGaAs), and if the thickness of the smaller band material is comparable to or smaller than the carrier de Broglie wavelength in the semiconductor, the quantum confined effects in QW become important and lead to a profound change in the properties of semiconductors. The changes result from the fact that carriers are confined along the growth direction z but are free in x and y directions. The confinement of the carriers leads to quantitization of the eigenenergies in the direction of restriction and to a parabolic dispersion due to the motion in all remaining directions. The energies and dispersion of such subbands in a 2D system are illustrated in Fig. 7.2 for the case of a GaAs/AlGaAs quantum well. The subband energies are shifted with respect to the bulk band gap by the confinement energies $E_{n_z,q}$ $(q = e, h)$ of electrons and holes, which are given for an infinite-height of the quantum well with width L_z by $E_{n_z,q} = \hbar^2 n_z^2 \pi^2 / 2m_q L_z^2$, where n_z is the quantum number and m_q is the electron or hole effective mass.

Because of the smaller effective mass of electrons, quantization in the conduction band is much more pronounced than in the valence band. The first light hole (lh) level lies typically farther from the valence band edge than the respective heavy hole (hh) level because of the lighter effective mass. If multiple QWs are stacked on top of each other, either a multiquantum well (MQW) or a superlattice (SL) structure is formed. The difference between these two types is that in the superlattice, the QWs are so close to each other such that the electron wavefunction is not localized into a single QW. The density of states distribution in QW is no longer a smooth parabolic curve as for the bulk semiconductor (3D) but it has a staircase structure shown in Fig. 7.3. More recently, great interest has been raising in even lower-dimensional heterostructures such as one-dimensional (1D) quantum wires

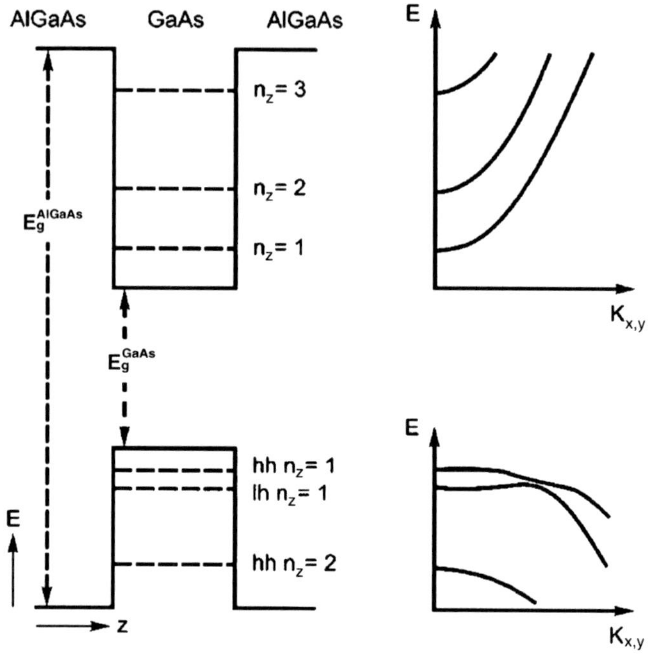

Fig. 7.2 Confined electron and hole energy levels and subband dispersion in AlGaAs/GaAs QW

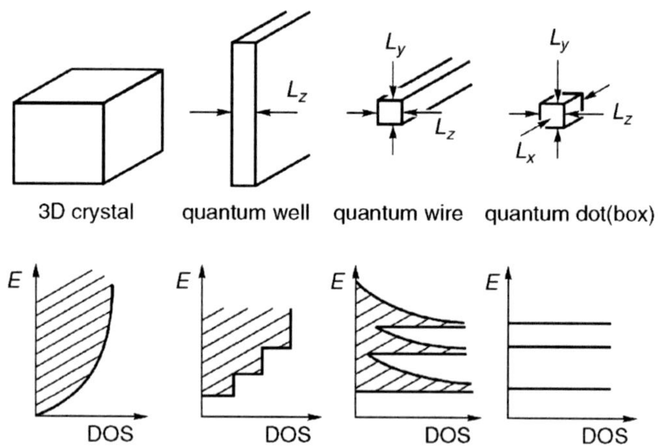

Fig. 7.3 Shape and density of states (DOS) in low dimensional (2D, 1D, 0D) semiconductor structures

(QWr) and zero-dimensional (0D) QD. However, the lack of technological capability in the fabrication of high-quality 1D and 0D heterostructures has so far limited the number of their possible applications. The electronic states in 1D and 0D structures are described in a similar way. The additional lateral confinement of the wave functions, along the y axis in QWR and along x and y axes in QD, results in a larger confinement energy for the discrete electronics states. The carrier confinement strongly modifies the properties of excitons (bound electron-hole pairs) in these low-dimensional structures with respect to the bulk.

During the last few years, there has been an intense theoretical interest in physics of 1D electron systems quantum wires, which are predicted to exhibit unique electronic and optical properties [13, 14].

Intensive investigations in QD structures have been motivated by promising QD properties such as δ-function-like density of states (Fig. 7.3) and by the perspective of realizing QD devices such as low-threshold QD lasers. QDs are typically 2–10 nm in diameter. Because of extremely small size of QDs, their optical and electrical properties are largely determined by quantum mechanics rather than by classical physics. The material can be made from just about any known semiconductor, and QD have been synthesized from silicon, germanium, gallium arsenide, indium phosphide, cadmium selenide, and zinc sulfide to name a few [15]. QDs are being analyzed for uses as LEDs, lasers, flat screen displays, memory, and solar sensors.

7.2.3 LEDs and OLEDs Basic Concept

LEDs are typical p–n junction devices used under a forward bias that emit incoherent light, when current passes through the semiconductor junction. The important property of a p–n junction is the current-voltage characteristic. The current under forward bias is given by $I = I_0 \exp[(eU / k_b T) - 1]$, where I_0 is the saturation current, U is the applied forward voltage, e is the electron charge, k_b is Boltzmann's constant, and T is the absolute temperature. When current passes through the junction, the injected minority charge carriers recombine radiatively with the majority charge carriers emitting photons within the active region. This is the basic light generation process in semiconductors. The wavelength of the light emitted λ_g [μm] is approximately determined by the band gap energy E_g [eV] given by $\lambda_g = 1.24/E_g$. Depending on the semiconductor material used in the active layer, the wavelength of the emitted light can be anywhere within the range from UV to infrared.

In general, LEDs can be realized on both homojunction and heterojunction structures. The homojunction LEDs have two drawbacks: the recombination process occurs over a relatively large distance because of the carrier diffusion length leading to increased nonradiative recombination, and reabsorption of the light produced in the active region becomes stronger with an increased material cap layer length as shown schematically in Fig. 7.4a.

For increasing the intensity of the output light, heterojunction LEDs use a double heterostructure (DH) or a quantum well in the active region. The DH device is based on two junctions between different semiconductor materials with different

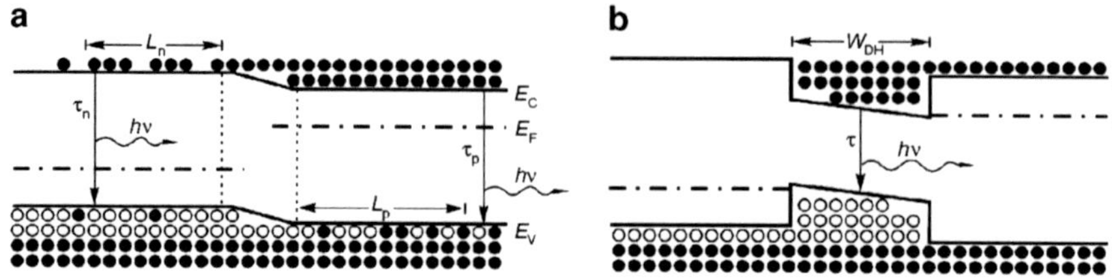

Fig. 7.4 Forward bias simplified energy band diagram for (**a**) homo and (**b**) double heterojunction

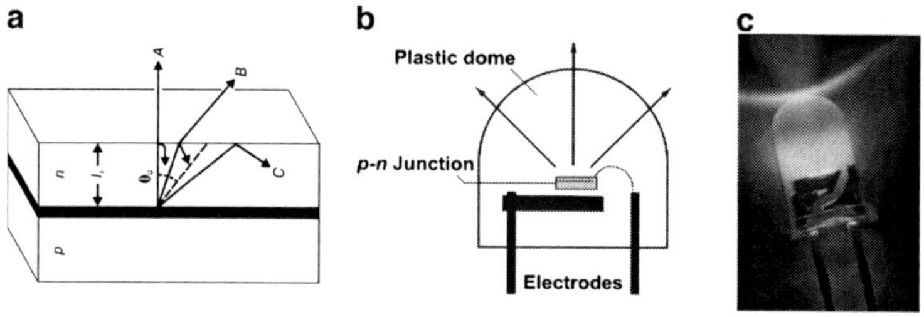

Fig. 7.5 (**a**) Light beam total internal reflection in the LED chip. (**b**) Cross section of a standard LED surrounded by plastic encapsulation. (**c**) Standard LED picture in plastic encapsulation

band gaps. The band structure and the distribution of injected carrier density under a forward bias in the DH device are shown schematically in Fig. 7.4b. There is a band gap change at the junction boundaries that results in a step-wise change in the bands creating an energy barrier that prevents diffusion of the injected carriers into the opposite region.

The performance of LEDs is best described by their ability to convert electrical energy to a luminous power. The power performance represents a value of the luminous power efficacy given by $F_V \Delta P_e$ [lm/W], where P_e is the power dissipated in the forward biased diode and F_v is the luminous flux [16]. The total light output power from the LED chip is never equal to the light power emitted from the active layer because light is emitted in random directions due to spontaneous emission as shown in Fig. 7.5a. The number of photons created by the spontaneous recombination in a unit volume of the active layer gives the internal quantum efficiency η_i. The internal quantum efficiencies of some direct band gap semiconductors can approach 100%. The reason of low external efficiencies η_e is the nontransparency of the substrate for visible light that results in photon absorption and mainly reflection of the emitted light at the chip surface as shown in Fig. 7.5a. Most of the emitted light strikes the material interface at an angle greater than the critical angle (C) and hence remains trapped. However, because of the large difference in the refractive index between the semiconductor and the air or surrounding encapsulant (n = 3.5 and 1 or 1.5, respectively), only light emitted in direction A, B within a cone of semiangle Θ_c (~16 for n = 3.5) will escape from the chip. So the external efficiencies of conventional LEDs are limited to about 4% for each available semiconductor surface [17].

The basic LED consists of a semiconductor diode chip mounted in the reflector cup of a lead frame that is connected to electrical (wire bond) wires and then encased in a solid epoxy lens (Fig. 7.5b). The colored light of a LED is determined exclusively by the semiconductor compound used to make the LED chip and is independent of the epoxy lens color. Solid state design allows LEDs to withstand shocks, vibration, frequent switching (electrical on and off shocks), and environmental (mechanical shocks) extremes without compromising their famous long life, typically 100,000 h or more.

7.2.4 OLEDs Technology

In the last decades, conducting organic materials have attracted a lot of interest as new materials for electronic applications because of their several advantages over the conventional semiconductors. They offer the potential for low fabrication cost, easy processing, and flexibility. OLEDs, organic field-effect transistors (OFET), photodiodes, and solar cells are applications under intense study, and the first products is already in commercial use [18]. There has been a growing research effort in "organic electronics" to improve the semiconducting, conducting, and light-emitting properties of organics [19, 20]. Nontraditional materials such as conjugated organic molecules, short-chain oligomers, and small molecules are being developed such that they emit light, conduct current, and act as semiconductors. The ability of these materials to transport charge (holes and electrons) due to the p-orbital overlap of neighboring molecules provides their semiconducting and conducting properties. Recombination of charge carriers under an applied field can lead to the formation of an exciton that decays radiatively giving rise to light emission. Two main technologies for OLEDs have emerged in the last decade, either based on conjugated polymers or sublimed films of small molecules [21]. With a commercial history of just 7 years, OLED manufacturing remains at an early stage, both in terms of technique and equipment. Small-molecule OLEDs are made using vapor deposition techniques, such as evaporation through a shadow or integrated shadow mask because OLED materials are too delicate for photolithography. Polymer PLEDs are made by solution processing, either spin-on techniques (for monochrome) or inkjet printing (for colour), although the latter has not yet been commercialized.

The principle of operation of bilayer OLEDs is similar to that of inorganic LEDs, as shown in Fig. 7.6. Holes and electrons are injected from opposite contacts into the organic layer sequence and transported to the emitter layer. Recombination leads to the formation of singlet excitons that decay radiatively. In detail, electrons are injected from a low work function metal contact (cathode), e.g., Ca or Mg. The last one is chosen for the sake of stability. A wide-gap transparent indium-tin-oxide (ITO) thin film is used for hole injection (anode). The efficiency of electron–hole recombination leading to the creation of singlet excitons is mainly influenced by

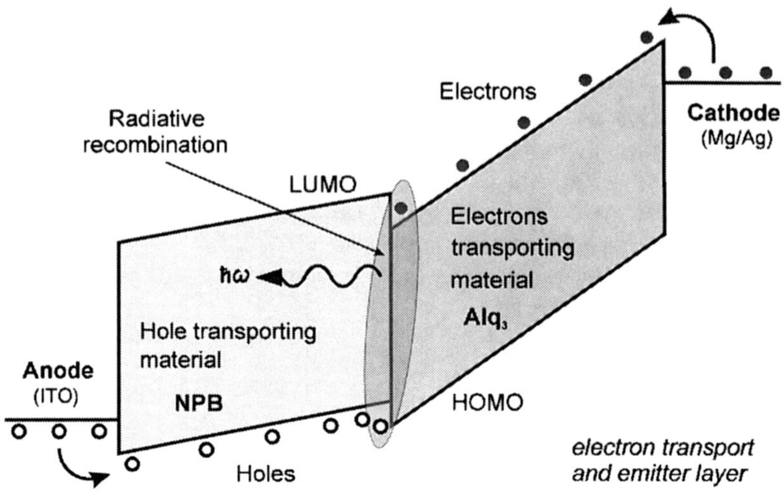

Fig. 7.6 Energy diagrams for bilayer OLED

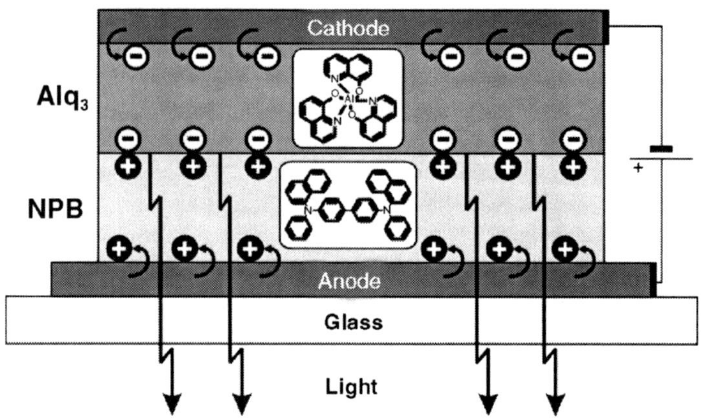

Fig. 7.7 NPB/Alq_3 bilayer OLED structure

the overlap of electron and hole densities that originate from carrier injection into the emitter layer. Schematic of a typical organic light-emitting diode, which uses Alq_3, tris(8-hydroxyquinolinato) aluminum, as the electron transport and emitting layer, and NPB, *N,N′*-di(naphthalene-1-yl)-*N,N′*diphenylbenzidine, as the hole transport layer, is depicted in Fig. 7.7.

As an example, the characteristic luminance, current density, voltage and spectral characteristics of bi-layer NPB/Alq_3 device are shown in Figs. 7.8 and 7.9 [22]. The luminance continuously increased with the increase in voltage and reached a value up to 1,000 cd/m^2 at 10 V showing typical Alq^3 spectral emission maxima at 530 nm. The turn-on voltage of about 4 V is mainly determined by the total layer thickness (60 nm NPB, 60 nm Alq^3) of green devices.

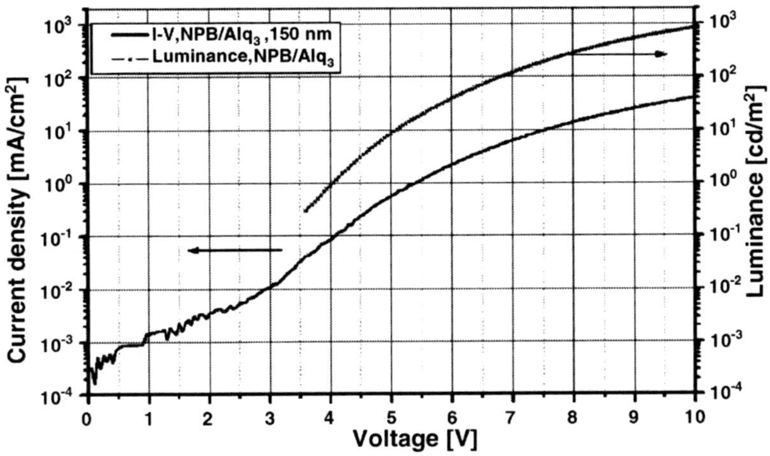

Fig. 7.8 Characteristic luminance-current density vs. voltage layer of bilayer NPB/Alq_3 device

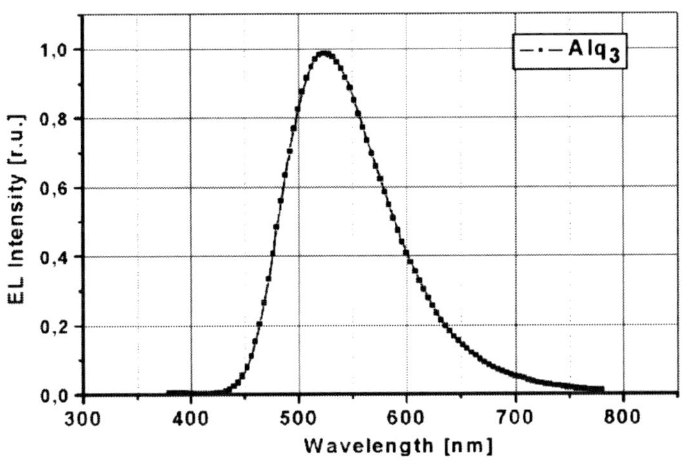

Fig. 7.9 Spectral characteristics of a bilayer NPB/Alq_3 device

7.2.5 *Research and Development of LEDs on Graded Buffer*

7.2.5.1 LED Simulation

The main engineering challenge in light generation efficiency of single source LED chips is now the extraction efficiency or the ability to get all the light out of the chip to where it is needed. This leads to a novel change, replacement of a gallium arsenide (GaAs) production wafer with transparent gallium phosphide (GaP) [23]. A new way to design LEDs can be the direct growth of InGaP or AlInGaP-based LEDs structure on a GaP substrate, using an $In_xGa_{1-x}P$ graded layer [24].

The LED structures grown on GaP substrate were designed and simulated using the APSYS -LED design software [25]. The multilayered $In_xGa_{1-x}P$ structure was designed as a sequentially grown graded buffer of 8 × 100 nm thick $In_xGa_{1-x}P$ layers with a change of $In_xGa_{1-x}P$ compositional value $D_x = 0.04$ per layer. The design of the graded region is necessary to bridge over the lattice mismatch between the GaP substrate and to reach composition over $x = 0.27$, where the ternary $In_xGa_{1-x}P$ is a direct band gap semiconductor as shown in Fig. 7.14. The design of a homojunction LED structure on a graded buffer consists of a 1-μm thick n-doped cladding layer with composition $In_{0.36}Ga_{0.64}P$ followed by an undoped 300-nm thick active layer and a 1.5 μm p-type layer. The simulated band diagram of the designed structure is illustrated in Fig. 7.10 for 0 and 2 V forward bias voltages. It is assumed that E_C, E_V, and E_g depend linearly on the composition x.

In the band energy diagram, one can distinguish the steps of the graded buffer as an interface between the wide band gap GaP substrate and the narrow-band gap $In_{0.36}Ga_{0.64}P$ layer.

In Fig. 7.11a, a simulated I-V characteristics of the designed LED structure is shown in linear and logarithmic scales. From the linear characteristics, the onset

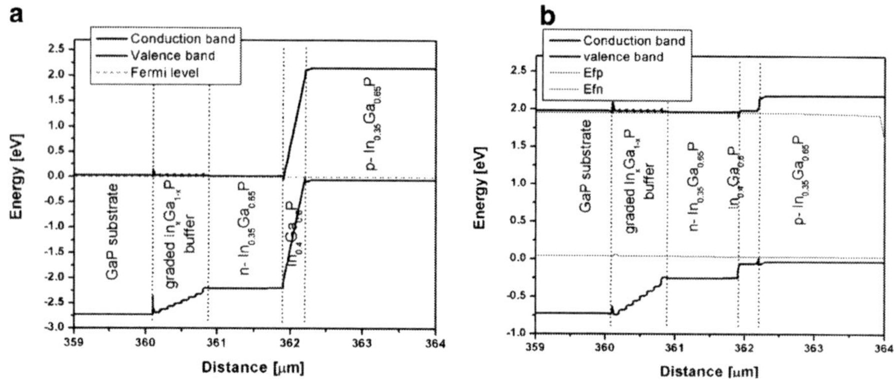

Fig. 7.10 Simulated LED structure band gap energy diagram (**a**) without and (**b**) with 2 V applied forward bias voltage

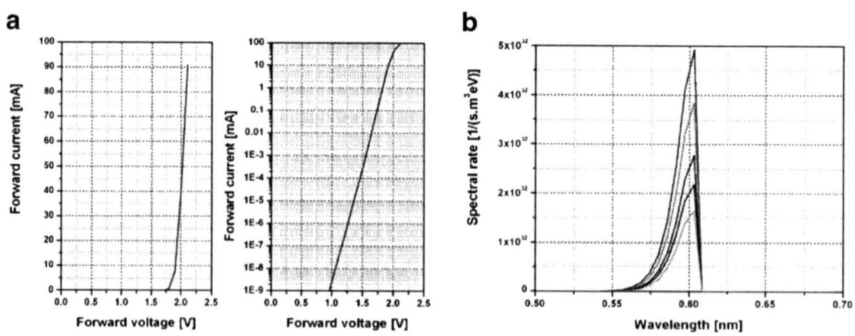

Fig. 7.11 Simulated (**a**) I–V and (**b**) spectral characteristics of the LED

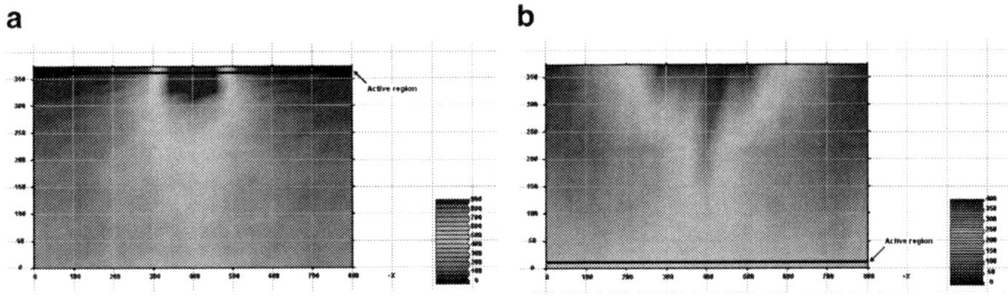

Fig. 7.12 Current spreading simulation of a LED structure, where a bottom full surface covering contact is placed (**a**) on the substrate and (**b**) on the active region side of the structure

voltage near 1.75 V needed for LED operation was determined. The simulated LED electroluminescence (EL) spectra for different driving currents in the range 10–50 mA (step 10 mA) are depicted in Fig. 7.11b. The EL spectra show the emission maxima at 605 nm for the designed homojunction LED containing an $In_{0.4}Ga_{0.6}P$ active layer. Variation of the active region thickness in the range of 300–700 nm has no significant influence on the emission intensity. For optimizing the LED structure contact, the current spreading in the LED structure was simulated, and the influence of the position of the contact covering the whole surface was analyzed. The device structure containing InGaP DH consisted of an undoped $In_{0.35}Ga_{0.65}P$ (500 nm) active layer inserted between n-$In_{0.15}Ga_{0.85}P$ and p-$In_{0.15}Ga_{0.85}P$ cladding layers with thickness 1 μm placed on the GaP substrate. The doping level of n and p-type cladding layers was 3×10^{18} cm^{-3}. Two simulations were made changing the position of the contact covering the whole surface, and the second shaped contact placed in the middle of the top side of a differently oriented LED chip. The width of the second contact was 200 μm. In Fig. 7.12a, b, the influence of the placement of the full surface contact is shown on the current spreading for LED chips in conventional and inverted structures, respectively. If the low area contact is on the active layer side of the chip, the current is flowing only through the narrow part of the active region as shown in Fig. 7.12a. Usually the substrate is the heaviest part of the LED structure, and in the case of its transparency, the substrate can act as a spreading layer. If the structure is inverted, which means that the top shaped contact is prepared on a GaP substrate, the current spreading problems in the narrow p-n part of device structures are eliminated because the full surface of the active layer is covered by an ohmic contact. In this case, the current density is equally distributed in the active region of the LED structure, which occurs at the bottom, as shown in Fig. 7.12b.

7.2.5.2 Growth and Characterization the LED Structure

An $In_xGa_{1-x}P$ graded buffer LED structure was grown by low-pressure IR-heated LP MO CVD on a sulfur-doped (100) GaP wafer. As precursors, phosphin (PH_3), trimethylgallium (TMGa), and trimethylindium (TMIn) were used. A 300-nm thick

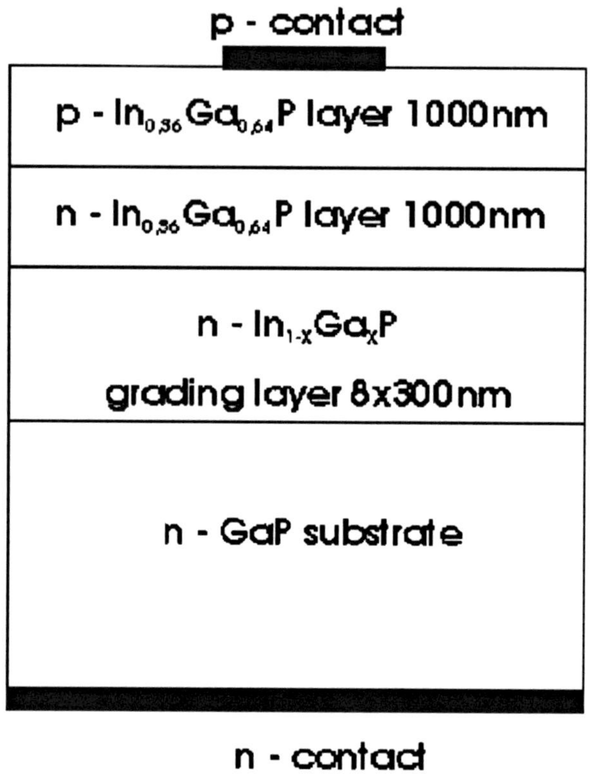

Fig. 7.13 InGaP-based homojunction LED directly grown on a GaP substrate

GaP Si-doped buffer layer ($n = 1.9 \times 10^{18}$ cm^{-3}) was deposited before the growth of the step-wise graded Si-doped $In_xGa_{1-x}P$ buffer. It consists of eight 300-nm thick $In_xGa_{1-x}P$ layers with compositional change $\Delta x_{In} \sim 0.04$ between the steps, as shown in Fig. 7.13. The role of this graded buffer is to change the lattice constant from the GaP substrate lattice constant value to the final active layer lattice constant while remaining able to filter dislocation propagation to the electroluminescent part of the LED structure, as depicted in Fig. 7.14. The growth conditions were described in detail in literature [26, 27].

On top of the graded buffer a homojunction LED structure was grown with equal 1-μm thick n and p layers of $In_{0.36}Ga_{0.64}P$ composition. The lattice parameter of the final layer in the buffer as determined by X-ray diffraction was 5.583 Å. The measured photoluminescence spectra at 300 K showed a peak position near 590 nm. Ohmic contacts to n-side (Ni/Au/Ge) and p-side (Pd/Be/Pd) of the LED structure were evaporated and afterwards annealed in a forming gas. A set of testing LED chips were prepared with surface area 500 × 500 μm^2 for measurements of electrical and optical properties.

Field emission SEM in the secondary electron mode was used to characterize the surface morphology of the prepared structures, and cross-sectional dislocation

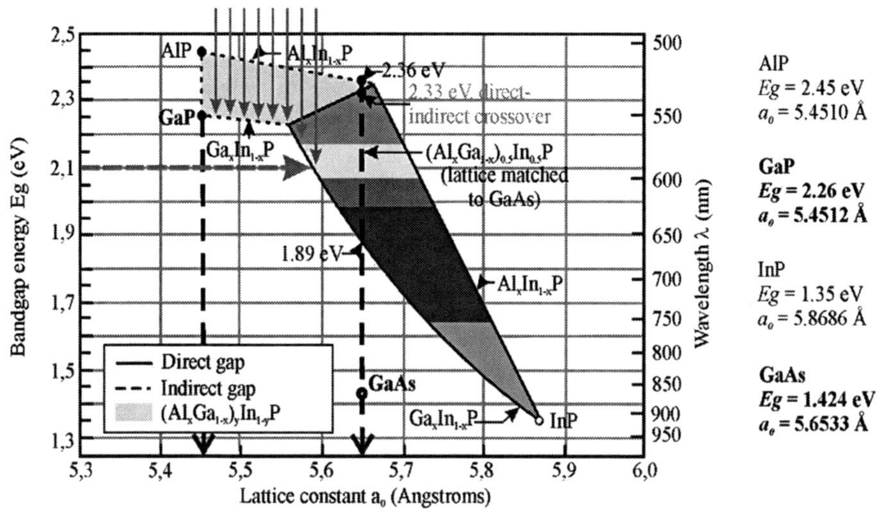

Fig. 7.14 Lattice constant and band gap energy variation with $In_xGa_{1-x}P$-graded buffer composition

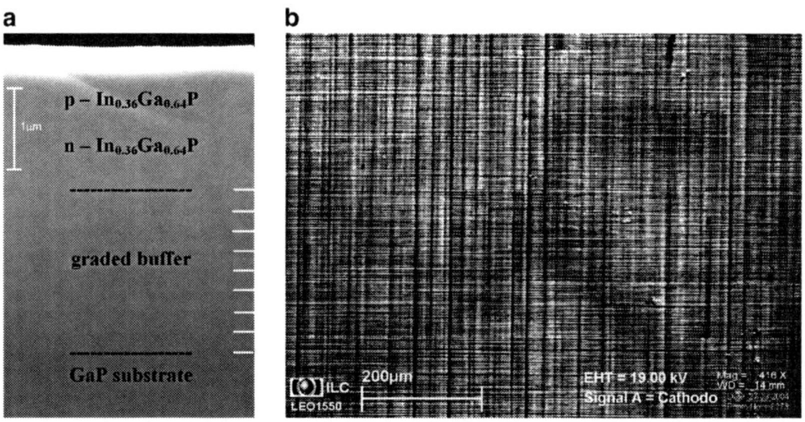

Fig. 7.15 (**a**) SEM image of the cleaved edge of the $In_xGa_{1-x}P$/GaP structure showing the graded buffer and the top homo-junction LED, (**b**) CL image of the top p-type $In_{0.36}Ga_{0.64}P$ layer

structure was examined by transmission electron microscopy. Figure 7.15a shows a SEM image of a cleaved edge of the $In_xGa_{1-x}P$/GaP structure. Starting points from the GaP substrate through individual grading, $In_xGa_{1-x}P$ interlayers and the homo-junction are clearly seen. No degradation in the cross-sectional SEM micrograph was revealed. The cathodoluminescence image plotted in Fig. 7.15b shows a top view of the sample, where crosshatch defects appeared due to the lattice mismatch and strain in the structure [28]. The total lattice mismatch is 2.37% between the

GaP substrate and top $In_{0.36}Ga_{0.64}P$ layer. The In content in the graded buffer was determined by secondary ion mass spectroscopy (SIMS) [29]. The SIMS depth profile of the graded buffer confirmed the expected compositional change $\Delta x_{In} \sim 0.04$ between the steps as documented in Fig. 7.16. The surface of the LED structure suffers from crosshatch defects due to the high lattice mismatch and strain in the graded layers. This effect is observable on EL emission from the LED chip under optical microscope, as shown in Fig. 7.17.

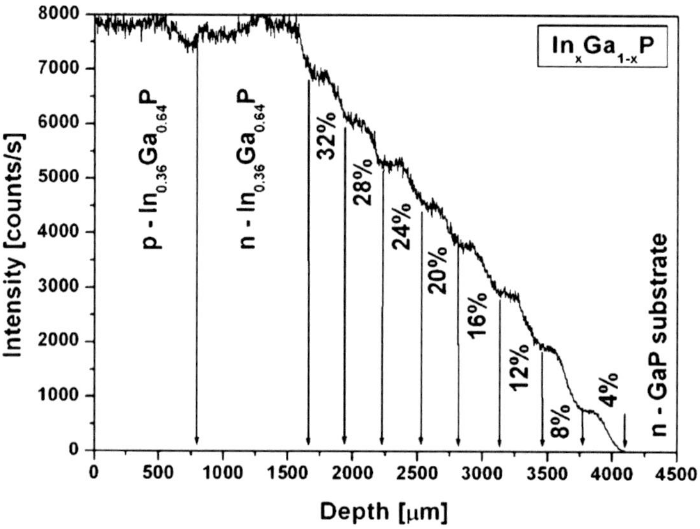

Fig. 7.16 SIMS depth profile of $In_xGa_{1-x}P$ from the graded buffer

Fig. 7.17 Image of LED chip light emission

The I-V characteristics of realized InGaP/GaP LEDs are depicted in Fig. 7.18a and compared with an LPE grown green GaP LED (dashed line). The I-V curve of the analyzed InGaP structure shows an identical behavior with a GaP LED corresponding to the decreased band gap and turn-on voltage in the range of 1.7 V [28]. Different behaviors appear in the low current region due to an increased leakage current probably through dislocations in the structure. For the inverted structure, which means that the top contact was prepared on the GaP substrate, the leakage current decreases at low voltages. Moreover, current spreading problems in the narrow p–n part of the device structures were also eliminated because the whole surface of the p-InGaP layer is covered by the ohmic contact. Hence, it results in total current spreading and homogenous emission from the active region of the diode structure as well as improvement of light emission generation, as clearly seen in Fig. 7.17. An improvement of light emission by a factor of 1.5 was reached in comparison with device configuration in which the full surface covering contact was placed on the GaP substrate. Figure 7.18b shows a typical electroluminescence (EL) spectrum of the investigated InGaP homojunction LED measured at 300 K. All the measured EL emission spectra peaks are located at around 593 nm (at 20 mA), and the full width at half maximum (FWHM) was 25 nm, which is typical for a high quality InGaP layer corresponding to the calculated $In_{0.36}Ga_{0.64}P$ composition [28] (2.091 eV). A small red shift 6 nm was obtained in the range of driving currents from 10 to 50 mA. This is due to the band gap shrinkage caused by Joule heating.

The results obtained from InGaP LED show a promising emission intensity comparable with conventional GaP LEDs and a single emission peak near 593 nm with FWHM value of 25 nm corresponding to $In_{0.36}Ga_{0.64}P$ composition. We have demonstrated the capability of growing an $In_{0.36}Ga_{0.64}P$ homojunction LED on a GaP substrate, using an $In_xGa_{1-x}P$-graded interlayer. The research work was supported by EC project VGF GaP LEDs – focused on the development of "New GaP substrate grown by VGF method for LEDs."

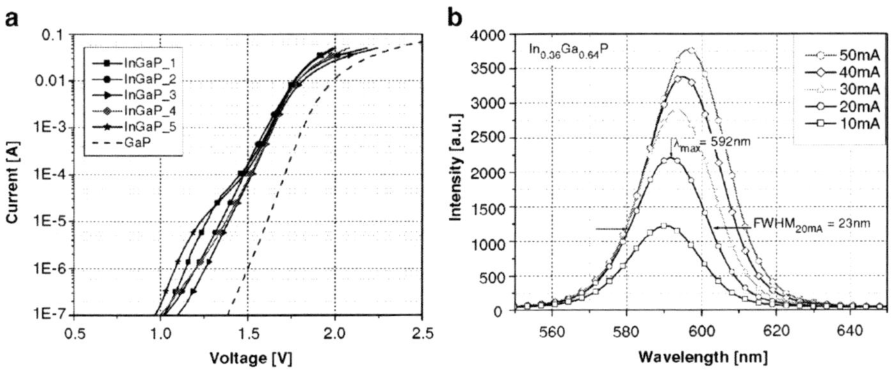

Fig. 7.18 Measured LED (**a**) I–V characteristics and (**b**) EL spectra at various forward currents

7.2.6 LEDs and OLEDs Recent Development and Trends

LEDs are among the most widely used semiconductor optoelectronic devices. Improvements in manufacturing the epitaxial material and packaging have enhanced the thermal conductivity and increased the overall light extraction efficiency of LEDs. Significantly, the light generation efficiency of these single-source LED chips has improved. Today's quantum efficiencies are very high so that the main engineering challenge is now the extraction efficiency or the ability to get all the light out of the chip to where it is needed. There even appears to be a LED version of the silicon industry's "Moore's Law" in which LED brightness doubles every 18 months [30]. In general, visible LEDs can be divided into two categories: "conventional" or low brightness, and new high brightness LEDs (HB LEDs). Conventional LEDs are found in any electronics, ranging from an indicator light up to an alpha numeric display. Most of the markets are well served by commercial LEDs, and they do not need to have high brightness. Conventional LEDs are based on homojunction devices that can be relatively inexpensively produced using older liquid phase epitaxy (LPE) and vapour phase epitaxy (VPE) growth techniques. The majority of HB LEDs are based on AlGaAs/GaAs, AlInGaP/GaAs, and InGaN/GaN heterojunction devices produced by MOCVD. Significant improvements in producing AlInGaP and InGaN structures have enhanced the brightness, specifically in green and blue LEDs. In Fig. 7.19, the LEDs evolution roadmap is illustrated showing the rate of improvement in performance of about tenfold increase in

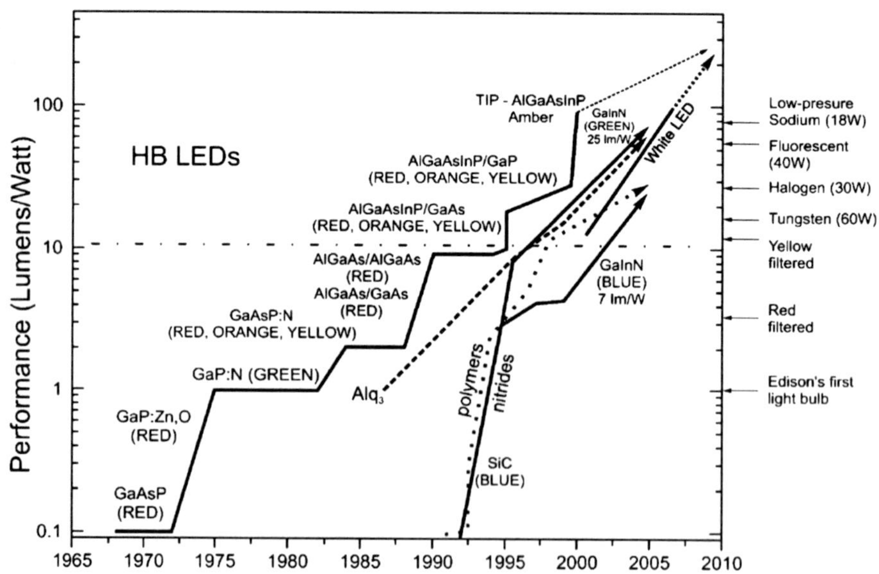

Fig. 7.19 Evolution of LEDs and OLEDs performance

light output per decade. The first LED was developed in 1960 based on GaAsP layers. Development of $A^{III}B^{V}$ compound semiconductors technology and their ternary and quaternary alloys incorporated in LED structures rapidly increased the performance. Recently, fabricated LEDs using double heterojunction (DH) or MQWs active layers increased their efficiency several times [21].

Development of HB LEDs is roughly divided to two main orientations: first using $(Al_xGa_{1-x})_{0.5}InP_{0.5}$ active layer lattice matched to GaAs (covering amber to red color), and the second, GaN/$In_xGa_{1-x}N$ nitride based technology (covering blue to green colors). LEDs using AlGaInP quaternary were developed in the mid nineties using a GaAs substrate, and their performance reached a value of ~12 lm/W. Disadvantage of this substrate is its nontransparency for visible light, which results in photon absorption by the substrate. To resolve this problem, the wafer bounding technique was used. In the wafer bonding technique, after selective etching of AlGaInP epilayers, the highly transparent GaP substrate is placed in contact with LED epilayers. Therefore, improving the light extraction is a major issue in fabricating high-efficiency LEDs [31]. Changing the absorbing substrate to a transparent one doubles the flux output from the same epitaxial structure. Using a different shape of the previously mentioned LED structure called a truncated-inverted-pyramid (TIP), the performance reaches 100 lm/W for amber light [32]. Recently, the major development is concentrated on increasing the brightness of nitride-based LED devices. The wider band gap of nitride semiconductors filled the wavelength gap between the common red-orange-yellow LEDs and the nonvisible near UV region [33]. GaN, AlGaN, and InGaN layers are grown by MOVPE on a sapphire or SiC substrate. The active $In_xGa_{1-x}N$ layer covers the band-gap energy in the range of 1.95–3.4 eV. All structures include a buffer layer because the interface between the surface and the nitride layer contains many dislocations (the lattice mismatch is 3.5% for SiC and 13% for sapphire) [6]. Sapphire is inexpensive and highly transparent while SiC gives many advantages from the epitaxial growth and device processing point of view. The main goal in nitride technologies is to increase the performance above 100 lm/W and to use blue HB LEDs as a basic light source in high brightness white light devices [34].

There are three basic approaches to making white light using LEDs. One is to combine the output from two or more LED chips; these could be blue and yellow, or more commonly red, green, and blue (RGB). The second approach is phosphor conversion, in which a blue LED chip is combined with a phosphor. Some of the blue photons are down-converted by the phosphor to produce a broad emission centered on yellow; this mixes with other blue photons to create white. The third main approach is to use a UV LED as a pump chip for a mixture of phosphors that emit across the visible spectrum [35].

Enhancements in performance have been seen in the development of OLEDs. Figure 7.19 shows the dramatic increase in the luminous efficiency of light-emitting molecular solids and polymers compared with typical inorganic LEDs over a 15-year time scale [7]. Pioneering work was done at Eastman Kodak on evaporated small molecules (Alq_3) in 1987 [36] and at Cambridge University on solution-processed semiconducting polymers in 1990 [37]. Currently, the highest observed luminous efficiencies of derivatives of these materials exceed that of incandescent light bulbs, thus eliminating the need for the backlight that is used in AMLCDs. The electronic and optical properties of these "active" organic materials are now suitable for some low performance, low-cost electronic products that can address the needs for lightweight portable devices. Through its application, a thin display that does not require backlight can be realized. The expectation is high, such as replacement for flat panel displays in mobiles. Development is in progress with the view toward using it in TV sets in the future. There is a hope for the organic EL elements to provide new forms in displays that did not exist before, such as forming them on plastic sheets and making them as flexible films. OLED, OLED displays, organic field effect transistors (OFET), photodiodes, and solar cells are applications under intense study, and the first products are already in commercial use. The steadily growing interest in OLEDs currently focuses on three major fields of applications: lightning/signature, passive-matrix displays for portable players and mobile phones, as well as active-matrix displays for computers and television screens [38].

In recent years, major advancements in LED technology have been achieved, such as developing a new technology that increases the LED light output by as much as 20 times over earlier generations, and allows production of daylight-visible LEDs virtually in any color of the spectrum. Even white light, long thought to be an impossibility, is now available in three different shades as a light-emitting diode. Over the past decade, LEDs performance has steadily improved until it exceeds that of incandescent lamps. If this progress continues, LEDs also will outperform fluorescent lamps and will be the primary sources for general illumination [39]. Recent improvements in the light output have enabled also new applications for LEDs such as automotive lightning, traffic signals, and more recently television displays [3, 40].

A very rapid progress was made over the last few years in developing organic semiconductor devices.

7.2.7 Photodetectors for Optical Communications

7.2.7.1 Photodetectors Basic Concept

If light of a proper wavelength is incident on the depletion region of a *p-n* junction diode while a reverse voltage is applied, the absorbed photons can produce additional electron–hole pairs. This yields a photocurrent in addition to the diode current so that the total diode current is given by $I = I_0 \exp[(eU / k_b T) - 1] - I_{ph}$, where the additional photocurrent, I_{ph}, is due to photogeneration of electrons and holes. The maximum photocurrent in a photodiode equals $I_{ph} = \eta_q (e\lambda / hc) P_{in}$, where η_q is the quantum efficiency and P_{in} is the incident optical power. The quantum efficiency is the ratio of the number of generated *e-h* pairs and incident photons.

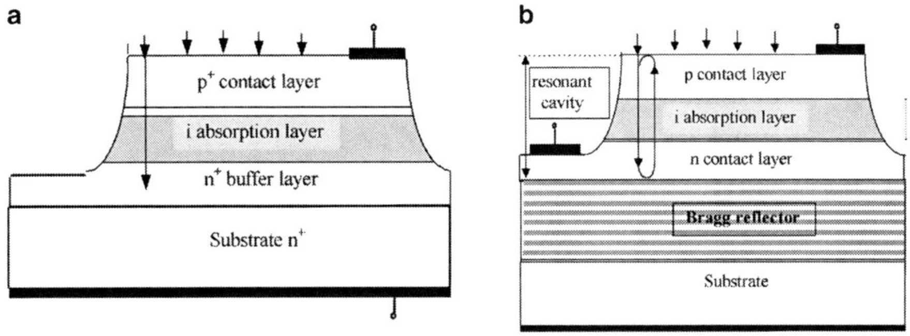

Fig. 7.20 (**a**) *p-i-n*, (**b**) RCE *p-i-n* photodetector

The responsivity (R) of a photodiode characterizes its performance in terms of the photocurrent per incident optical power at a given wavelength and is given by $R = I_{ph}/P_{in}$ (A/W).

Reverse-biased p–i–n diodes are probably the most widely used photodetectors for optical communications [8, 9]. Schematic structure of a p–i–n photodetector is shown in Fig. 7.20a. In such a structure, the electrons and holes are generated by absorbed light in the undoped region and separated by the built-in electric field toward the electrodes. Efficient detection of the incident optical power requires that maximum of the light is absorbed within the intrinsic region. The p–i–n photodiodes can achieve quantum efficiencies of up to 80%. The response time of photodiodes is limited by two factors: the finite transit time of carriers across the depletion region, and the RC time constant associated with electrical parameters of the diode and its external circuitry.

RCE photodetectors are optoelectronic devices whose performance is enhanced by placing the active device structure inside a Fabry-Perot resonant microcavity as shown in Fig. 7.20b. Such RCE devices benefit from the wavelength selectivity and the large increase of the resonant optical field introduced by the cavity. The increased optical field allows photodetectors to be made thinner and therefore faster, while simultaneously increasing the quantum efficiency at the resonant wavelength.

Avalanche photodiodes (APDs) are used in optical communications because of their high speed and internal gain. An avalanche photodiode is a specially designed reverse biased photodiode in which an incoming light signal initiates an avalanche breakdown. When the reverse bias is increased sufficiently, avalanche occurs and the optically generated photocurrent is amplified. Hence, carriers generated by light create other carriers via impact ionization, providing an internal gain. APDs use special design features to prevent edge breakdown. They exhibit a high gain and high speed, with response times as short as several tens of picoseconds.

7.2.8 RCE Photodetectors

7.2.8.1 Design of RCE Photodetectors

RCE photodetector consists of a photodetector enclosed in an optical resonator, a resonant cavity of the Fabry-Perot type (Fig. 7.20b) [41, 42]. On the one hand, photons with wavelength corresponding to the resonance of the cavity are reflected back by the mirrors and pass the cavity multiple times, increasing the probability that they will be absorbed in the active layer of the detector. The sensitivity of the detector at the resonant wavelength is hereby increased compared with a similar structure without a resonator. On the other hand, photons with wavelengths different from the resonant wavelength are reflected back from the surface. As a consequence, sensitivity of the detector is suppressed at these wavelengths compared with a similar structure without a resonator. The resonant wavelength depends on the optical length of the cavity

$$\lambda = \frac{L_{\text{eff}} n}{k}, \tag{7.1}$$

where L_{eff} is the effective length of the cavity including the penetration depth to the mirrors, n is the refractive index of the cavity material, and $k = 1, 2, 3$ is the order of resonance. The spacing between adjacent resonant wavelengths is

$$\Delta\lambda = \frac{\lambda^2}{2nL_{\text{eff}} + \lambda}. \tag{7.2}$$

With an increasing length of the resonant cavity this spacing decreases, and for long cavities more than one resonant peak can occur within the reflectivity window of the Bragg reflector. Usually the material of the resonant cavity is chosen so that there is no significant absorption in the cavity except in the active layer.

7.2.8.2 Semiconductor Bragg Reflectors for 855 and 1300 nm Wavelength Ranges

The structure of RCE photodetectors consists of layers of different materials grown on each other by MBE or MO CVD. All materials should be able to grow on each other; usually lattice matched to one of the common substrate materials (GaAs or InP). From the electrical point of view, the materials should be dopable to both types in required concentrations, and they should have a low defect rate. Materials for intrinsic layers should have low background doping concentrations, and the contact layer materials should be able to create ohmic contacts with metals. The material for the cavity itself and for the Bragg reflector should have the lowest possible absorption at the design wavelength, while the material of the active layer should be absorbing at the same wavelength. In other words, materials for the cavity and for the Bragg reflector should have the band gap energy higher than the energy corresponding to the working wavelength, while the active layer material band gap energy should be below that. Additionally, the refractive index contrast for the Bragg reflector materials should be as high as possible [41].

AlGaAs/GaAs is the most popular material system for investigation of the heterostructures. This is due to the following factors: historically, GaAs was one of the first $A^{III}B^{V}$ compound semiconductors studied, and the commercial success of the GaAs/AlGaAs heterojunction-based transistors and the first semiconductor LDs boosted the research and development of this material system. GaAs substrates are available in high quality (low defect density), at low cost and with large diameters (up to 6"). AlGaAs can be grown on GaAs substrates by MBE as well as by MO CVD almost lattice matched at any fraction of Al, providing materials with band gap ranging from 1.424 eV for GaAs (direct band gap) up to 2.17 eV for AlAs (indirect band gap), in terms of absorption edge, 870–570 nm, respectively. The AlAs/GaAs material combination exhibits the highest refractive index contrast amidst practical semiconductors for a wide range of wavelengths (n_{GaAs}/n_{AlAs} = 3.5/2.95 at $\lambda = 1$ µm, 3.4/2.9 at $\lambda = 1.3$ µm, 3.37/2.89 at $\lambda = 1.55$ µm), making this material combination attractive for Bragg reflector design. A high reflectivity Bragg reflector can be obtained using a small number of layers, and the high reflectivity wavelength band is relatively wide.

The band gap range limits the utilization of this material system to the wavelength range below 870 nm. This is a serious drawback in optical communications, as the absorption of the glass fiber is relatively high at these wavelengths. However, for shorter interconnections, optical local area networks (OLAN), and for very short interconnections (on-chip, chip-to-chip, board-to-board), the 850-nm wavelength region still remains the choice because of the advantages mentioned earlier.

The most commonly used material systems for $\lambda \sim 850$ nm are AlAs/GaAs and $Al_xGa_{1-x}As$/AlAs. The highest refractive indices difference $\Delta n \sim 0.6$ is between AlAs and GaAs. For modeling the optical properties, such as reflection, transmission, and absorption in structures consisting of stacked planar layers of optically homogeneous materials, the transfer matrix technique was used [43]. However, GaAs is absorbing for the wavelength of 855 nm and a nonabsorbing mirror could be made of an $Al_{0.3}Ga_{0.7}As$/AlAs quarter wave stack. Figure 7.21 shows the dependence of the maximum mirror reflectivity on the number of $Al_xGa_{1-x}As$/AlAs pairs. With 20 pairs a reflectivity of 99% could be achieved, and it remains constant in the wavelength range 830–880 nm. Figure 7.22 shows the calculated reflectivity spectra of 20 pairs of $Al_{0.3}Ga_{0.7}As$/AlAs for the 855 nm wavelength range [44].

The most important material system for optical communications is the heterostructure based on InP/InGaAsP quaternary material system. The invention of MO CVD greatly enhanced the development of epitaxially grown phosphide compounds, mainly on InP substrates. InP substrates are substantially more difficult to prepare, hence more expensive and of lower quality than GaAs substrates. However, the variety of materials that can be epitaxially grown lattice matched on InP gives it a superior position for optoelectronic devices operating at 1.3 and 1.55 µm

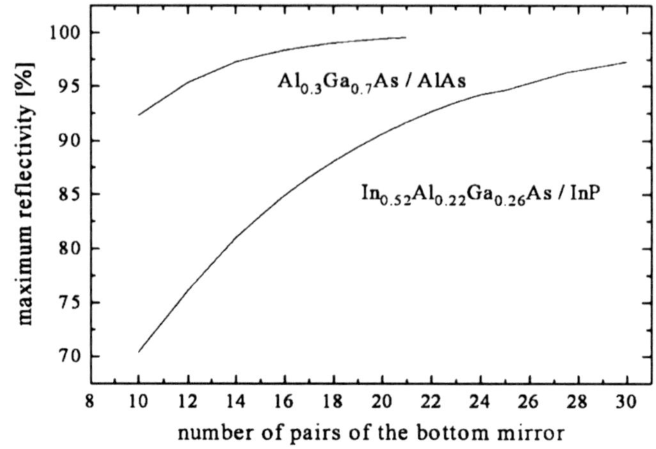

Fig. 7.21 Maximum reflectivity vs. the number of pairs of mirrors

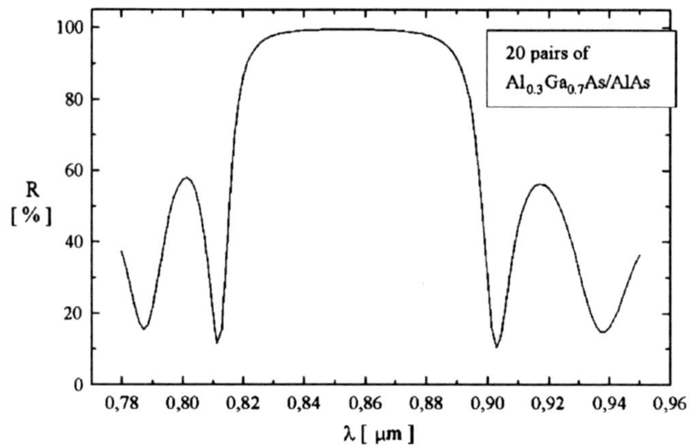

Fig. 7.22 Simulated reflectivity spectra of an AlGaAs/GaAs mirror

wavelength ranges. Ternary and quaternary compounds of In, Ga, Al (Group III), As, and P (Group V) elements provide an infinite variation of materials with a varying refraction index and absorption edge corresponding to the band gap energy. Although quaternary compounds such as InGaAsP or InGaAlAs provide more degrees of freedom in choosing materials for a given working wavelength, generally they are more difficult to prepare, as the ratio of more components should be controlled carefully. Materials lattice matched to InP exhibit a rather small refractive index contrast (e.g., $n(In_{0.63}Ga_{0.37}As_{0.8}P_{0.2})/n(InP) = 3.42/3.17$ at 1.55 μm). Consequently, Bragg reflectors require a big number of layers, making their growth expensive and time consuming.

For the wavelength range over 1.3 μm, a typical material system is $In_{1-x-y}Al_xGa_yAs$/InP. Since the refractive indices difference is smaller than for AlGaAs/AlAs system, more pairs must be grown to get the high reflectivity. For comparison, the maximum reflectivity vs. the number of pairs of $In_{0.52}Al_{0.22}Ga_{0.26}As$/InP mirror designed for 1.3 μm is plotted in Fig. 7.21 as well. In this case up to 30 pairs are necessary to give a reflectivity of 97%. The difference in refractive indices is increasing with the increasing content of Al but at the same time the energy gap is decreasing and the mirror would be absorbing.

7.2.8.3 Tunable MQW RCE Photodetector Design and Characterization

In many systems, it is required to tune the components to a particular wavelength. Tunability of the RCE photodetector can be obtained by replacing the (wide spectrum) absorbing layer of the detector by a multiple-quantum well (MQW) structure [45].

In RCE MQW structures, the excitonic peak enhanced by the resonator provides wavelength selectivity and the quantum confined stark effect (QCSE) provides tunability. The tunnability based on QCSE can be utilized because the excitonic absorption peaks shift to red with an increasing field in MQW region. As exciton absorption decreases in amplitude with an increasing electric field, this effect is compensated by the optical gain of the resonant cavity. The optical resonator cavity should, therefore, be of relatively low quality so that the enhancement peak is relatively wide and a single Bragg reflector structure is sufficient [44].

An electrically tunable RCE photodiode based on the QCSE in GaAs/AlGaAs MQWstructure for the 855 nm wavelength region was grown by MOCVD [46]. The optical resonator in the designed structure of RCE MQW is formed by 15.5 pairs of the bottom AlAs/$Al_{0.2}Ga_{0.8}As$ Quarter-Wavelength Stack Bragg reflector on the substrate side and the natural semiconductor-air interface on the top of structure (Fig. 7.23). The Bragg reflector broad reflectivity maximum is centered at 855 nm. The absorber is an undoped GaAs/$Al_{0.3}Ga_{0.7}As$ MQW structure with 20 GaAs wells of 9.5-nm thickness. The thickness of the adjacent $Al_{0.3}Ga_{0.7}As$ layers is calculated so that the enhancement maximum occurs at 855 nm. The top GaAs cap layer should be kept very thin to avoid significant absorption. Test diodes were fabricated from the designed sample. The measured spectral photoresponse of the prepared RCE MQW PIN photodiodes for different reverse voltages is shown in Fig. 7.24.

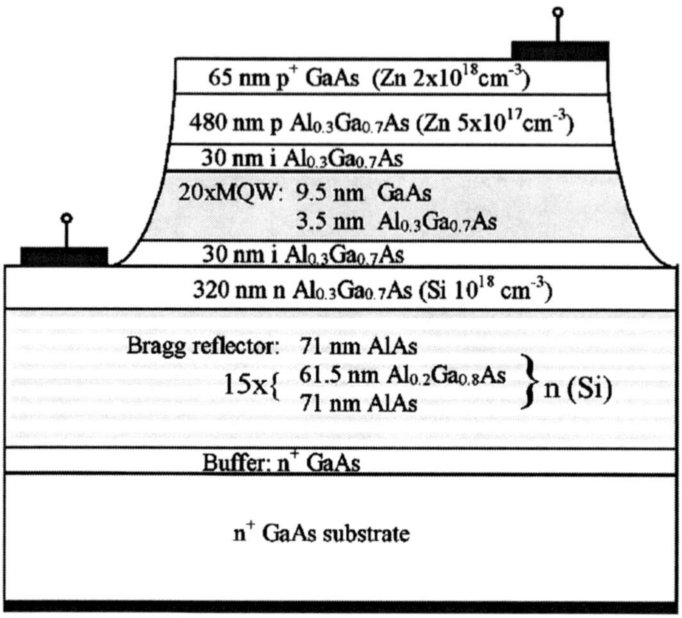

Fig. 7.23 Structure design of RCE MQW *p-i-n*

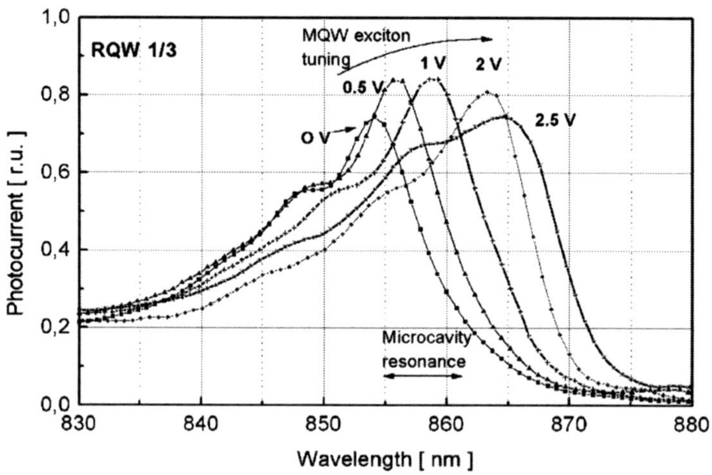

Fig. 7.24 Spectral photoresponse of tunable RCE

The tuning range of 12 nm at 2.5 V of the applied reverse bias was obtained for a structure with one bottom Bragg reflector and the physical principle of tuning based on the Quantum-Confined Stark Effect in the GaAs/AlGaAs MQW absorber was verified [44, 46].

7.2.9 Design and Characterization of Separated Absorption, Charge, and Multiplication (SACM) InGaAs/InGaAsP/InP APD Structure

An InGaAs/InP avalanche photodiode (APD) with a sectional InGaAsP/InP charge layer at the heterointerface between the InGaAs absorption and InP multiplication region has been designed, fabricated, and tested [47, 48]. Cross-sectional view of the designed SACM APD structure is depicted in Fig. 7.25a. In this design, long wavelength infrared light is absorbed in a narrow band gap material ($In_{0.47}Ga_{0.53}As$), while the photogenerated carriers are transported to and multiplied in a wider band gap (InP) material capable of sustaining high electric fields for providing an internal avalanche gain. The design was focused on the investigations of the electric field distribution in separated absorption, charge, and multiplication (SACM) APD layers in dependence on their doping concentration and thickness. The APSYS software was employed for simulations of the electric field distribution, carrier concentration, thickness of layers, and current–voltage characteristics of the designed APD structure at low voltages. The electric field distributions in the structure under different reverse bias voltages are depicted in Fig. 7.25b. The electric field in the central active region of the junction is enhanced by a selectively increased charge density under that region. The simulation results revealed that the sectional charge layer could be used to control the electric field profile in the APD structure.

The designed structure was grown by MOCVD in a single step on n-InP substrate. The test samples have been fabricated using standard photolithography, wet chemical etching, and lift-off ohmic contacts processing. For connecting the mesa diode (40 μm in diameter) top ohmic contact with a bounding pad, a gold air bridge

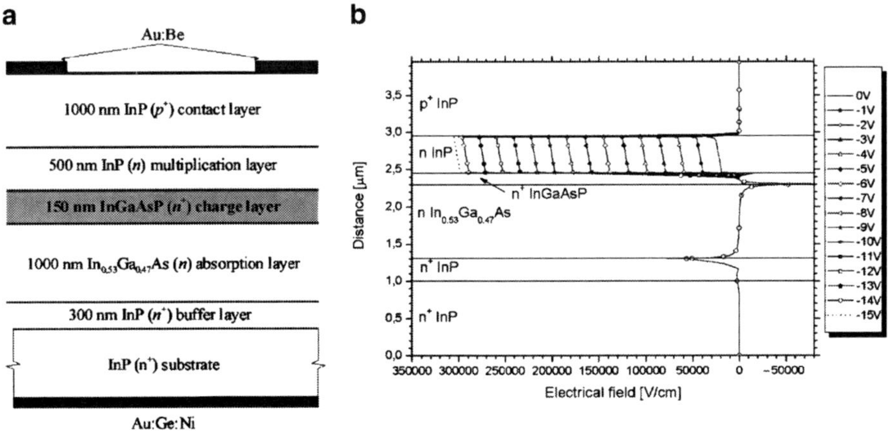

Fig. 7.25 SACM APD device: (**a**) Cross-sectional view, (**b**) Simulated electrical field distribution in the structure for different reverse bias voltages

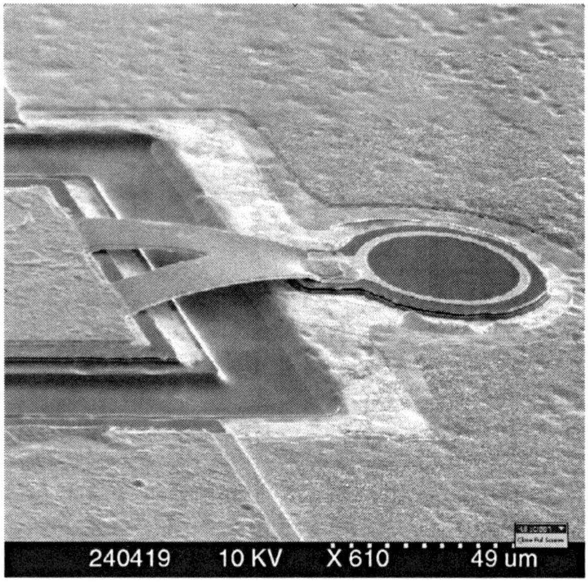

Fig. 7.26 *Top view* SEM image of processed photodiode

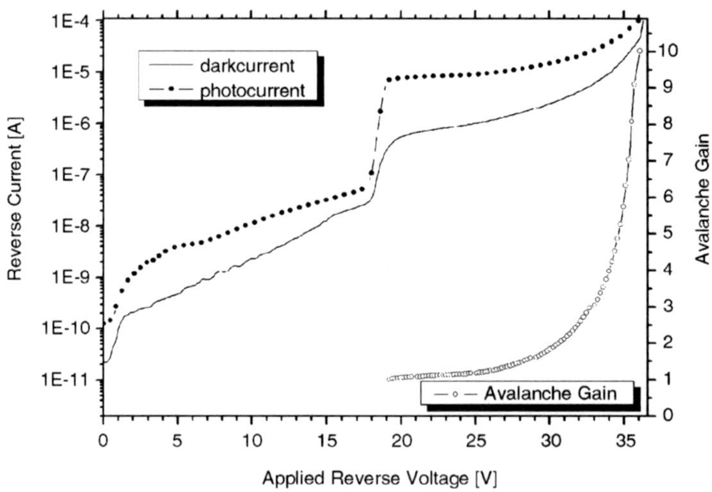

Fig. 7.27 I–V dark current and photocurrent characteristics

was formed as shown in Fig. 7.26. Dark current and photocurrent I–V characteristics under 1310 nm calibrated source illumination are shown in Fig. 7.27. The measured step in the I–V characteristics near 19 V is due to an extension of the electric field into the absorption layer.

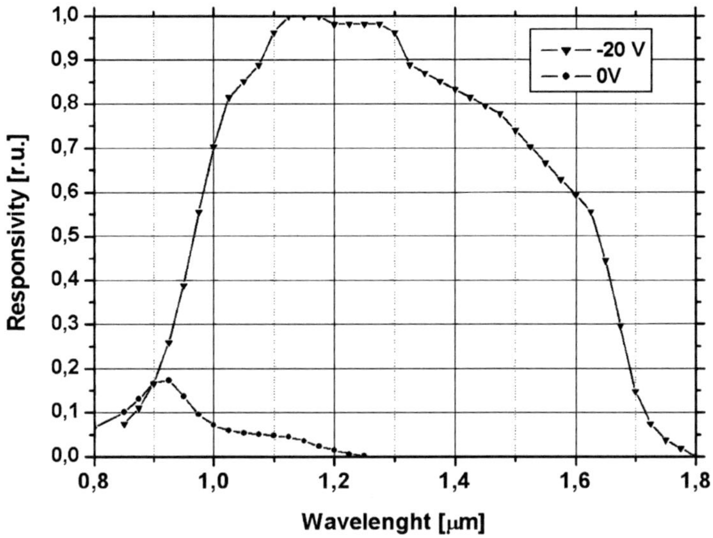

Fig. 7.28 Spectral characteristics of SACM APD at different reverse bias voltages

Within this mode of operation, the entire charge sheet and absorption layer are depleted. By increasing the bias voltage over 25 V, the carriers generated within the InGaAs absorption layer can drift into the multiplication region with high electric field thus providing an internal gain. Figure 7.28 shows the photodiode spectral characteristics at applied reverse bias voltages of 0 and 20 V. The SACM APD spectral sensitivity at reverse bias up to 15 V is low and corresponds to generation of carriers in the InP multiplication and InGaAsP charge regions. For the reverse bias voltage over 18 V, the depletion region is extended into the absorption layer and the device operates similarly as a p-i-n photodetector. At bias voltages higher than 25 V the sensitivity increases providing an internal avalanche gain. The measured photoresponse of the photodiodes without antireflection coating at 1310 nm was evaluated as 0.85 A/W at reverse voltage 20 V. Avalanche gain up to 10 was reached near the breakdown voltage of 35 V for designed SACM APD by utilizing an InGaAsP sectional charge layer.

7.2.10 Recent Developments in Optical Communications

In particular, microcavity devices such as VCSELs, RCE LEDs, and RCE photodetectors have recently received increased interest in optical interconnect technology. An ingenious concept for increasing the transfer rates of existing and future optical links is the WDM, where data are transmitted by light of different wavelengths simultaneously via the same optical fiber. RCE photodetectors represent a convenient solution for simple WDM receivers because of their inherent wavelength selectivity and potential for very high speeds [41].

Emerging optical interconnection technologies offer also a new capability. These novel technologies include parallel optics, high bandwidth plastic optical fibers, and new VCSELs, which cover an increasing range of wavelengths. Parallel optics offers low-cost interconnection with perhaps the best use of the backplane space. Plastic optical fibers (POF) have been demonstrated with high bandwidths above 1 GHz km and low loss between 850 and 1300 nm [1].

GaAs-based VCSELs are the dominant technology in transmitters for high performance, short-range data links, and optical network switches because of their low cost, ease of fiber coupling, and straightforward fabrication of large multielement 2D arrays. Unfortunately, these lasers operate only at 850 nm because of materials limitations. This short wavelength operation limits the system performance due to the wavelength-dependent dispersion and loss properties of silica fibers. The cost structure for high-density metro systems will require long-wavelength VCSELs and VCSEL arrays, which can be directly modulated, operated uncooled in ambient conditions, and produced at low cost. Such long-wavelength VCSELs will be the key elements in the next generation high-bandwidth data communication systems [49].

Avalanche photodetectors are preferred in long-haul and high bit rate transmission systems since they offer an improvement of the receiver sensitivity by several decibels [50, 51]. For APDs based on InGaAs/InP heterostructures, a higher carrier multiplication can be achieved in wider band gap InP layer. When the absorbing ternary layer is separated from the binary multiplication layer by an intermediate band gap or a graded quaternary layer, this structure is known as separate absorption and multiplication (SAM) or separate absorption, grading, and multiplication (SAGM) APD [52, 53]. To obtain a high gain-bandwidth product, the doping level and thickness of the multiplication layer must be very carefully controlled [54].

7.3 Silicon Photonics

The rapid development of microelectronic circuits over the past 50 years had a significant impact upon modern society. This is now the basis of complex microprocessors, large memory circuits, and other digital and analog electronics. Silicon as the most important material in the field of microelectronics is associated primarily with electronic devices. Silicon photonics in general means the use of silicon-based materials for generation, guidance, control, and detection of light to communicate information [55]. As silicon has proven to be cost-effective for the fabrication of integrated circuits, building optical devices on a silicon platform is considered a major thrust for the next generation of optoelectronic-integrated circuits. Implementation of silicon-based photonic devices, maybe even electrically pumped silicon lasers and silicon amplifiers [56, 57], could possibly lead to much smaller and cheaper photonic devices, making accessible a range of applications which so far have been impossible already for the reasons of too high costs [58].

Currently, a question of high interest is whether the silicon technology platform can also be effectively utilized in the area of photonics, leading to silicon-based photonic integrated circuits (PICs). Such circuits could be used, e.g., to establish very fast communication between circuit boards, between chips on a board, or even within single chips. There is a strong need for such fast communication links because the rapid progress of microprocessors may soon be severely limited by the transmission bandwidth capabilities of electronic connections made, e.g., of copper. Optical data transmission allows much higher data rates and would at the same time eliminate problems resulting from electromagnetic interference. The technology may also be useful for other areas of optical communications, such as fiber to the home (FTTH or generally FTTx).

Microphotonics is a related technology involving manipulation of light on a microscopic scale. It can be viewed upon as an equivalent to microelectronics, the electrical signal being replaced by an optical one. The analogy may become clearer when we consider the propagation of the signal. An electrical signal resides in the region of high electrical conductivity, whereas an optical signal propagates along the region of a high refractive index. The equivalent to an electrical wire in microphotonics is an optical waveguide, the refractive index of which is higher than that of the surrounding material.

Similarly, PICs correspond to electrical integrated circuits. The basic functions in PICs are the generation, guiding, splitting, multiplexing, amplification, switching, and detection of the light signal. It is possible to fabricate hybrid photonic devices, where the photonic functions are provided by structures made of III–V semiconductors (with a direct bandgap and electrooptic properties), such as, e.g., indium phosphide, and these are placed on a silicon chip containing the bulk of electronic components. There are some techniques based on epitaxial regrowth procedures or applying a sophisticated bonding process to combine a silicon chip containing waveguides with an indium phosphide-based laser chip providing the efficient coupling of the laser output light into other silicon photonic devices [59]. For that reason, hybrid devices tend to be expensive and are strongly limited in complexity. Still, all-silicon solutions, arising from the "siliconization of photonics," would probably be more suitable for widespread applications.

7.3.1 Silicon-Photonic Circuits

Waveguides, splitters, couplers, mode transformers, tapers, and arrayed waveguide gratings (AWGs) belong to the basic building blocks of PICs. Optical splitters and couplers are used to split or couple light between different waveguides, mode transformers and tapers are used to couple the light from/to input/output fiber or between planar waveguides with different cross-sections, whereas AWGs are used for spatial combining or separation of different wavelengths. All these devices are constructed from combinations of various waveguide structures that create the basic elements of photonic devices.

7.3.2 *Planar Silicon Waveguides*

In optical applications, silicon is known in its oxidized form, silicon dioxide, as the basic material of the optical fiber. It offers a combination of a low material cost and relatively simple processing, and a useful temperature range makes it currently the best compromise among the various competing materials. Intense development work has resulted in the maturity and cost-efficiency of silicon-based fabrication technology. In silicon photonics devices, light is confined to the silicon material by a top and bottom cladding of silicon dioxide.

There are several waveguide structures that can be used to realize a waveguide on silicon, as shown in Fig. 7.29. The simplest is the slab waveguide (Fig. 7.29a), which has a uniform high-index core layer on a low-index cladding layer. When silicon is used as a waveguide core and is surrounded, for example by a SiO_2 cladding, a large index step is obtained. In such waveguides, light is highly confined in the core, which can have cross-sections as small as few hundreds of nanometers and the bending radius can be reduced to a few micrometers. The slab waveguide offers only vertical confinement, and cannot principally confine light horizontally. From the waveguide types offering full confinement, a channel waveguide (Fig. 7.29b) can be formed from the slab structure by photolithography and wet or dry etching techniques. After etching of the waveguide structure, a top cladding layer is typically deposited to provide a uniform low-index surrounding for the core. A variation of the strip waveguide is a ridge waveguide (Fig. 7.29c) in which the high-index core layer is not fully etched. However, the difference between the effective indices of the partially etched slab and the ridge is sufficient to confine light within the ridge section.

As there are several different waveguide structures available, there are also many materials that can be used to build a waveguide on silicon. These include, e.g., silicon dioxide, silicon oxynitride, polymers, germanium, structuring III–V semiconductors, and silicon itself. From the different material systems available, siliconon-on-insulator (SOI) is becoming an important platform for realizing PICs and for integrating them monolithically with control electronics. SOI technology takes advantage of commercially available SOI wafers that have a single-crystal Si layer (SOI layer) on top of a buried oxide (BOX) layer made of SiO_2. This structure readily acts as a slab waveguide. From this, the ridge waveguides fabrication is

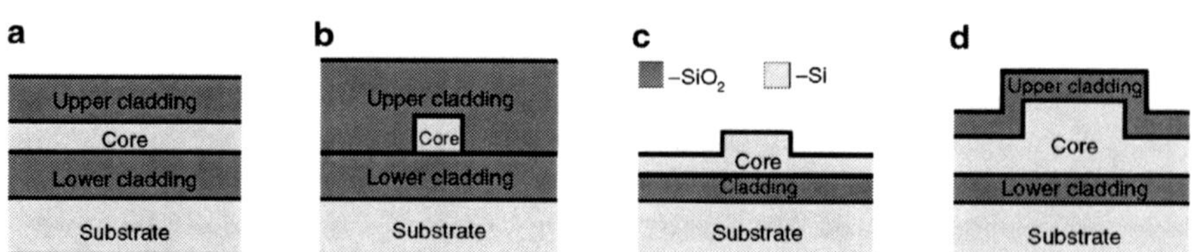

Fig. 7.29 Different waveguide structures: (**a**) slab waveguide, (**b**) channel waveguide, (**c**) ridge waveguide, and (**d**) ridge waveguide in SOI

straightforward as shown in Fig. 7.29d. Because of the ultra-high refractive index contrast between Si and SiO_2, SOI technology allows large-scale integration of optical circuits. Ultra-compact planar waveguide devices can thus be made in Si, and the transparency range includes the 1.3 and 1.55 μm telecommunication windows.

There are a number of techniques that can be used to grow the core and cladding materials for a waveguide on a silicon wafer. In principle, any of the deposition techniques known in microfabrication can be applied. These include thermal oxidation, sputtering, chemical vapor deposition, flame hydrolysis deposition, ion-assisted deposition, spin–coating, sol–gel technique, etc. [60]. The given application and chosen material system determine the requirements for the deposited thin film. However, there are some common requirements that the deposition technique must fulfill. First, the thin film should enable low optical losses. This means that the surface roughness should be low, and the material should have a negligible amount of impurities that can absorb light in the operating wavelength. For example, a high hydrogen content in a glass film grown with plasma enhanced chemical vapor deposition results in strong absorption at telecommunication wavelengths [61]. It is important to be able to control the film thickness accurately since the waveguide thickness is one of the main design parameters.

7.3.3 Moore's Law for Photonics and Beyond

The current state of photonic technology is such that a variety of different materials platforms are needed for different device functionalities. Thus, not only active devices such as lasers and modulators are needed, but also detectors, filters, multiplexers, wavelength converters, optical switches, and isolators. To do all this, we need to put photonics on the path of its own Moore's Law, when we could put photonics and electronics on a single chip [62].

This can be achieved by making the devices smaller, more efficient with less power consumption. Unlike electronics, where there is a single building block in the form of a complementary metal-oxide semiconductor (CMOS) transistor, photonics needs multiple functionalities. A most exciting prospect is that as the CMOS foundries decrease the feature size from 65 to 45 nm and, ultimately, 10 nm, the photonic devices can ride this progress and continue to become smaller, faster, and less power hungry. Once high performance photonic circuits can be created in a CMOS-compatible manner, they can coexist on the same of chip of silicon with electronic circuits [63].

Extending this materials capability to photonics and also merging the two primary device types, photonics and electronics, into one integrated dual functional device brings the benefits of large economy and wide applications, far beyond a simple combination of separate devices. With research growing in numerous industrial laboratories, one expects 100 Gbps devices for optical communication and novel devices for sensors and other application to be soon commercially available [62].

Acknowledgments The authors are grateful to Kouchi Zhang for stimulating and continuous encouragement of this work. **This chapter** is a review of the design and development of selected optoelectronics devices, and we are indebted to all of the researchers who have been contributing to this field. Only a few of these researchers are cited here, so we also thank those whose work has not been referenced. The authors are grateful to colleagues who have worked with them in this research since the early 1990s at the Microelectronics Department of the Slovak University of Technology in Bratislava in close collaboration with Slovak Academy of Science in Bratislava, University of Leipzig in Germany and City University of Hong Kong-COSDAF. The assistance of J. Breza and PhD students L. Peternai, J. Waclawek, and D. Hasko is greatly appreciated also.

This work was done in Center of Excellence CENAMOST (Slovak Research and Development Agency Contract. No. VVCE-0049–07) where support of European Commission projects, CP 93, No. 12 283, "DEMACOMINT" and IST-2001–32793-VGF GaP-LEDs as well as projects of the Slovak Research and Development Agency under the contracts APVV-20–055405 and APVV-0290–06 are gratefully acknowledged.

References

1. Suhir, E.: Microelectronics and photonics – the future. Microelectronics Journal **31,** 839–851 (2000)
2. Hecht, G.: Long-haul DWDM systems go to the distance. Laser Focus World, October, 125–132 (2000)
3. Mills, A.: Strategies in Light 2006: record LED sales but price erosion. III-V Review **29** (3), 35–39 (2006)
4. Türk, V., Bimberg, D.: Nanotechnology in optoelectronics - trends and prospects. Mstnews 3, 17–19 (1999)
5. Diff. authors: Things to watch in 2001, "High Brightness LEDs for Lightning", "Convergence of Optics and Electronics", "Explosive Market Growth for VCSELs", "Indium Phosphide Planar integrated Optics Comes of Age". Compound Semiconductor **7,** 45–84 (2001)
6. Nakamura, S.: III-V Nitride based light emitting devices. Solid State Communications **102 (2–3),** 237–242 (1997)
7. Forrest, S.R.: The road to high efficiency organic light emitting devices. Organic Electronics **4,** 45–4 (2003)
8. Kasap, S.O.: Optoelectronics and photonics, principles and practices. Prentice-Hall, Upper Saddle River, New Jersey (2001)
9. Suematsu, Y., Adams, A.R.: Semiconductor lasers and photonic integrated circuits. Chapman & Hall, London (1994)
10. Herman, M.A., Sitter, H.: Molecular beam epitaxy; Fundamentals and current status. Springer-Verlag, Berlin (1989)
11. Moon, R.L.: MOVPE: is there any other technology for optoelectronics? Journal of Crystal Growth **170,** 1–10 (1997)
12. Davies, J.H.: The physics of low-dimensional semiconductors: an introduction. Cambridge University Press, Cambridge, UK (1998)
13. Cingolani, R.: Semiconductor superlattices and interfaces. In: Proceedings of the International School of Physics "Enrico Fermi", Course CXVII, North-Holland (1993)
14. Kalt, H.: Optical properties of III-V semiconductors, Springer-Verlag, Berlin (1996)
15. Sullivan, S.C.: The anatomy of a quantum dot. Photonics Spectra 63 (2007)
16. Bergh, A.A., Dean, P.J.: Light-emitting diodes. Oxford University Press, Ely House, London (1976)
17. Wilkson, J., Hawkes, J.: Optoelectronics an introduction. Prentice Hall, London (1998)
18. Shaw, M.J., Seidler, P.F.: Organic electronics: introduction. IBM Journal of Research and Development **45(1),** 3–9 (2001)
19. Sheats, J.R., Antoniadis, H., Heuschen, M., Leonard, W., Miller, J., Moon, R., Roitman, D., Stocking, A.: Organic electroluminiscent devices. Science **273,** 913 (1996)

20. Dodapalabur, A., Torsi, L., Katz, H.E.: Organic transistors: two-dimensional transport and improved electrical characteristics. Science **268,** 270 (1996)
21. Kováč, J., Peternai L., Lengyel, O.: Advanced light emitting diodes structures for optoelectronic applications. Thin Solid Films **433,** 22–26 (2003)
22. Kováč, J., Wong, T.C., Fung, M.K., Liu, M.W., Kremnican, V., Bello, I., Lee, S.T.: Transient electroluminescence of single and multilayer organic light emitting devices. Materials Science and Engineering **B85,** 172–176 (2001)
23. Gardner, N.F., Chui, H.C., Chen, E.I., Kramers, M.R., Huang, J.-W., Kish, F.A., Stockman, S.A., Kocot, C.P., Tan, T.S., Moll, N.: 1Δ4§ efficiency improvement in transparent- substrate (AlGa)InP light -emitting diodes with thin active regions. Applied Physics Letters **74 (15),** 2230–2232 (1999)
24. Wallart, X., Deresmes, D., Mollot, F.: Growth of strained $Ga_{1-x}In_xP$ layers on GaP (100) by gas source molecular beam epitaxy: similarities and differences with the growth of strained arsenides. Journal of Grystal Growth 227–228, 255–259 (2001)
25. Peternai, L., Kováč, J., Jakabovič, J., Gottschalch V.: Numerical simulation and analysis of GaP/GaNP/GaP double heterostructure light emitting diode. In: WOCSDICE 2004 Proceedings, Smolenice Castle, Slovak Republic, 101–102 (2004)
26. Hasenöhrl, S., Novák, J., Vávra, I., Šatka, A.: Material properties of graded composition In_xGa1-xP buffer layers grown on GaP by OMVPE. Journal of Crystal Growth **272,** 633–641 (2004)
27. Novák, J, Hasenöhrl, S., Kúdela, R., Kučera, M.: Growth and characterisation of $In_xGa_{1-x}P$ layers with composition close to crossover from direct to indirect band gap. Journal of Crystal Growth 275, e1287–e1292 (2005)
28. Kováč, J., Peternai, L., Jakabovič, J., Šatka, A., Hasenöhrl, S., Novák, J., Gottschalch, V., Rheinläender, B.: New development of LED structures directly grown on GaP substrate. In: Proceedings "Electroluminiscence 2004", Toronto, Canada (2004)
29. Vincze, A., Šatka, A., Peternai, L., KováÜJ., Hasenöhrl, S., Veselý, M.: SIMS and SEM analysis of $In_{1-x-y}Al_xGa_yP$ LED structure grown on $In_xGa_{1-x}P$ graded buffer. Applied Surface Science **252,** 7279–7282 (2006)
30. Mills, A.: High-brightness LEDs lightning up the future. III-Vs Review **14 (2),** 32–37 (2001)
31. Streubel, K., Linder, N., Wirth, R., Jaeger, A.: High brightness AlGaInP light-emitting diodes. IEE Journal of Selected Topics in Quantum Electronics **8 (2),** 321–331 (2002)
32. Kramers, M.R., Ochai-Holcomb, M., Hofler, G.E., Carter-Coman, C., Chen, E.I., Tan, I.H., Grillot, P., Gardner, N.F., Chui, H.C., Huang, J.W., Stockman, S.A., Kish, F.A., Tan, T.S., Kocot, C.P., Hueschen, M., Posselt, J., Loh, B., Sasser, G., Collins, D.: High-power truncate-dinverted-pyramid (Al_xGa_{1-x})0:5In0:5P)/GaP light-emitting diodes exhibiting >50% external quantum efficiency. Applied Physics Letters **75 (16),** 2365–2367 (1999)
33. Cook, M.: Excitonic prospects for UV. III-V Review **29 (5),** 22–26 (2006)
34. Akasaki, I.: Nitride semiconductors-impact on the future world. Journal of Crystal Growth 237–239, 905–911 (2002)
35. Whitaker, T.: Innovations push white LEDs towards new applications. Compound Semiconductor **75** (2003)
36. Tang, C.W., Van Slyke, S.A.: Organic Electroluminescent Diodes. Applied Physics Letters **51**, 913 (1988)
37. Burroughes, J.H., Bradley, D.D.C., Brown, A.R., Marks, R.N., Mackay, K., Friend, R.H., Burns, P.L., Holmes, A.B.: Light-emitting diodes based on conjugated polymers. Nature **347,** 539 (1990)
38. Rentberger, S.: Doped OLEDs overcome limitations of traditional devices. Laser Focus World May, 79–84 (2007)
39. Bergh, A.: Commercial applications of optoelectronics. Photonics Spectra February, 54–61
40. Segler, D.J.: LEDs in high-definition television. Photonics Spectra, October, 62–65 (2007)
41. Ünlü, M.S., Strite, S.: Resonant cavity enhanced photonic devices. Journal of Applied Physics **78 (2),** 607–639 (1995)
42. Kishino, K., Ünlü, M. S., Chyi, J.-I., Reed, J., Arsenault, L., Morkoc, H.: Resonant cavity enhanced (RCE) photodetectors. IEEE Journal of Quantum Electron **27 (8),** 2025–2034 (1991)

43. Borgulová, J., Uherek, F., Kováč, J., Šatka, A.: Design of Multilayer Optoelectronic Devices. IEEE, SPIE Proc. Ser., 3820 (0277–786X/99), 239–247 (1999)
44. Waclawek, J.: Resonant cavity enhanced photodetectors. PhD thesis, Slovak University of Technology, Faculty of Electrical Engineering and Information Technology, Bratislava, Slovakia (1997)
45. Lai, K., Campbell, J.C.: Design of a tunable GaAs/AlGaAs multiple-quantum-well resonant-cavity photodetector. IEEE Journal of Quantum Electron **30 (1),** 108–114 (1994)
46. Waclawek, J., Kováč, J., Rheinländer, B., Gottschalch, V., Škriniarová, J.: Electrically tunable GaAs/AlGaAs MQW RCE Photodetector. Electronics Letters **33 (1),** 71–72 (1997)
47. Haško, D.: Lavínová fotodióda s oddelenou absorpčnou, nábojovou, a násobiacou vrstvou na báze InGaAs/InP (in Slovak language), PhD thesis, Slovak University of Technology, Faculty of Electrical Engineering and Information Technology, Bratislava, Slovakia (2005)
48. Haško, D., Kováč, J., Uherek, F., Škriniarová, J., Jakabovič, J., Peternai, L.: Avalanche photodiode with sectional InGaAsP/InP charge layer. Journal of Electrical Engineering **57 (6),** 369–372 (2006)
49. Harris, J.S. Jr: GaInNAs long-wavelength lasers: progress and challenges. Semiconductor Science and Technology **17,** 880–891 (2002)
50. Brennan, K.F., Haralson, J., Parks, J.W. jr., Salem, A.: Review of reliability issues of metal-semiconductor-metal and avalanche photodiode photonic detectors. Microelectronics Reliability **39 (12),** 1873–1883 (1999)
51. Budianu, E., Purica, M., Rusu, E.: Heterostructures on InP substrate for high-speed detection devices over a large spectral range (0:8–1:6mm). Microelectronics Engineering 51–52 (5), 393–400 (2000)
52. Saleh, M.A., Hayat, M.M., Kwon, O-H., Holmes, A.L., Campbell, J.C., Saleh, B.E.A., Teich, M.C.: Breakdown voltage in thin III-V avalanche photodiodes. Applied Physics Letters **79 (24),** 4037–4039 (2001)
53. Cho, S.R., Yang, S.K., Ma, J.S., Lee, S.D., Yu, J.S., Choo, A.G., Kim, T.I., Burm, J.: Suppression of avalanche multiplication at the periphery of diffused junction by floating guard rings in a planar InGaAs-InP avalanche photodiode. IEEE Photonics Technology Letters **12,** 534–536 (2000)
54. Watanabe, I., Nakata, T., Tsuji, M., Makita, K., Torikai, T., Taguchi, K.: High-speed, highreliability planar-structure superlattice avalanche photodiodes for 10-Gb/s optical receivers. Journal of Lightwave Technology **18,** 2200–2207 (2000)
55. Hecht, J.: Development of silicon photonics focuses on high speed optical interconnects. Laser Focus World, April, 86–89 (2007)
56. Boyraz, O., Jalali, B.: Demonstration of a silicon Raman laser. Optics Express **12 (21),** 5269 (2004)
57. Rong, H., et al.: Low-threshold continuous-wave Raman silicon laser. Nature Photonics **1 (4),** 232 (2007)
58. Greene, K.: Intel speeds up silicon photonics. In: Technology Review, Infotech, January 22 (2007)
59. Jones, R., Cohen, O., Paniccia, M., Fang, A.W., Bowers, J.: The first electrically pumped, hybrid silicon laser overcomes one of the major hurdles to photonic chips. Photonics Spectra, January, 55–63 (2007)
60. Cheben, P.: Wavelength dispersive planar waveguide devices: Echelle gratings and arrayed waveguide gratings. In: Optical waveguides: from theory to applied technologies, M. L. Calvo and V. Laksminarayanan, Eds., Chapter 5, Taylor & Francis, London, 174–217 (2007)
61. Osinsky, A.V., Bellman, R.A., Akwani, I.A., Sachenik, P.A., Logunov, S.L., McCamy, J.W.: Optical loss mechanisms in GeSiON planar waveguides. Applied Physics Letters 81 (11), 2002–2004 (2002)
62. Melle, S.: Network capacity growth demands new solutions. Laser Focus World, October, 113–115 (2006)
63. Salib, M., at al.: Silicon photonics. Intel Technology Journal 8 (2), 143–160 (2004)

Chapter 8
Semiconductor Image Sensing

N. Blanc, P. Giffard, P. Seitz, P. Buchschacher, V. Nguyen, and M. Hoheisel

Abstract Silicon is an excellent detector material for electromagnetic radiation in the wavelength range of 0.1–1,000 nm. In the visible spectral range (400–700 nm), external quantum efficiencies approaching 100% are obtained. When combined with the amazing miniaturization capabilities of the semiconductor industry, this fact explains why silicon is the material of choice for very efficient, highly integrated, cost-effective image sensors: In 2007 about one billion image sensors were produced and employed in camera systems. Profiting from the unrelenting progress of semiconductor technology, silicon-based image sensors with astounding performance have been demonstrated, in terms of resolution, pixel size, data rate, sensitivity, time resolution, and functionality: 111 million pixels on a single CCD chip were produced; pixels with a period of 1.2 μm were fabricated; sustainable image acquisition and readout rates of four billion pixels per second were realized; single-photon sensitivity at room temperature and at video rates was achieved; timing resolution of the pixels in lock-in image sensors below 5 ps was obtained, and the processing complexity of "smart pixels" was raised to several ten thousand transistor functions per pixel. The future of semiconductor image sensing lies in the extension of the accessible wavelength range to the infrared spectrum (1.5–10 μm), the development of affordable, high-performance X-ray image sensors, in particular, for the medical energy range (20–120 keV), the realization of sensitive and cost-effective sensors for Terahertz imaging (100–500 μm), as well as the integration of an increasing amount of analog and digital functionality on single-chip custom camera systems. The Holy Grail is the "seeing chip," capable of analyzing the contents of a scene and of recognizing individual objects of interest.

Keywords Solid-state image sensing • Electronic imaging • X-ray imaging • Infrared sensing • Seeing chips

---

N. Blanc (✉), P. Giffard, P. Seitz, P. Buchschacher, V. Nguyen, and M. Hoheisel
CSEM SA, Photonics Division, Technoparkstrasse 1, CH-8005, Zurich, Switzerland

G.Q. Zhang and A.J. van Roosmalen (eds.), *More than Moore: Creating High Value Micro/Nanoelectronics Systems*,
DOI 10.1007/978-0-387-75593-9_8, © Springer Science+Business Media, LLC 2009

8.1 Introduction

Over the last decade, electronic imaging has made tremendous progresses, basically replacing film-based cameras in almost all application fields, including consumer, security, industrial, and scientific applications. These cameras rely on the excellent optoelectronic properties of semiconductor materials, with silicon being in most cases the material of choice. High quantum efficiency over the visible range and the cost-effective manufacturing capability of the semiconductor industry have enabled the production of image sensors at affordable prices with excellent performances, in particular in terms of spatial resolution, pixel size, dynamic range, and sensitivity. Until the year 2000 the large majority of commercially available electronic cameras and camcorders were based on charge coupled devices (CCDs). Today CCDs still play an important role in electronic imaging, due to their high performances providing for example spatial resolution of up to 111M pixels, dynamic range in excess of 80dB and sensitivity down to single-photon detection. However, complementary metal-oxide-semiconductor (CMOS) image sensors have been since the 1990s the subject of significant development. They take advantage of the capability to integrate analog and digital circuits on chip for the control and readout electronics. This on one hand enables the manufacture of cost-effective image sensors and on the other hand allows the realization of imagers with added functionalities, leading to increasingly smarter and more compact imaging devices. As a result, CMOS image sensors have today become worldwide the dominating technology both in terms of units sold as well as in terms of revenues. CMOS image sensors are now widely used for consumer applications, such as in compact digital still picture cameras (DSCs), mobile phones and camcorders, as well as in numerous other fields, including surveillance, security, robotics, automotive, industrial control and even medical applications.

CMOS image sensors clearly profit from the progresses realized in the semiconductor industry and the continuous trend toward miniaturization, smaller minimal feature size, and the increased numbers of transistors/pixels per unit area, following in this regard Moore's law as often referred to for other integrated circuits such as processors and memories. On the other hand the trend in the development of image sensors goes beyond just a race for more pixels and a higher level of integration. Image sensors are more and more used for applications covering a significant broader spectral range, from X-ray up to Terahertz imaging. Moreover, the development of smart imagers opens up completely new possibilities, for example, in three-dimensional imaging, object tracking, smart intrusion detection, or spectral imaging.

The present chapter gives an overview on the fundamentals of semiconductor image sensing and technology, giving insights in current technological trends and the state of the art in high-sensitivity electronic imaging. This is followed by application examples in high-speed imaging, color imaging, and optical 3D time-of-flight imaging. While many applications and products are focused on the visible range of the electromagnetic spectrum, the extension toward shorter wavelength (e.g., X-ray imaging) and longer wavelength (such as infrared and Terahertz imaging) is gaining in importance. Examples of IR sensors include quantum well IR photodetectors, bolometers, and HgCdTe sensors. Magnetic resonance and ultrasound imaging in turn are not addressed within the scope of this chapter.

8.2 Fundamentals of Semiconductor Image Sensing

8.2.1 Interaction of Light and Semiconductor

The capability of solid-state image sensors to detect light is related to their semiconducting properties. The interaction of light and semiconductors has two principal components. First, light waves travel at a reduced speed $c_{mat} = c_{vac}/n$, where n is the refractive index of the material and $c_{vac} = 3 \times 10^8$ m/s is the speed of light in vacuum. Second, the incident light intensity I_0 (W/m^2) is absorbed exponentially with the thickness x of the material through which it traveled, according to Beer's law:

$$\lambda_{.} = \frac{L_{\text{eff}}\, n}{\quad} \tag{8.1}$$

The absorption coefficient α (in units of cm^{-1}) depends strongly on the wavelength and thus the energy of the incident light: For energies below the bandgap E_g, the material is essentially transparent. For energies larger than the bandgap, the material absorbs the incident radiation. The higher the energy (the lower the wavelength), the smaller is the penetration depth. For direct-bandgap semiconductor materials such as GaAs the $\alpha(E)$ curve is even steeper than for indirect-bandgap materials such as silicon (Fig. 8.1). The bandgap of the material used has thus a direct impact on the spectral range that can be addressed. Silicon with a bandgap of $E_g = 1.1$ eV at room temperature has a cut-off frequency of 1.1 μm, whereas germanium with a lower bandgap of 0.67 eV leads to a cut-off frequency of 1.8 μm enabling to address applications in the near infrared part of the spectrum.

$$\lambda_c(\mu\text{m}) = \frac{1.24}{E_g(e\text{V})} \tag{8.2}$$

8.2.2 Quantum Efficiency

The processes of photon absorption and thermalization of charge pairs happen in sequence, contrary to the generation of light via a recombination process, requiring the simultaneous presence of a suitable phonon in indirect-bandgap materials to satisfy momentum conservation. For this reason, all semiconductor materials are excellent photodetectors, independent of their type of bandgap, direct or indirect. Virtually 100% of the incident photons with energy above the bandgap could therefore be detected, in principle. In actual devices, the external quantum efficiency (QE),

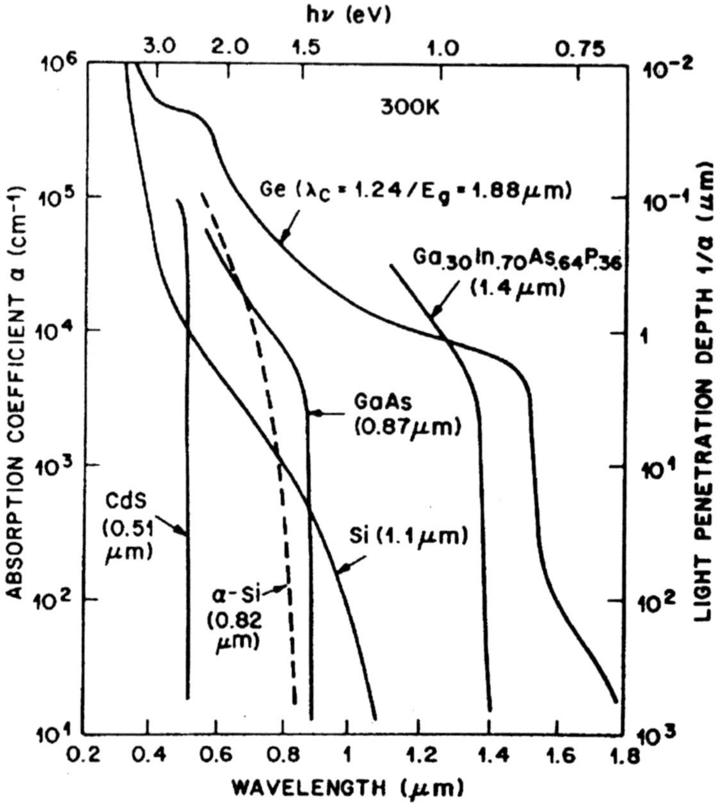

Fig. 8.1 Optical absorption coefficients and light penetration depths for various semiconductor materials [1]

defined as η = number of photogenerated charge pairs/number of incident photons, is smaller than 100% for the following reasons:

1. Fresnel reflection of the incident light at the surface of the device.
2. Multiple reflections in thin layers covering the device. This causes the characteristic thin film interference oscillations in the spectrum.
3. Absorption of the incident light either in the covering layers or in the electrically inactive part of the semiconductor near the surface.
4. Absorption deep in the semiconductor, at a distance greater than the diffusion length L, where charge carrier pairs recombine instead of being able to diffuse to the depletion region of the photodetector near the surface.
5. The semiconductor is too thin (too transparent) so that not all the incident light is absorbed and parts of it (the longer wavelengths) are transmitted.

Figure 8.2 shows the measured spectral QE of an n^- p-substrate silicon photodiode, realized in a 2-μm CMOS process (Orbit Inc., Sunnyvale, USA) that can be explained with a model taking into account the effects mentioned earlier [2].

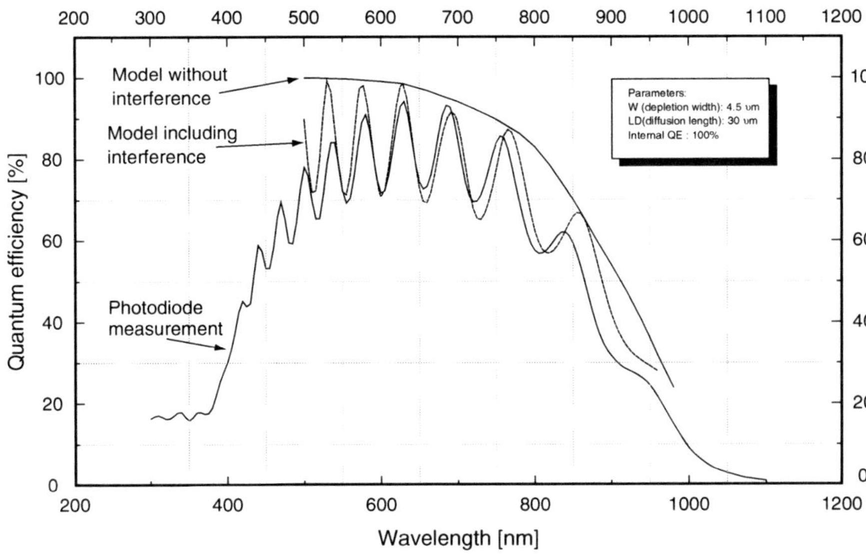

Fig. 8.2 Comparison of a model of the quantum efficiency (*dashed line*) and actual measurements (*solid line*) of a photodiode, realized with a standard 2.0-μm CMOS process [2]

The QE η is a measure of the number of photogenerated electronic charges per incident photon. If, instead, the generated photocurrent I_{ph} (A) is measured as a function of the incident light power P (W) at the wavelength λ, the corresponding responsivity R (in units of A/W) can be defined as

$$R = \frac{I_{ph}}{P}. \tag{8.3}$$

R is obviously related to the quantum efficiency η:

$$R(\lambda) = \frac{q\lambda}{hc}\eta(\lambda), \tag{8.4}$$

with Planck's constant $h = 6.63 \times 10^{-34}$ J s, and the speed of light in vacuum $c = 3 \times 10^{8}$ m/s.

8.2.3 Temperature Effects: Dark Current

The strongest impact of temperature is undoubtedly related to leakage currents. The so-called dark current (i.e., current measured without any applied optical signal) significantly varies from one technology to another with CCD still providing the best figures (below 1 pA/cm²). Optimized CMOS processes for image sensing technologies report figures of a few 10 s of pA (30–50 pA/cm^2) at room temperature. It is important to note that the dark current depends strongly on temperature, rendering high-temperature applications very challenging. This dependence is essentially given by that of the intrinsic carrier concentration n_i in the semiconductor:

$$n_i \propto T^{\frac{3}{2}}\, e^{-\frac{E_g}{2kT}}. \quad (8.5)$$

With $E_g = 1.11$ eV for silicon at room temperature the dark current is found to double roughly every 10°C. Experimentally this doubling occurs even every 7–8°C. Thus, cooling the sensor by 25°C leads to a decrease of the dark current by almost a decade, which is significant and explains why most scientific cameras targeting applications with long exposure time (e.g., in astronomy) still need to be cooled. The current trend toward higher integration and smaller design features leads to typically higher doping levels and increased leakage current and dark current densities. This renders the realization of imagers providing high performances especially under low-light conditions even more challenging. In this specific case the higher integration/continuous trend toward miniaturization basically works against key performance figures of imaging devices.

8.2.4 Photosensor Principles: Photodiode and CCD

In image sensors the photogeneration of an electrical signal takes place in most cases either in a photodiode (i.e., a p–n junction) or in a MOS capacitance. Other approaches involve the use of specialized devices such as avalanche photodiodes (APDs) and phototransistors. Generally speaking photodiodes are used in CMOS image sensors, whereas MOS capacitances are preferably implemented in CCD imagers. However, interline transfer CCDs as mostly used in consumer cameras are typically based on photodiodes as photosensitive elements, whereas some CMOS imagers take advantage of photogates, a structure that is basically derived from CCD imagers.

Assuming a pixel size of 10 × 10 μm^2, a fill-factor of 100%, and a quantum efficiency of 100%, an illumination level of 1 lux (which corresponds to about 10^{16} photons/s m^2 for white light) leads to a photocurrent of the order of 160 fA. To effectively read out such low signals special care must be taken. Different detection circuits have been developed over the last decades to address this challenge, the simplest and also most often implemented one being a source follower. In this approach the photogenerated current is first integrated during a given integration time T_{int} onto a capacitor C. After this integration/exposure time, the charge ΔQ accumulated is converted to an output voltage ΔV_{out} according to

$$\Delta Q = \int_0^{T_{int}} i_{photo}(t)\,dt, \quad (8.6)$$

$$\Delta V_{out} = (\Delta Q/C) g_a, \tag{8.7}$$

with $g_a \approx 0.6 - 0.8 \approx 1$ being the "gain" of the source follower.

For a capacitance of C = 10fF, the conversion gain amounts to q/C =16μV/e leading to a measurable signal for typical charge packet of a few thousands of electrons. Higher conversion gain can be achieved by lowering the capacitance, or using current amplifier with very small feedback capacitances. This becomes mandatory if one wants to detect small signals, ultimately reaching for single electron and photon detection.

8.3 Semiconductor Technology for Imaging

8.3.1 Silicon Sensors

The unrelenting progress of semiconductor technology made it possible in the early 1960s to fabricate reliably several ten thousand transistors on the same chip. This capability was quickly used for the production of semiconductor image sensors using the then-available MOS process technology. These image sensors were based on the photodiode pixel architecture illustrated in Fig. 8.3. Each pixel consists of a photodiode that can be connected individually to an output amplifier through a row-select transistor connecting the photodiode to a column electrode, and through a column-select transistor connecting the column to the video output line.

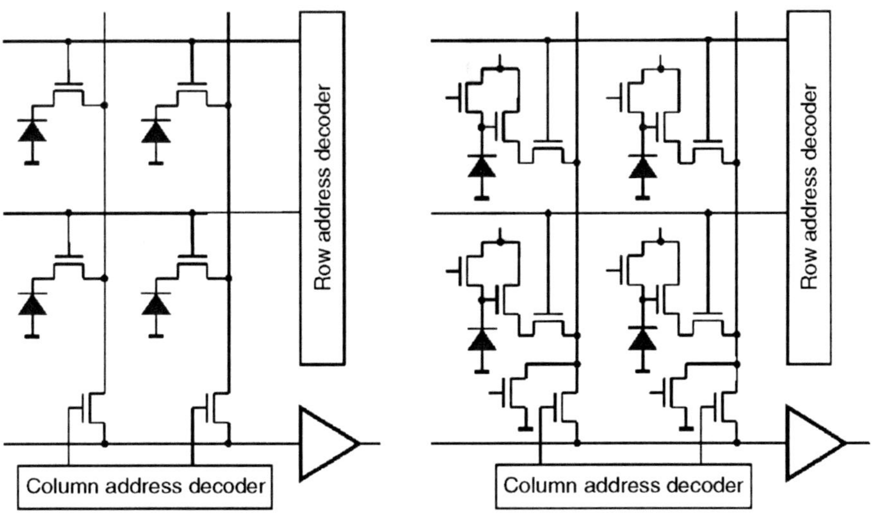

Fig. 8.3 Architecture of photodiode-based image sensors: (**a**) MOS or photodiode array, (**b**) CMOS-compatible active pixel sensor (APS) imager

The charge detector measures the stored photocharge on the reverse-biased photodiode's depletion capacitance, and at the same time it performs a reset operation on the photodiode. These long signal lines result in effective capacitances at the input of the charge detector circuit of several pF, causing significant readout noise, typically in excess of 1,000 electrons.

It was realized as early as 1968 that the invention of the active pixel sensor (APS) principle makes it possible to utilize a standard MOS or CMOS process for the fabrication of image sensors offering a similar imaging performance as CCDs [3]. The basic idea, illustrated in Fig. 8.3, consists of supplying each pixel with its own source follower transistor, thus reducing the effective capacitance at its gate to a few tens of fF or less. Since each pixel still requires a row-select transistor and a means of resetting the photodiode – implemented with a reset transistor – a CMOS pixel with three transistors, a so-called 3T APS pixel, results. Surprisingly enough, this insight was only picked up commercially 15 years later, once the urge to cointegrate analog and digital functionality on the same image sensor chip became significant [4]. For almost a decade afterward, industrial CMOS processes were employed for the fabrication of CMOS image sensors and single-chip cameras of growing complexity, increasing pixel number and rising imaging performance.

In the late 1990s it became clear that the roadmap of mainstream CMOS technology, in order to satisfy the device scaling according to Moore's law, foresees technological changes that are partially detrimental to image sensing [5]. This is summarized in Table 8.1, making it evident that the increasing demands in semiconductor imaging cannot be met with mainstream CMOS technology long after 2000.

As a result, the semiconductor industry started to develop variants of their established CMOS processes under the acronym CIS (CMOS image sensing) processes. Such CIS processes continue to profit from the enduring reduction in the feature size of the mainstream CMOS processes. This permits the integration of a growing number of increasingly smaller pixels. The largest semiconductor image sensor fabricated to date contains the impressive number of 111 million pixels, while the smallest pixel pitch reported to date is 1.2 μm [6]. The reduction of the threshold voltage to a current value of about 0.3 V is also beneficial for semiconductor image

Table 8.1 Technological trends on the roadmap of mainstream CMOS technology, an evaluation (Eval) of their positive or negative impact on semiconductor imaging with these processes, and the main reason for this particular evaluation

CMOS technology trend	Eval	Reason
Reduction of feature size	++	More pixels and functionality on chip
Reduction of threshold voltage	+	Higher signal levels and dynamic range
Reduction of power supply voltage	+	Reduced power consumption
	–	Lower signal levels and dynamic range
Gate material: poly-Si → silicides	–	Opaque gates in the visible spectrum
SOI or epi-Si technology	–	Low red response and reduced QE
Increase of substrate doping	–	Larger dark current
Increase of number of metal layers	–	"Tunnel vision"; low QE of small pixels
Reduction of gate oxide thickness	– –	Larger dark current

sensors because the available voltage swing is increased. The reduction of the power supply voltage, however, is a mixed blessing: On one hand, power consumption is lowered appreciably, since it is proportional to the square of the supply voltage; on the other hand, this lowers also the voltage swing for representing photosignals and, as a consequence, also the dynamic range of the image sensor is reduced.

The replacement of doped polysilicon as the gate electrode material with silicides ($CoSi_2$ or $TiSi_2$) is beneficial since the gate's electrical conductivity is improved but since silicides are essentially opaque to visible radiation, photosensitive devices employing photogates lose much of their sensitivity. The replacement of bulk silicon with epitaxial silicon layers of a few micrometers thickness [as in silicon-on-insulator (SOI)] technology also reduces the sensitivity of photosensitive devices, in particular, in the red and near infrared spectral region where the penetration depth of light in silicon is particularly large. To enhance the electrical conductivity of the silicon substrate, doping levels of the implantations are increased to typical levels of about 10^{18} cm^3. This implies a significant increase of the dark current density to values beyond 1 nA/cm^2 at room temperature. Such values are clearly unacceptable in many imaging applications, in view of the low dark current densities of a few 10 pA/cm^2 in commercially offered CIS processes and the record value of 0.6 pA/cm^2 for the best CCD process.

The combination of shrinking pixel period and increasing number of metallization and via layers results in "chimney pixels," as illustrated in Fig. 8.4, which at the same time reduces the optical sensitivity of the pixels and increases the angular

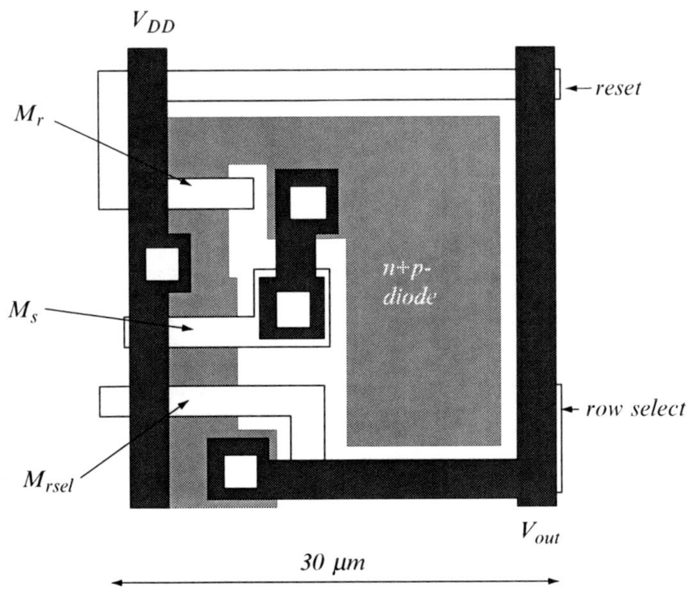

Fig. 8.4 Example of pixel layout for an active pixel sensor

dependence of their photoresponse. Finally, the reduction of the gate oxide thickness to today's typical values of less than 2 nm increases the tunnel density current through the oxide to values exceeding 1 mA/cm^2, which is clearly unacceptable in image sensing applications.

For these reasons, CIS processes are offered today that are derivatives of industrial CMOS processes, where all the problems mentioned earlier are addressed appropriately, and where great care is taken that the photosensitive devices can fully profit from the excellent optoelectronic properties of the silicon base material.

An example of pixel layout for a CMOS APS is given in Fig. 8.4. Because of the presence of circuitry within the pixel the area that is effectively sensitive to light is reduced to typically 25–65% of the total pixel area. This corresponds to the so-called fill factor. In comparison, CCDs typically have fill factors of 70–100% for full-frame CCD (FF-CCD) and frame transfer CCD (FT-CCD), whereas interline transfer CCDs (IT-CCD) achieve fill factors in the range of 25–30%.

A typical cross section of a pixel structure fabricated with such a commercially available CIS process is illustrated in Fig. 8.5. In the silicon substrate photosensitive regions are created using suitable implants: As detailed earlier, the preferred devices are photodiodes, buried (pinned) photodiodes, or CCD structures.

Oxide material is deposited on top of the silicon, containing the copper interconnects and vias, as well as the gate structures for the transistors. Different types of color filters are produced over the individual pixels, followed by microlens structures to enhance the effective light collection efficiency of the pixel. Their focusing effect is illustrated with the broken lines in Fig. 8.5, indicating schematically the ray tracing modeling that is necessary to optimize the optoelectronic performance of the pixels. Typical dimensions of state-of-the art CIS pixels as employed, for example, in

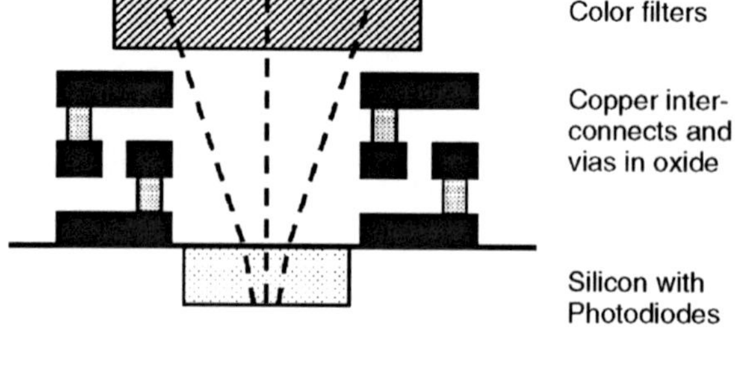

Fig. 8.5 Cross section of a typical CIS pixel, consisting of a silicon substrate with implanted photodiode regions, on top of which oxide material is deposited, where the copper interconnects, vias, gate structures, and color filters are fabricated. To improve the light collection efficiency, a microlens array is produced

cost-effective image sensors used in mobile phone cameras are periods of 2–3 μm and a total thickness of 3–4 μm.

To produce more pixels for less money, the CIS industry will not cease to invest in the continuing reduction of pixel size. This involves solving the problem of reduced capacitance and reduced photosensitive surface of the ever-shrinking pixels: As an example, consider a pixel with a storage capacitance of 1.6 fF; assuming a maximum voltage swing of 1 V, such a pixel cannot store more than 10,000 electrons.

Since the photogenerated and stored charges show Poisson noise statistics, perceived picture quality deteriorates as a function of the stored photocharge numbers [7]. For this reason, good quality images require photocharge storage capacities of a few 10,000 electrons in each pixel. This problem is quite similar to the one in memory circuits: Despite the continuing reduction of the unit cell size, the capacitance of the unit storage cell must be kept at a certain value to assure reliable operation of the storage function. The solution adopted in CIS technology is, therefore, also related: The third dimension is exploited to increase the specific capacitance of pixel storage devices. An example of this development is the recently developed "stratified pinned photodiode," realized by corrugating the outer shape of the buried photodiode implant [8].

The other problem of smaller pixels with increased "tunnel vision," as illustrated in Fig. 8.5, is the reduced photosensitive surface and the increased angular dependence of the response. While microlenses can help to alleviate the problem, there will always remain an angular dependence of the pixel's response, which reduces the pixel's overall sensitivity, in particular when imaging lenses with small *f#* are employed. A promising technological approach to solve this problem is the separation of the optoelectronic transduction and the electronic circuitry into different layers. This is achieved by depositing a thin film of amorphous or microcrystalline silicon (or another suitable semiconductor) on top of an ASIC structure (TFA = thin film on ASIC technology). Photodiodes are fabricated in the top layer of TFA image sensors, and electrical connections to the charge sensing circuits on the ASIC are produced. In this way, an effective geometrical fill factor approaching 100% can be achieved, and the photoelectric conversion properties, including, for example, very low dark current, can be optimized separately from the ASIC in the thin film [9].

8.4 Examples and Applications of Imagers

8.4.1 Electronic Imaging in the Visible Spectrum

In the past, the main drivers of electronic imaging have been cost, performance, and functionality; without doubt, these drivers will stay unaffected in the foreseeable future. The challenge, therefore, is to take advantage of the continuing progress of semiconductor physics to satisfy the demands of the market in terms of cost, performance, and functionality of the realized image sensors and camera systems.

8.4.1.1 Challenges and Opportunities

The key to meeting present-day challenge is integration: More electronic circuitry will be placed on each image sensor, and the pixels themselves will be supplied with all the analog and digital elements to improve their performance and to increase their functionality. Whenever economically justified, an application-specific single-chip digital camera (SoC=system-on-chip) will result that just needs to be complemented with inexpensive optics to arrive at a highly cost-effective solution.

A first huge, fast-growing, and highly contested market is electronic imaging for cell phones. It is estimated that close to one billion cell phone cameras will be sold in 2007. Consequentially, the price of such a complete single-chip digital camera has dropped to less than \$5 in volume, while its resolution is already exceeding 1 megapixel. From this development, other mass markets such as automotive applications, game console interfaces, personal computer peripherals, and videoconferencing will surely profit. Typical single-chip camera developments for these and related applications where price is the main driver are described later.

Two other fast-growing markets with considerably higher margins are digital still cameras (DSC) and security/surveillance. Although both markets combined require only about 100 million image sensors in 2007 – slightly more than 10% of the cell phone camera market – the performance expectations of the customer are much higher: In the case of DSCs, multi-megapixel resolution (already exceeding 10 megapixels even for amateur cameras), high uniformity, and low noise are demanded. The performance expectations of the security/surveillance cameras are similar; however, the emphasis is not on the number of pixels but on the low-noise performance for low-light applications. The ultimate physical performance in this respect is single-photon resolution, and as soon as cost-effective solutions will be available, most other electronic imaging applications will demand such a performance, as well. The most promising approaches for low-light semiconductor image sensors are detailed later.

The third driver of electronic imaging is functionality: In conventional image sensors, the available functionality is restricted to the four basic tasks symbolically illustrated in Fig. 8.6: conversion of incident light intensity into a linearly related photocurrent, integration and storage of the resulting photogenerated charge packets, sequential access to these charge packets, and amplification/buffering of the signals

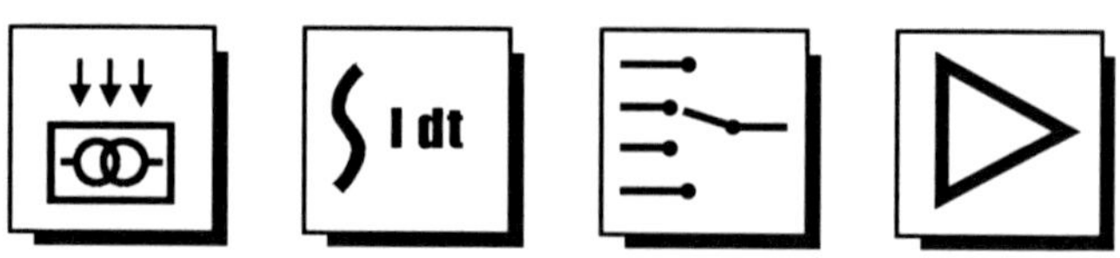

Fig. 8.6 Symbolic functionality of conventional imagers

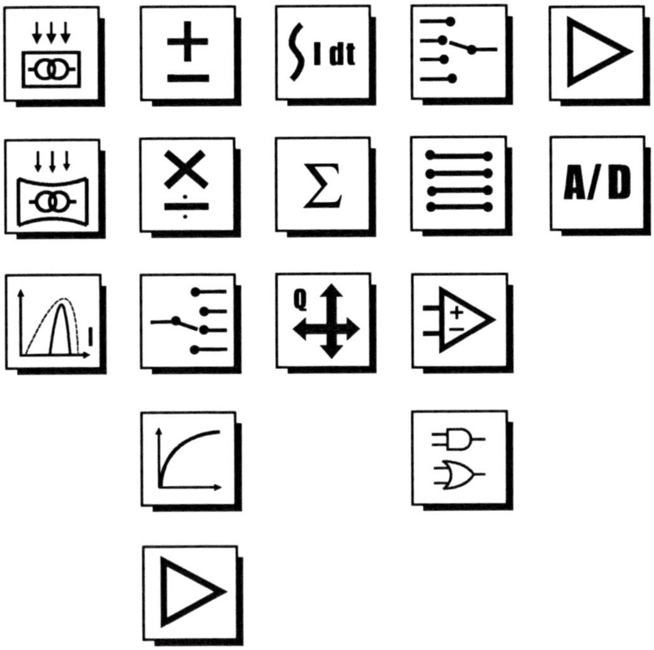

Fig. 8.7 Symbolic illustration of a few examples of functional building blocks available for smart pixels and custom image sensors

to make them available off chip. By complementing this basic optoelectronic functionality with analog and digital circuit building blocks, in each pixel or adjacent to the pixel matrix, a very rich toolbox for the realization of application-specific "smart image sensors" is created. Some of this additional functionality, all fabricated with the technological capabilities of the same CIS process, is schematically illustrated in Fig. 8.7, indicating the broad range of building blocks available for the design of custom imagers. This includes, among other things, smart pixels with nonlinear response for a dynamic range approaching 200 dB, color pixels making use of the wavelength-dependent absorption of silicon, pixels with unusual geometries implementing selected linear or nonlinear transforms, in-pixel analog processing circuits for basic arithmetic operations in the photocharge domain, as well as conventional analog and digital circuits in the pixels or as complementary building blocks beside the photosensor area on the image sensor [10].

The availability of this toolbox of functionality is the key for improving the performance of novel, smart pixels over previous types of pixels, and for providing the custom functionality required for the successful integration of special image sensors for a large variety of applications.

As detailed later, this development is expected to culminate in the realization of complete single-chip machine vision systems, so-called "seeing chips."

8.4.1.2 Single-Chip Digital Low-Power Camera

The continuous trend toward miniaturization and higher density in CMOS mixed-signal technologies has a strong impact on the development and progress in solid-state image sensors and related applications. Whereas image sensors based on a passive MOS (metal oxide semiconductor) array architecture were already demonstrated in the seventies, their noise figures and overall performances turned out to be by far not sufficient to compete with dedicated CCD-based imaging products. It is only with the introduction of the APS concept in the nineties that CMOS imagers have been able to step by step close the gap with CCD imagers. The main improvement came with the capability to integrate active transistors within every pixel of the imager in order to locally amplify the typically very small photogenerated signals. This approach basically reduces significantly the large stray capacitances that characterize passive MOS array imager to a few tens of fF, i.e., values similar to the input capacitance of CCDs' output stages. The use of more aggressive technologies and submicron processes was thus instrumental for the performance improvements of CMOS imagers. Furthermore, the use of standard CMOS technologies opens up the possibility of manufacturing imaging devices that can be monolithically integrated: functions such as timing, exposure control, and analog to digital conversion (ADC) can be implemented on one single piece of silicon, enabling the production of single-chip digital image sensors. With ever decreasing transistor size, it also becomes technically and economically possible to combine the image sensor with functions such as digital signal processor, microcontroller, or further interfaces (e.g., USB). Clearly the higher number of pixels that can be integrated on ever decreasing areas have fuelled the success of CMOS imagers. As such, to a large part, the success of single-chip digital cameras can be seen as a direct result of "more Moore."

Today solid-state cameras have become pervasive tools in many markets. It is forecast already that next year more than a billion solid-state imagers will be sold worldwide, the dominating applications and markets being mobile phones, digital still photography, security, automotive, and toys. CMOS technologies and in particular CMOS imagers have also intrinsically the potential for lower power consumption. CMOS imagers can be driven from one single supply voltage (typically <3.3 V) and with power consumption below a few tens of mW making these devices particularly attractive for mobile applications, such as the mobile phone market. The latter market corresponds today to roughly 50% of the units sold worldwide.

Low power consumption also helps to reduce the temperature (or temperature gradient) of the sensor chip and camera head, leading in general to improved performances. Figure 8.8 shows a single-chip digital camera with a spatial resolution of 176 × 144 pixels. It operates at 1.5 V and consumes less than 1 mW [14]. The low power consumption is achieved by the reduction in the power supply voltage and by applying special techniques for the analog design. Sensors operating at less than 1 mW have also been demonstrated [15]. Video cameras in mobile phone have become a reality today. Undoubtedly further applications, for example,

Fig. 8.8 Micrograph of a digital single-chip camera with a power consumption of less than 2 mW

in medicine (e.g., minimally invasive medicine/endoscopy) or surveillance and security, will profit from the progress made in low-power digital imaging.

8.4.1.3 High-Sensitivity Electronic Imaging, Single-Chip Digital Low-Power Camera

As we have seen in Sect. 2, the external quantum efficiency of silicon can be close to 100% in the visible spectral domain (for monochrome imaging). Since this means that almost each incident photon is converted into an electron–hole pair, the physical problem of high-sensitivity image sensing is really an electronic problem of high-sensitivity charge sensing. The electronic photocharge detection circuit used almost exclusively in semiconductor image sensing is illustrated in Fig. 8.9: In a p substrate, an n^+ output diffusion (OD) is implanted. Using the reset transistor R, this diffusion is reverse-biased to a certain reset potential, and it is left floating afterward. The photocharge Q that should be measured is then placed on OD, lowering its potential by the amount $V = Q/C$, where C indicates the total effective capacitance at the gate of the source follower transistor SF. The voltage at C in each pixel can be measured individually by connecting SF with the row-select transistor RS to the column signal line Col, which terminates in a load transistor for SF.

Two noise sources dominate the charge measurement process in the source follower circuit shown in Fig. 8.9: (1) Reset noise – also called kTC noise – introduced

Fig. 8.9 Typical readout structure for semiconductor image sensors

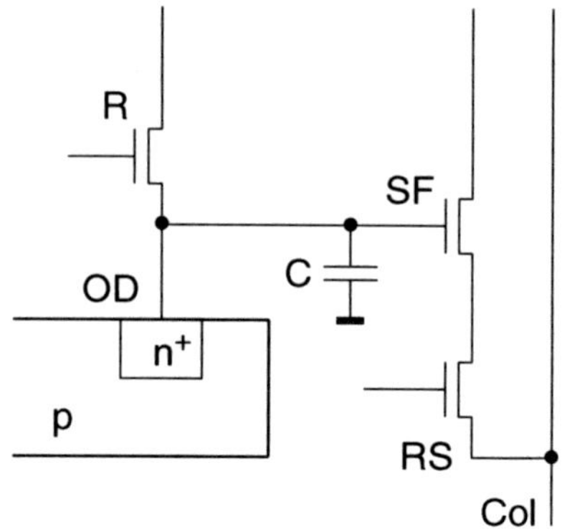

by the reset process effectuated by transistor R, and (2) Johnson noise in the channel of the source follower transistor SF. Both effects cause a statistical uncertainty ΔQ (standard deviation) in the measurement of charge packet Q:

$$\Delta Q_{\mathrm{reset}} = \sqrt{kTC} \tag{8.8}$$

and

$$\Delta Q_{\mathrm{Johnson}} = C\sqrt{\frac{4kTB\alpha}{g_{\mathrm{m}}}}, \tag{8.9}$$

where k denotes the Boltzmann constant ($k = 1.381 \times 10^{-23}$ J/K), T indicates the absolute temperature, B is the bandwidth of the measurement process, α is a constant – with a value not too far from unity – that depends on the way the source follower transistor is operated, and g_{m} denotes the transconductance of the transistor SF.

For the ultimate high-sensitivity photodetection performance of an image sensor, both of these charge measurement uncertainties must be minimized. Although it has already been demonstrated that single-photon imaging performance can be achieved with several approaches, even at room temperature and at video readout frequencies (corresponding to an output circuit bandwidth of a few tens of MHz), the affordable CIS-compatible single-photon image sensor with at least megapixel resolution is still elusive. It is not clear, therefore, which of the various high-sensitivity charge detection approaches will be the winner in this race:

In conventional photodiode-based CIS image sensors, where the output diffusion OD is identical with the photodiode (as illustrated, for example, in Fig. 8.9), the reset noise described by (8.8) is typically one order of magnitude larger than the Johnson noise in (8.9). With the invention of the CCD principle, it became possible to completely eliminate reset noise by adopting the following measurement sequence: OD is reset to a certain reset voltage and then left floating. The voltage on OD is measured and the value is stored. The photocharge Q is physically transferred on OD, employing the CCD principle, and the resulting voltage drop is measured.

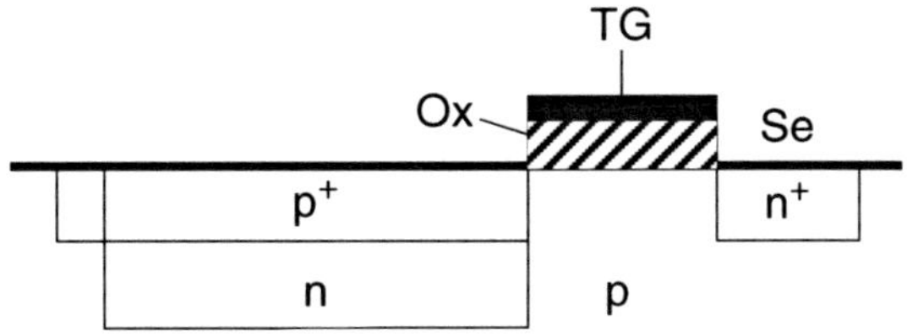

Fig. 8.10 Cross section through a pinned (or buried) photodiode

The difference of these two measurements results in a voltage value that is linearly related to the photocharge Q, and its statistical uncertainty ΔQ is reduced to the Johnson noise described in (8.9).

Since it was believed that it is not possible to transfer a stored amount of charge from one diffusion (e.g., a photodiode) to another one (e.g., the OD in Fig. 8.9), without introducing additional noise, CCD image sensors were considered superior in terms of noise performance than imagers fabricated with CMOS-compatible processes. A conceptual breakthrough occurred with the invention and the fabrication technology of the pinned (or buried) photodiode (PPD), illustrated in Fig. 8.10.

As shown in this figure, a buried n-type volume is entirely surrounded by p-doped silicon. This PPD can be fully depleted, so that CCD-type operations such as complete charge transfer using the transfer gate TG become possible. This implies that image sensors fabricated with CCD as well as CIS technology can reach the Johnson noise limit described by (8.9). Using this approach, a charge measurement noise of less than 10 electrons r.m.s. can be obtained in practice.

For the past two decades, researchers have grappled with the Johnson noise equation, with the aim of attaining single-electron photocharge detection noise in an image sensor at room temperature and at video frequencies (corresponding to a readout bandwidth of 10–20 MHz):

- The effective capacitance C can be reduced to values below 1 fF using double-gate field-effect transistor (DG-FET) technology. Although single-electron noise is achieved, it comes at the expense of complex technology modifications and voltages above 20 V.
- A similar approach with buried charge modulation devices (BCMD) results in charge noise values of a few electrons, again requiring complex technology modifications.
- CCD image sensors used for astronomical applications are read out at very low frequencies of around 50 kHz, and they are cooled down to about $-100°C$. This results in single-electron noise performance but at the expense of very long readout times.
- The so-called Skipper CCD is operated at room temperature, using floating gate readout structures to measure the same photocharge repeatedly in a nondestructive way. The averaged measurements show the desired single-electron charge noise but the necessary averaging over more than 100 individual measurements per pixel slows down readout significantly.

- To circumvent the problem of noisy electronic charge detection, physical amplification mechanisms with low excess noise factors are actively pursued. Image sensors with APDs, either working in the linear or in the Geiger mode, have been realized, exhibiting the wanted single-electron charge noise at room temperatures and at video frequency readouts. Several products such as the Impactron™ CCD image sensor with avalanche amplification section, developed by Texas Instruments, are offered commercially. Unfortunately, the high electric fields required to cause avalanche amplification still necessitate voltages of 20 V and more, making the use of avalanche-based single-electron imagers somewhat unpractical.
- Probably the most promising approach for cost-effective single-electron image sensors is "bandwidth engineering": The measurement bandwidth B in (8.9) is significantly reduced, either at each pixel site or in each column amplifier. Suitable imager architectures with parallel readout capabilities provide for data readout at unimpeded speeds, despite the low-pass filtering effect; for example, see [11]. The large advantage of this approach is its entire compatibility with standard CIS technology, as well as low-power, low-voltage operation.

None of these approaches has yet resulted in the commercial availability of cost-effective, easy-to-use image sensors with close to single-photon performance. Without doubt, such devices, based on one of the described principles, will become available in the next few years.

Pixel dimension reduction is justified not only to decrease the costs, following the Moore's law trend, but also to reduce the volume and thickness of the imager including its optical part. The reason is that for a given optical *f#*, a reduction of the lens diameter results in a reduction of the focal length and hence the total thickness of the imager. The main drawbacks of this trend concern the optics, with diffraction effects close to the pixel pitch, and the pixel collected charge possibilities, reduced as the pixel size is lowered. The drastic size reduction of the well collecting the photogenerated electrons, able to collect typically 40,000 e^- with 16 μm in pitch pixel, 10,000 e^- with 2.2 μm, and 6,000 e^- with 1.45 μm, is the strong motivation to decrease the total noise coming with each pixel reading: the dynamic range being reduced with its upper limit, the lower limit must improve drastically. The best read noise results are now in the range of 1–2 electrons RMS. Nevertheless, even with a low noise reading, the fluctuations of number of photogenerated electrons are correlated with a Poisson distribution, which predicts a RMS value given by the square root of the total electrons number. This phenomenon comes from the photon number statistical fluctuation and cannot be avoided without increasing the total number of collected electrons. To obtain a SNR (signal-to-noise ratio) of 10, 100 photogenerated electrons are necessary with a perfect nonnoisy reading circuit. To increase this collection number, several directions are to be considered.

SNR limitations come primarily from the optics size; a lower *f#* is necessary to both lower the diffraction and increase the collected photons. This leads to large incident angles not compatible with the path through interconnection layers, as illustrated in Fig. 8.11.

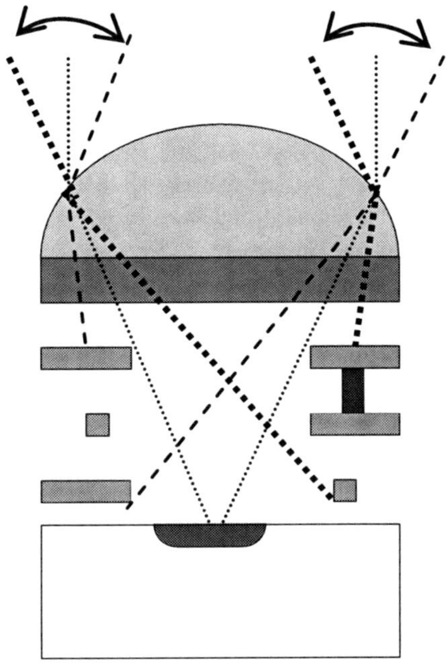

Fig. 8.11 Light path through interconnects for large optical aperture (low *f*#)

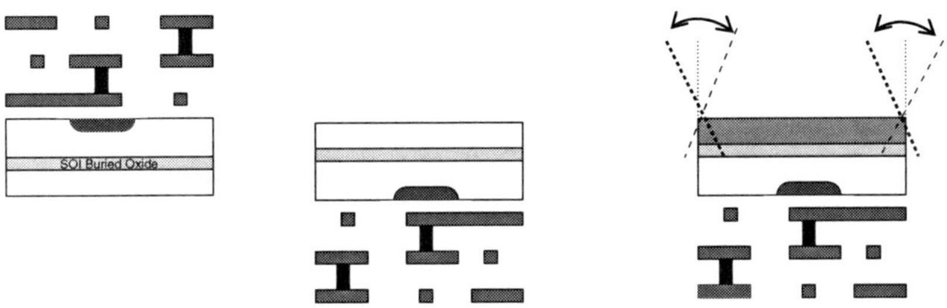

Fig. 8.12 Back side technique to overcome the light path screening through interconnects for large optical aperture (low *f*#)

To overcome this limitation, large-angle collection is provided by backside and thinning approaches. This technique, yet used for high-end CCD products for space applications, provides high fill factor, high photon collection angle, and improved pixel crosstalk. This technique is based on wafer bonding and thinning, and the thin top silicon layer, necessary to collect efficiently the photogenerated electrons, is difficult to obtain and passivate. One way is to use a SOI substrate, allowing to keep the initial buried oxide as the top final passivation layer, for high-quality interface and controlled silicon thickness: this backside technique is illustrated in Fig. 8.12. Another way is to improve the thinning control to reach the desired value and to passivate to the final silicon surface by the use of laser anneal and low-temperature deposition.

8.4.1.4 Color Imaging

Appropriate color restitution requires that the color recording is as reliable as possible. Human sense of color is based on triplet cells in the eye's retina: the spectral sensitivity of these three types of cells is the base of the color recording. Color science, as investigated by the CIE (Commission Internationale de l'Eclairage), gives the rules to achieve good restitution of the available colors in the world, as seen by the human eye. The main result is as follows: the reconstruction of all colors is possible if for each image point three different light intensity measurements are acquired – these measurements having relative spectral sensitivities given in Fig. 8.13, or linear combinations of them.

The classical way to obtain these spectral sensitivities is the use of color filters. They are made of photoresist layers deposited on top of the imager circuits, as previously shown in Fig. 8.13. Three types of photoresists, e.g., red, green, and blue are necessary; complementary colors could be used as well; they are generally deposited in a 2 × 2 pixels pattern known as the Bayer pattern (Fig. 8.14).

The limitations of this technique are as follows:

- The thickness of the resist, around 800 nm, becomes high with respect to the width of the pixel. Thinner filters would lead to reduced optical aberrations due to lateral optical path from the top surface to the silicon.
- The maximum temperature allowed by the resist is much lower than any other layers in the circuit: no high-temperature process step can be done after color filter deposition.

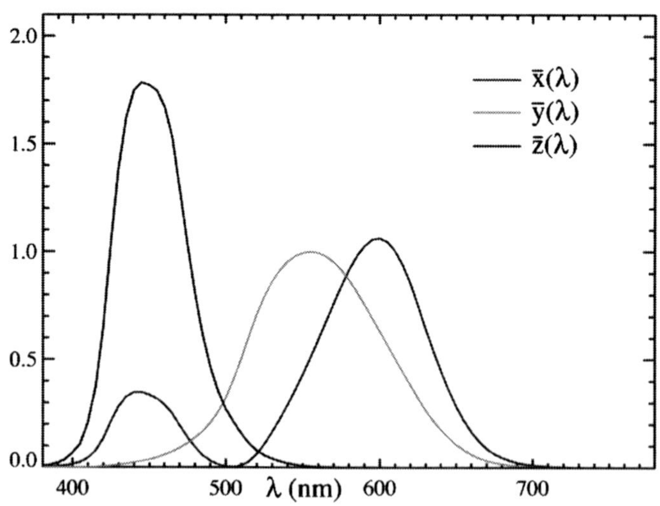

Fig. 8.13 Normalized spectral sensitivities allowing reliable reconstruction of all colors

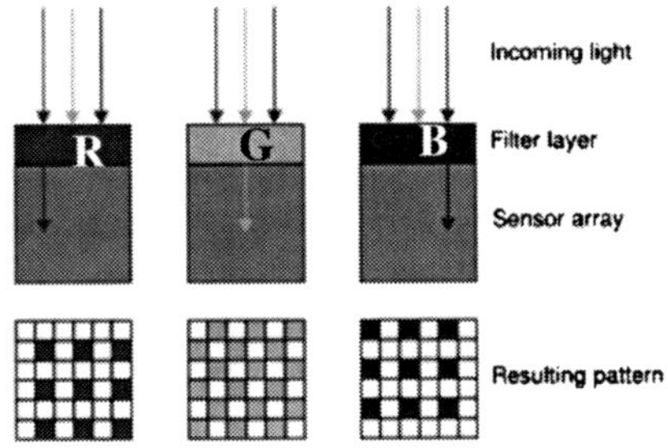

Fig. 8.14 Bayer pattern for color filters

- The filters absorb around two-third of the incoming light; this is a main cause of loss of photons and thus sensitivity.
- The three types of color sensors are not at the same location, resulting in some color reconstruction difficulties and aliasing/moiré effects.

To overcome these limitations, several techniques have been investigated. The three types of sensors can be arranged vertically, i.e., using stacked photodiodes as demonstrated by Foveon: it uses the light absorbance variation in silicon to detect the red signal deep into the silicon, while red and green signals are absorbed closer to the surface and finally red + green + blue signals are detected in the top p–n junction that is located closest to the surface of the silicon. An alternative technique consists in using mineral filters integrated in the CMOS process. This leads to a higher temperature resistance and a thinner total stack thickness, thus lowering the lateral shift of photons causing diaphoty. More recently the use of the filtering behavior of submicron metal patterns creating plasmons when impacted by light has been envisioned and is currently been investigated.

8.4.1.5 High-Speed Image Sensing

High-speed imaging allows motion analysis in various application fields such as scientific measurements, machine vision, or holographic data storage to mention a few. Impressive application examples are available from various suppliers [16].

In a broad sense, image sensors can be considered as "high speed" as soon as they exceed video rates [over some 60 frames per second (fps)]. The "speed" (maximum frame rate) of an image sensor is closely related to its spatial resolution and more particularly to its number of rows. For example, a sensor capable of 1,000 fps while scanning 1,000 rows (row rate = 1 μs) is able to deliver 10,000 fps when scanning a partial image of 100 rows (row rate = 1 μs). The reason why the number of columns has almost no impact on the sensor frame rate is that the majority of today's high-speed sensors use column parallel readout circuitry.

The highly parallel architecture of CMOS sensors has led within the last decade to a technology shift from CCDs toward CMOS imagers for high-speed imaging. For low spatial resolution in turn, CCD sensors are still capable of recording high speeds, which have by far not been equaled by CMOS. As a comparison, a 320 × 260 pixel CCD sensor using in situ storage is capable of 1 M (10^6) fps [17], while a 352 × 288 pixel CMOS sensor using innovative in situ ADCs reaches 10k (10^4) fps [18]. For higher spatial resolution (and lower speed) CMOS image sensors (CIS) have, however, overtaken the market place. This success is mainly attributable to the system integration capability of CIS, which leads to lower cost cameras. Up to the 1990s, CCD and early CIS [19] delivered analog output and thus A/D conversion had to be performed in separate ICs. With the advent in the late 1990s of the first on-chip column-parallel ADCs [20], the way was paved for implementing high-speed/high-resolution digital imagers [21]. Today, a typical 1.3-megapixel high-speed sensor delivers 500 fps at full resolution [22].

To speed-up image acquisition, high-speed sensors use snap-shoot pixels operating in the "integrate-while-read" (IWR) mode of operation. IWR-capable pixels do per definition require in-situ storage, a property that was interestingly also required in the very first CMOS photodiode array published in the early 1990s [23]. Therefore, the same 5T pixels were reused [19] and fine-tuned for improved responsivity [24], one of the key requirements to the sensing part of a high-speed sensor, since high frame rates imply short integration time. To improve the sensitivity at pixel level, the conversion factor (CF) must be maximized, e.g., the capacitance of the photodiode and floating diffusion must be minimized. With the 3-μm process in the early 1990s, C_d and C_{FD} were in the order of 50 fF [23], yielding a poor CF of ~1.6 μV/e. With a modern 0.25-μm process, a CF of 13 μV/e is typical [22], meaning that C_d and C_{FD} have been lowered to some 6 fF. Various pixels with improved sensitivity have been proposed by either incorporating an amplifier in front of the shutter [25–27] or an in-pixel charge amplifier [28]. The first approach improves CF by a factor of 2 (~20 μV/e), while the second allows boosting the CF to some 38 μV/e. With the advent of pinned photodiode, the CF of traditional 5T pixels improves by a factor of 2, which allows smaller pixels [29] to rival with more complicate ones in term of CF. In addition to CF improvement at pixel level, almost all sensors further increase the overall sensitivity by electronic gain boosting inside the column amplifier (prior to A/D conversion).

Three types of column-parallel ADCs are used in today's CIS: single-slope, successive-approximation (SA), and cyclic/pipeline converters. Single slope converters are widely used in low-speed imagers. Recently, however, such an ADC has been pushed to a remarkable speed of 180 fps at 1,440 rows (row-rate = 3.86 μs) for 10 bits by using a 300-MHz clock [30]. Sensors with higher throughput usually use SA-ADCs, which typically reach 5,000 fps at 512 rows (row rate = 0.39 μs) [24] or 440 fps at 1,728 rows (row rate = 1.3 μs) [31] for 10 bits. It is difficult to achieve more than 10-bit gray-scale resolution with SA-ADCs on CIS. To increase the ADC resolution, the authors of [28] have developed cyclic converters capable of 3,500 fps at 512 rows (row rate = 0.56 μs) for 12 bits. Other publications have reported the use of multiplexed pipeline converters [27] [32] with similar aggregate performances as early-day SA-ADCs (500 fps at 1,024 rows/2 μs row rate for 10 bits [21, 32]).

High-speed digital CIS produce tremendous amounts of data: for example, the 440-fps 4.1-megapixel sensor of [31] delivers 18 Gbit/s (4.1 megapixels × 440 fps × 10 bit), which are dumped into some external IC over 160 pins toggling at 120 MHz. Such I/O bandwidths are indeed quite challenging to handle both at IC level (large power consumption, on-chip noise) and at board level (large buses, signal integrity). Some newer designs thus propose LVDS outputs [27, 32, 33] to solve those issues. Recently, incorporation of on-chip parallel image compression circuits for reducing the data traffic has been proposed [34]. In a prototype 3,000-fps sensor, a compression ratio of 4.5 has been demonstrated.

In future, further functionality will be integrated onto CIS dies. An example of such a sensor using programmable single-instruction multiple data-stream (SIMD) parallel digital processing has been published in [35]. The IC, aimed at high-speed machine vision, combines image acquisition, digitalization, and column-parallel SIMD processing with a computational power of 100 GOPS. With the availability of sub 100 nm mixed DRAM/imager technology, integration of a frame memory will also become economically viable [34].

8.4.1.6 Optical Time-of-Flight 3D Cameras

Humans are capable of perceiving their environment in three dimensions. This key capability represents in turn a major challenge to film-based and electronic camera systems. In fact, so far, cameras have been typically optimized for, and also limited to, the acquisition of two-dimensional images, providing basically only monochrome or color images without any direct information on depth. The race for higher (lateral) spatial resolution continues unabatedly; today's still picture cameras for the consumer market offer resolution in excess of 10 megapixels, whereas scientific and professional cameras already count more than a few tens of megapixels. This race for more pixels is very reminiscent of the continuous trends toward a higher degree of integration, a higher number of transistors as described by "more Moore." In contrast the shift from 2D imaging toward real-time 3D imaging represents a major technological step and paradigm change. A new type of information is available, based on sensors and pixels with added functionality – thanks to processing capabilities (at the pixel level). These smart sensors and pixels can be seen as vivid examples of "more than Moore." It also opens up completely new application opportunities in numerous fields such as in security, in automotive, in industrial control and machine vision, in the medical sector as well as in consumer electronics and mobile robots. All these applications can very directly benefit from camera systems that are able to capture the world in real time and in all three dimensions. This is particularly true if such 3D systems can be made compact, robust, and cost-effective. The technological requirements for such 3D systems are nevertheless stringent: To a large extent obtaining reliable real-time distance information over an entire scene is indeed very challenging and originally distance measuring systems have been limited to point measurements or to scanning systems, the latter being rather slow and/or expensive. Over the last few years however, technological progress in semiconductor devices and microtechnologies has lead to sensors that can "see distances" within complete scenes at affordable prices [36–39]. The breakthrough came with a new type of optical 3D camera based on the time-of-flight principle that uses the finite velocity of light ($c = 3 \times 10^8$ m/s) to measure distances. Either a pulse or a continuously modulated light wave is sent out by an illumination module. Correspondingly, either the "time of flight" t, which is the time the light needs to travel from the illumination module to the target and back to the sensor, or the phase delay ϕ of the continuously modulated light wave for this round trip is used to calculate distances within a scene. The latter method, referred as continuous wave (CW) modulation presents several advantages over the light pulse technique. In particular, the requirements to the illumination module and the electronics are not so demanding. Short optical pulses require a high optical power and a high bandwidth of the electronic components. In the CW modulation scheme, light from a LED or laser diode array is modulated to a frequency f_{mod} of a few tens of MHz and illuminates the scene. The light that is reflected back by the objects and/or persons in the scene is imaged with a lens onto a custom solid-state image sensor. Each of the pixels of this sensor is capable of synchronous demodulation of the incident modulated light for the precise local determination of the phase delay. From the phase delay ϕ, distances can be computed directly as given by the following equation.

$$I(x) = I_0\, \mathrm{e}^{-\alpha x}. \tag{8.10}$$

Current sensors achieve distance resolution in the millimeter range for distance up to a few meters in real time. Camera systems providing lateral resolutions of up to 176 × 144 pixels are commercially available today [40]. Higher resolutions up to 360 × 240 pixels have been demonstrated [41] and products with spatial resolution of up to VGA are expected in the near future. Figure 8.15 shows a sample image of a person entering a room. The information on the depth allows typically distinguishing reliably and effectively objects and persons from the background of the scene. In fact, the acquired 3D image data enables the very simple extraction of important information, such as the position and distance of objects and/or persons in the scene. This type of information turns out to be often key, in particular to human beings in solving many day-to-day tasks as well as more complex tasks. Thanks to 3D cameras – many applications in the field of automation, security, etc. are expected to profit as well from this 3D seeing capability.

8.4.2 Beyond the Visible Spectrum

Since imaging is performed in most cases by means of electromagnetic waves (magnetic resonance and ultrasound imaging are ignored within this document), the key parameter is the wavelength used. Visible imaging enables to cover all applications

Fig. 8.15 Sample image of 3D data (with permission of Mesa Imaging) [40]

where the information needs to be closed to the human vision. Beyond the visible spectrum, X-ray, infrared, and Terahertz electromagnetic waves are of most interest in many applications.

8.4.2.1 Challenges and Opportunities

Medical imaging, for example, is accomplished by means of X-ray, gamma rays (in nuclear medicine), visible light (ocular fundus, skin), infrared light (mamma transillumination), or even Terahertz waves, which has been proposed recently. In X-ray imaging, the gold standard is the X-ray-based mammography, where a special radiographic unit is used. Optical imaging of the retina is the only way to examine veins directly and noninvasively. Tele-ophthalmology has been demonstrated in a large international project.

As for medical imaging, industrial control and surveillance applications really take the benefit from detector innovations for X-ray range (e.g., CdTe material developments and new hybridization technics), infrared technologies (continuous need for improvement to decrease price), and new terahertz technologies

8.4.2.2 X-ray Imaging

Today X-ray imaging is used in many domains: medical, nondestructive testing, and security check. Most of the new developments are currently driven by medical requirements. This paragraph aims at giving a specific focus on medical image sensors.

First based on films, in the 1970s analog fluoroscopy became widely used. This modality was realized by an X-ray image-intensifier tube coupled to a TV camera. Sometimes, this kind of system was also used to take single images where the image resolution, however, was limited by the camera performance. In the 1980s, X-ray radiography became digital by introducing the so-called computed radiography where a storage phosphor in a cassette is irradiated and later on read out by a laser scanner.

State-of-the-art all-digital X-ray imaging is accomplished by flat-panel detectors [43]. These detectors are based on a large-area semiconductor film, i.e., hydrogenated amorphous silicon (a-Si), forming an array of thin-film transistors (TFT) managing readout of the picture elements. The fundamental interaction of the electromagnetic radiation with the material is the absorption. It is dependent on the energy. Secondary effects are called reflection scattering diffraction and refraction. All these interactions can be studied with Monte Carlo simulations. Details of interest differ from the surrounding tissues. This results in the contrast. To "see" its amplitude it needs to be compared with image noise.

The total image noise includes the detector noise and it is still of most interest to work on the performance of those X-ray image sensors.

Current Digital X-ray Image Sensors

Current digital X-ray imagers are FD detectors (flat panel solid-state detectors), and they are suitable for most X-ray imaging applications. They are based on an amorphous silicon a-Si readout matrix; each pixel comprises an a-Si switch and a sensing element. The restricted space limits the resolution to a minimum pitch of 70 μm.

For better resolution, the most advanced X-ray image sensors are CCD or CMOS based. The pixel size can be as low as 10 μm × 10 μm. These image sensors are responsive in the visible range. For X-ray imaging, the conversion from X-rays to light is performed by a scintillator bound to a fiberoptic face plate. This configuration leads to the best resolution (better than 20 lp/mm).

Large surfaces up to 49 mm × 86 mm are currently commercialized. However, CCDs and CMOS are there at their limits and increasing the surface in such technologies is hardly compatible with reasonable production cost.

For large field of view X-ray imaging such as mammography, chest… it is important to work on new large sensors.

New Image Sensor Concepts

In recent years, several advanced imaging solutions have been realized or currently are under development. In the following, some of these technologies are presented.

CCD-based detectors for very high spatial resolution: As mentioned earlier, FD detectors are suitable for most X-ray imaging applications. They are based on an a-Si readout matrix, each pixel comprising an a-Si switch and a sensing element. The restricted space available leads to a minimum pixel pitch of some 70 μm; otherwise, the spatial fill factor of the pixels would become too low. Especially for mammographic biopsies imaging with very high resolution is desirable. Therefore, a CCD-based detector has been developed with a pixel pitch of 12 μm. It can be also used in a 2 × 2 or 4 × 4 binned mode, resulting in 24 or 48-μm pixels, respectively. The detector uses a CsI scintillator that is coupled via a fiberoptic plate to the sensor with 4k × 7k pixels. The sensitive area is 49 mm × 86 mm being the largest CCD in serial production in the world [49].

CdTe is also a material of choice. Up to now, price versus performance was questionable, but current R&D works performed on both the material growth (in order to improve the yield and size) and on hybridization technics (in order to decrease the pitch and hybridization cost) give good results and open new perspectives for this material (such as X-ray intraoral application) for medium-size imaging. This technology should increase drastically the contrast on the image and aim at being compatible with market price.

Organic semiconductor-based detectors: Current FD detectors are based on a-Si technology that has proven to be suitable for radiological, fluoroscopic, and angiographic applications. These detectors are built on glass substrates, which are rigid, heavy, and fragile. The a-Si layers and electrodes are patterned by photolithography, a process enabling the production of fine structures. Depositing a-Si requires elevated temperatures in the order of 250°C. In summary, there is also a demand for cheaper alternatives to a-Si technology.

Some investigations are going on into all-organic detectors [50]. Plastic substrates can be used that are flexible, lightweight, and unbreakable. The organic semiconductor and electrode layers can be deposited in the desired pattern, e.g., by jet printing. This results in cheap detectors that have the potential for new applications. Since an organic detector is not as heavy as an a-Si-based detector, it lends itself to use in a portable bedside device.

It has been shown that organic photodiodes and transistors are feasible, but further work is necessary to improve their performance.

8.4.2.3 Infrared Image Sensors

Today thousands of cooled IR detectors are produced taking advantage of the good know-how regarding this domain [57–60]. However, these current IR detectors have some limitations regarding their ability to operate in all weather conditions, and in terms of compactness and reliability. Therefore, research for moving to the next generation of detectors is in progress to overcome these limitations as well as to offer more performance. In parallel, the production cost reduction is one of the main challenges for cooled IR detectors, and new technologies are developed to answer this need.

These new researches and developments are dedicated to smaller pixel pitches, large formats, APDs, multicolor detectors, and active imagery detector needs as well as optimization of microbolometer technology, one of the best choices for compactness/very low price compromise. This technology still requires miniaturization but is very promising and should open new applications in volume markets such as automobile and energy monitoring in buildings.

Infrared Detector Development

A lot of different technologies and sensitive materials are well mastered and available for IR detector production. As a matter of fact, quantum well IR photodetectors (QWIP) have been developed [ex Thales Research and Technologies (TRT)] and are at mass production level in cooperation with Sofradir. Then, InSb technologies are also available.

As to HgCdTe technologies, they have been widely used for high-performance IR detectors and are at mass level production for years. Finally uncooled technology was developed based on amorphous silicon microbolometers at CEA Leti. Then it was transferred to ULIS (the subsidiary of Sofradir) and is at mass production level since 2003. In addition, other developments are running like InGaAs technology in Xenics and III–V lab.

The mastering of all these technologies in Europe shows leading position in Europe for the present generation but also for the next generation in development. These different technologies are complementary and are used depending on the needs of the applications mainly concerning the detection range needs as well as their ability to detect in bad weather environmental conditions. They can be classified as follows (see Fig. 8.16):

- High-performance IR detectors for long-range detection systems: Long-range detection systems are dealing with detection ranges ranging from 6 km to tens of kilometers. High-performance detectors are necessary for long detection

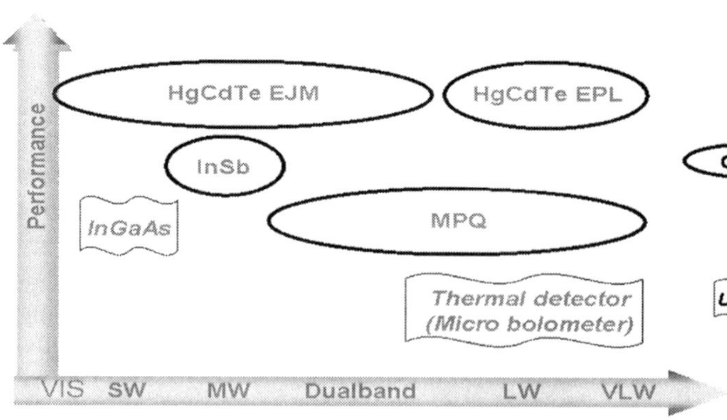

Fig. 8.16 Different IR detector technologies

range and will also be necessary for scientific applications including spectrometry, where you have very small signal (or emitted flux) to detect. For these applications high-performance cooled technologies are mandatory, and mainly both HgCdTe and InSb can reach today these detection ranges.

- HgCdTe material: Based on the unique characteristics of this semiconductor, MCT IR detectors can be sensitive in a very large range of wavelengths, starting from visible up to about 18 μm. These detectors exhibit high quantum efficiency coupled with high signal-to-noise ratio and can be operated at a rather high operating temperature compared with concurrent cooled technologies.
- In addition high production capacity and low cost are possible with the new growth method [molecular beam epitaxy (MBE)] for short wave (SW) and medium wave (MW) bands as well as for dual-band SW/MW and MW/MW. For long wave and very long wave bands, the classic liquid phase epitaxy (LPE) will continue to be used in the coming years. Finally this material offers a large range of improvements regarding APD and dual-band devices, which confirms that it is dedicated to high-performance systems.
- InSb: This semiconductor is just sensitive in the MW with a fixed cut-off band at 5.5 μm at 77 K, and there are presently some technological limitations in operating temperatures as well as in pixel pitch reduction. Consequently, performance obtained may be limited in some demanding applications.

Quantum well photoconductors (QWIP) may also be used for long range in some specific cases but with very low detector operating temperatures and low imaging frame rates.

- IR detector for medium-range detection systems: Medium detection ranges are dealing with few kilometers to about 6 km maximum. For these applications medium- to high-performance cooled technologies are also mandatory. As a matter of fact, in some cases, the use of high-performance detector can allow an IR system cost reduction (reduction of optics size, simplification of signal processing, relaxation of reliability constraints, etc.). Consequently the different candidates could be the same than for long range but in some cases, QWIP for LW and InGaAs for SW can offer a good quality/price ratio.

  - QWIP: there are mainly used for LW range but they have a limited efficiency as well as a higher dark current than concurrent high performance technologies, which limits their performance.
  - InGaAs: it is mainly used in short wavelength range and it is in production up to about 1.9 μm cut-off. For very low input signal they may be limited by the readout circuit noise performance.

- IR detectors for short-range detection system: Short detection ranges are dealing with few hundreds of meters to 1 or 2 km at maximum, and uncooled technologies answer these needs with uncooled operation. The most successful technology offering the best quality/price ratio is the microbolometer based on amorphous silicon that is fully compatible to CMOS silicon technology.

Focus on Microbolometer Technology

The principle of the microbolometers relies on the heating produced by the incident infrared radiation.

Thus, the measurement is indirect and involves a temperature variation-sensitive element: the thermometer. The scene at a temperature T_{sc} emits an infrared light through an optics and is converted by an absorber to heat the thermometer.

Radiation power is, however, very low and thus thermal insulation of the thermometer with respect to the readout circuit is of most importance.

Absorption can be performed directly within the thermometer itself and it can be further increased – thanks to a quarter wave cavity arrangement. At this point the bolometer is built as a thermometer with infrared antennas.

The thermometer can be realized using different principles: pyroelectricity, ferroelectricity, thermopiles, or variation of resistors.

The readout circuit then translates heating into voltage or current (Fig. 8.17).

From a device point of view, the evaluation of the quality of a bolometer is provided by the noise equivalent temperature difference (NETD).

The response is the variation of the output signal ΔI_{signal} with respect to the scene temperature variation ΔT_{sc}:

$$\text{Response} = \frac{\Delta I_{signal}}{\Delta T_{sc}}.$$

The NETD is the noise equivalent at the input of the device:

$$\text{NETD} = \frac{I_{noise}}{\text{response}} = I_{noise}\frac{\Delta T_{sc}}{\Delta I_{signal}}.$$

Considering now the particular technology developed at LETI, the pixel features a microbridge structure, which affords a very high thermal insulation of the sensitive part of the sensor from the readout circuit. This sensitive part is made from a thin

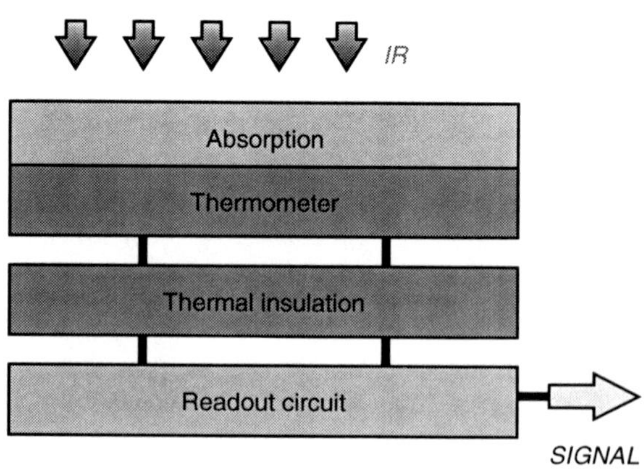

Fig. 8.17 Bolometer principle

film of amorphous silicon. The proper design of the support legs provides the required thermal insulation, while it also ensures the mechanical strength and electric connection of the thermometer. To further improve the thermal insulation, the sensor is packaged under vacuum. Although the elementary module represented here measures 50 µm × 50 µm, IRFPA with 25-µm pixel are now totally affordable, and latest developments are now focused on 17-µm pixel achievement.

On top of the readout circuit a reflector is deposited. It forms with the electrodes a quarter wavelength cavity for a wavelength of 10 µm and it boosts the absorption by creating a maximum intensity at the detector level (Figs. 8.18–8.20).

Absorption of such a structure while considering that the reflector and the electrodes are separated by vacuum is as follows:

$$\varepsilon(\lambda)=\frac{4\dfrac{R_{\square}^{\text{vide}}}{R_{\square}}}{\left(1+\dfrac{R_{\square}^{\text{vide}}}{R_{\square}}\right)^{2}+\cot^{2}\left(\dfrac{2\pi d}{\lambda}\right)},$$

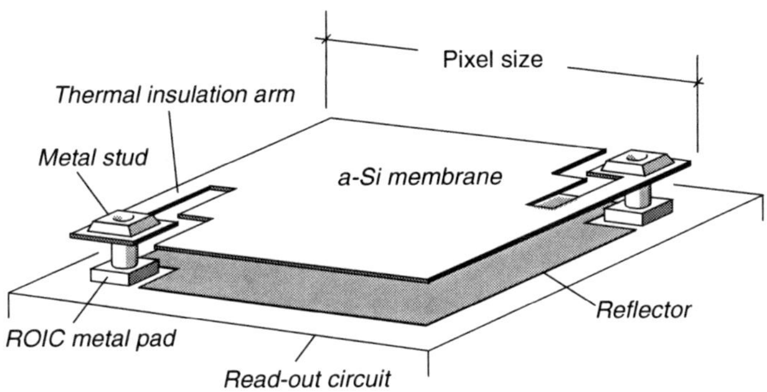

Fig. 8.18 Pixel architecture

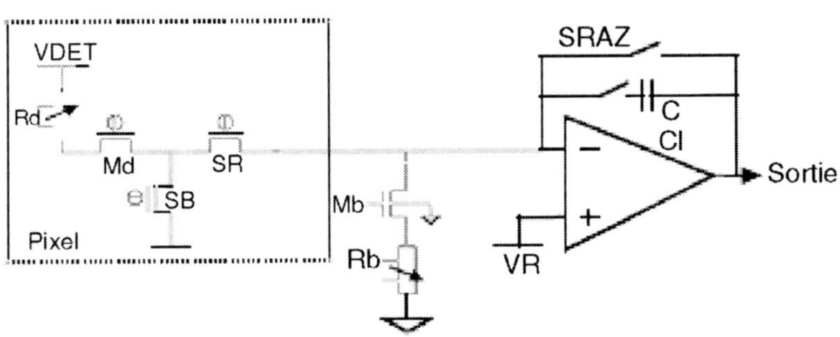

Fig. 8.19 Zoom on the contact between readout circuit and thermometer

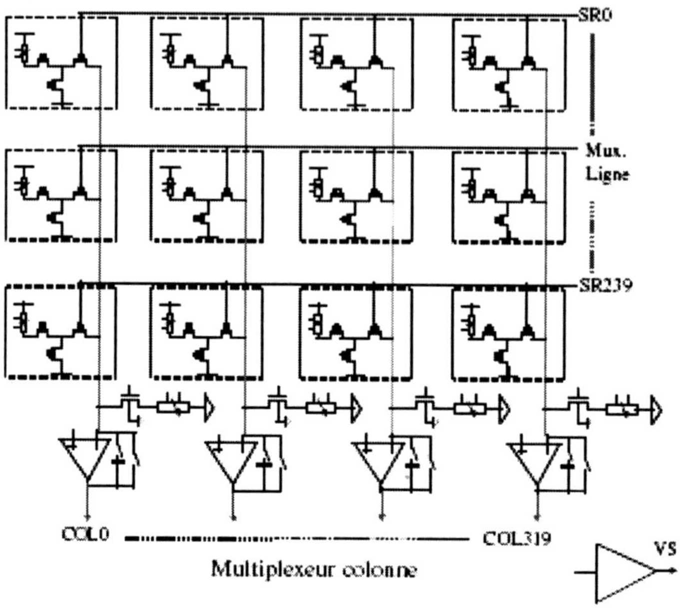

Fig. 8.20 2D microbolometer readout circuit architecture

where R_π is the resistivity of the electrodes, R_π^{vide} is the resistivity of the vacuum layer, i.e., 377 $\Omega/\square$, d is the distance between the reflector and the electrodes, and λ is the wavelength. The maximum value of ε is obtained for a resistivity of the electrode layers equal to the vacuum. In these conditions the device is called *adapted*.

Absorption modelization of such Fabry Perrot cavity is widely described in the literature.

Focus on MCT

$Hg(1-x)Cd_xTe$ presents unique properties that make it an ideal candidate for most of the needs in infrared detection in all the IR spectral bands.

First, the bandgap can be tuned by just controlling the *x*Cd/*x*Hg ratio of the alloy, making possible the entire infrared band to be covered from visible to very long wavelengths (zero bandgap for *x*Cd:0.15) as presented in Fig. 8.21a.

Second, the lattice parameter does not change at the first order from CdTe to HgTe. This allows the use of lattice-matched CdTe substrates. $Cd(1-y)Zn_yTe$ is more extensively used today for its better quality and because it leads to a perfect lattice matching by adjusting at the second order with *y*Zn composition to the *x*Cd composition. Very high quality epitaxial layers of $Hg(1-x)Cd_xTe$ can be grown today on very high quality $Cd(1-y)Zn_yTe$ oriented substrates with a crystallographic quality similar to the substrate.

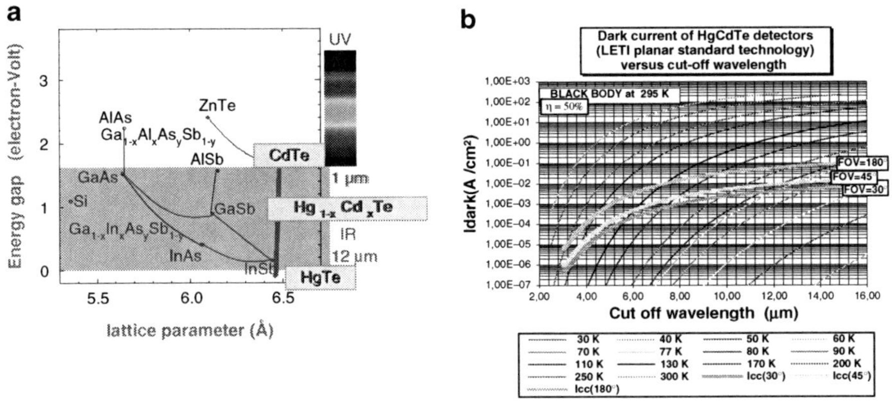

Fig. 8.21 (**a**) Lattice parameter of HgCdTe versus composition, (**b**) Dark current model for n on p photodiodes for cut-off wavelengths from 2 to 16 μm and operating temperature from 30 and 300 K

The very small lattice parameter variation versus composition makes also possible the fabrication of lattice-matched multilayer epitaxies with only a very small amount of defects and dislocations. Today, very high quality Hg(1−x)Cd$_x$Te/Cd(1−y)Zn$_y$Te epitaxies can be grown either by liquid phase epitaxy or by vapor phase epitaxy (molecular beam epitaxy for example).

This semiconductor can be doped n-type (e.g., with indium) and p-type (by mercury vacancies or arsenic) making possible n on p or p on n junctions to be fabricated.

Fundamental parameters such as lifetime are relatively high leading to low dark current and large quantum efficiency (near 1) photodiodes. Moreover the performances of detectors are only limited by the physics and predictable by validated models in a large domain of temperatures and compositions (see Fig. 8.21b).

This very large flexibility of these alloys makes possible the following:

- The fabrication of a large variety of infrared detectors, in particular, multicolor detectors that need multilayer epitaxies.
- To adjust the composition of the alloys to the useful infrared band to get the ideal composition xCd/operating temperature couple to get the optimum performance for the focal plane array.

Among the other advantages recently pointed out we can mention the following:

- An optimum light collection in backside illumination from the cut-off of the detector to the near UV, including the entire visible spectrum, without any decrease of quantum efficiency.
- A unique specificity in the semiconductor field for MCT avalanche photodiodes that can exhibit very large gain at moderate bias, without any excess of noise [$F(K)$ strictly equal to 1].

All these properties allow this semiconductor to answer all the high performance needs and classes of detectors in infrared bands up to at least 20 μm.

8.4.2.4 Terahertz Image Sensing: New Concepts

The terahertz (THz) part of the spectrum has recently been investigated. The ability to detect metallic parts under clothes without an ionization source is seen as being the answer to increased screening airport controls after 01/09/11.

Currently, no cost-effective industrial solution exists and thus detectors are still at early development of research and development stage.

Up to now, several kinds of detectors have been studied worldwide for terahertz sensing. The following table summarizes the available devices: it was displayed by VTT in Finland and shows a THz detector classification of existing technologies. This classification stresses on three major issues: sensitivity, frequency range, and price (Fig. 8.22).

Since last 10 years, all R&D efforts spent in Europe were performed on monolithic microwave integrated circuits (MMIC), heterodyne detection, and cryogenic bolometer. These technologies exhibit a major drawback: they can not provide large matrix for THz imaging. Moreover, cost of these technologies is not compatible with industrial system applications described in the previous introduction.

From this table it is straightforward that antenna-coupled microbolometer operating at room temperature is a technology able to fulfill price constraint to deliver THz system for mass market application. Raytheon in the USA started research on this topic since 2002 supported by homeland security effort and they already displayed significant results. In fact, at the SPIE Defense and Security conference in Orlando in 2005, they presented a 22 × 28 THz focal plane with a CMOS reading circuit (SPIE vol. 5778). The performances of this device

Technology	Sensitivity	Price
Coherent heterodyne	Good	Huge
Coherent direct (with MMIC preamplification)	Good	Large
Cryogenic Microbolometers	Excellent	Moderate
Incoherent direct (i.e. diodes, no preamplification)	Moderate	Small
Antenna coupled microbolometers	Poor (active only)	Tiny

Maximum frequency

~200 GHz	600 GHz	> 1THz

Fig. 8.22 Existing THz technologies (VTT)

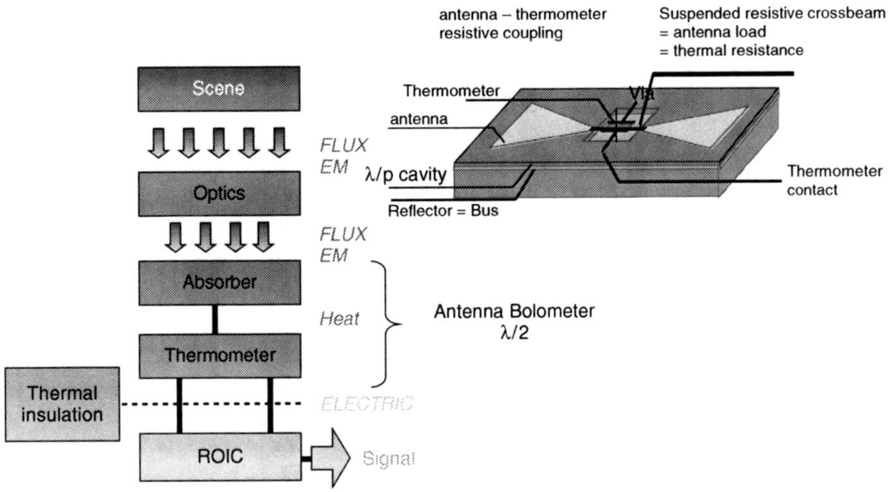

Fig. 8.23 Operating principle of the antenna-coupled microbolometer existing THz technologies

were presented in a qualitative manner only and there was no discussion of the electro-optical performances of the component in terms of sensitivity or equivalent thermal resolution (Fig. 8.23).

If we compare it with the European approach, based mainly on *direct detection*, the following technical limitations exist in addition to high delivery cost. Direct detection includes signal reception on an antenna and its amplification in the same (large) bandwidth, by means of several HEMT (high electronic mobility transistor) stages using GaAs or InP devices. These needs contributed to the development of integration, hybridization and, finally, the appearance of specific integrated circuits: MMICs. These decrease losses by coupling and provide miniaturization, suggesting that integration into a focal plane will be possible. Also, it should be noted that this technology is limited in terms of frequency (spectroscopy for explosives, above 1 THz, is not feasible) and currently the cost of the technological lines will not allow mass production.

New terahertz image sensor concepts still need to be found. These should need to match image quality, production cost, power consumption, and volume production.

A possible technology for terahertz detection is based on CMOS-compatible processing leading to sensors operating at room temperature and capable of low-cost production. THz detector principle is slightly different from the IR microbolometer technology due to the fact that absorber and thermometer are physically separated. Metallic antenna enforced with quarter wavelength cavity will provide the coupling with the electromagnetic wave, producing a current flowing into a matched load resistance. Then, the thermometer material will sense like a calorimeter the heat coming from THz radiation Joule effect on the load. Schematic presentation of the operating principle is displayed on the picture given later.

8.5 Outlook: The Future of Semiconductor Image Sensing

Thanks to the amazing, relentless progress of semiconductor technology, in particular, the specialized CIS processes that are derived from mainstream CMOS processes; we will soon have more cameras than people on earth. This fact has a rather simple consequence, namely that many more pictures will be acquired than can be looked at by human beings, which is particularly true and worrying in the case of security and surveillance cameras. For this reason, experts agree on the direction electronic imaging will progressively take in the future: Image sensors need to be equipped with an ever-increasing amount of digital processing power, enabling these imagers not only to acquire pictures but also to improve the pictures' quality with suitable postprocessing and to extract meaningful information about the contents of a scene using powerful object recognition and classification hardware. The goal will be single-chip machine vision systems that "understand" what they see.

Impressed by the demonstration of image sensors with more and more integrated functionality, a prominent researcher in the field proclaimed the imminence of such "seeing chips," already two decades ago [68]. It is true that for certain tasks, a few examples of image sensors with complete, integrated image processing hardware have been reported, such as single-chip fingerprint recognition and identification systems. Other types of smart image sensors have been demonstrated that are capable of carrying out a few important, but still only rather basic functions for the vision process on a single chip. The more research results are reported, however, the more obvious it becomes that "vision is difficult," as suspected in [68]. It is clear now that early expectations of monolithically integrated single-chip vision systems were too high. While it is possible today to cointegrate an image sensor and all the necessary processing circuitry on a single chip for the solution of a given, not too complex, machine vision problem, such an accomplishment is far removed, though, from the original idea of a seeing chip, capable of visually perceiving important aspects of an imaged scene.

The main obstacle is not a technological one: today, a lot of digital processing power and memory can be packed onto a single chip. The problem is rather our still very limited understanding of the natural vision process and how to translate it into reliably working algorithms, for which dedicated circuitry can be fabricated on a "smart" image sensor. Despite almost 50 years of research in machine vision, the man-made, robust machine vision system performing as well as natural vision systems, such as the visual sense of insects, are still elusive. Recent research is increasingly taking clues from the successful natural vision systems, where evidence can be found that all kinds of cognitive functions in humans are implemented with the same basic functionality and system architecture, the so-called *Mountcastle observation* [69]. This is an exciting prospect, of course, that may not only lead the way to "seeing chips" but also to "hearing microphones" and – more generally – to "aware sensors." In any case, all these developments clearly require more insight and for its implementation more microelectronic functionality on each chip – more Moore, and more than Moore!

References

1. S.M. Sze, *"Semiconductor Devices – Physics and Technology,"* 2nd edition, Wiley, New York, 1992.
2. T. Spirig, "Smart CCD/CMOS Based Image Sensors with Programmable, Real-time, Temporal and Spatial Convolution Capabilities for Applications in Machine Vision and Optical Metrology," Thesis, ETH Zurich, No. 11993, 1997, Chapter 3.
3. P.J.W. Noble, "Self-Scanned Silicon Image Detector Array," *IEEE Trans. Electron Devices*, Vol. 15, pp. 202–209, 1968.
4. E.R. Fossum, "Active Pixel Sensors: Are CCD's Dinosaurs?," *Proc. SPIE*, Vol. 1900, pp. 2–14, 1993.
5. H.S. Wong, "Technology and Device Scaling Considerations for CMOS Imagers," *IEEE Trans. Electron. Devices*, Vol. 43, pp. 2131–2142, 1996.
6. G. Agranov et al., "Super Small, Sub 2μm Pixels for Novel CMOS Image Sensors," *Proc. 2007 International Image Sensor Workshop*, pp. 307–310, 2007.
7. P. Seitz, "Solid-State Image Sensing," in *Computer Vision and Applications – A Guide for Students and Practitioners*, B. Jähne, and H. Haussecker (Eds.), pp. 111–152, Academic, San Diego, 2000.
8. Y. Lim et al., "Stratified Photodiode – A New Concept for Small Size High Performance CMOS Image Sensor Pixels," *Proc. 2007 International Image Sensor Workshop*, pp. 311–315, 2007.
9. N. Wyrsch et al., "Vertical Integration of Hydrogenated Amorphous Silicon Devices on CMOS Circuits," *Mater. Res. Soc. Symp. Proc.*, Vol. 869, pp. D1.1.1–D1.1.12, 2005.
10. P. Seitz, "Smart Pixel Arrays," in *Encyclopaedia of Modern Optics*, B. Guenther and D. Steel (Eds.), Elsevier, Oxford, 2005.
11. A. Krimsky, N. Khaliullin and H. Rhodes, "A 2 e– Noise 1.3 Megapixel CMOS Sensor," *Proc. 2003 IEEE Workshop on Charge-Coupled Devices and Advanced Image Sensors*, May 15–17, 2003.
12. C. Koch, "Seeing Chips: Analog VLSI Circuits for Computer Vision," *Neural Comput.*, Vol. 1, pp. 184–200, 1989.
13. J. Hawkins and S. Blakeslee, *On Intelligence*, Henry Holt, New York, 2004.
14. S. Tanner, S.C. Lauxtermann, M. Waeny, M. Willemin, N. Blanc, J. Grupp, R. Dinger, E. Doering, M. Ansorge, P. Seitz and F. Pellandini, "Low-Power Digital Image Sensor for Still-Picture Image Acquisition," *Proc. SPIE Int. Soc. Opt. Eng.*, Vol. 4306, p. 358, 2001.
15. K.B. Cho, A Krymski and E.R. Fossum, "A Micropower Self-Clocked Camera-on-a-Chip," *IEEE Workshop on Charge-Coupled Devices and Advanced Image Sensors*, Cal-Neva Resort, NV, USA, pp. 12–13, June 7–9, 2001; K.B. Cho et al., "A 1.2V Micropower CMOS Active Pixel Image Sensor for Portable Applications," *ISSCC Digest of Tech. Papers*, pp. 114–115, Feb. 2000.
16. http://www.weinbergervision.com; http://www.photron.com; http://www.aostechnologies.com; http://www.visionresearch.com; http://www.redlake.com; www.pco.de; www.dalsa.com.
17. T.G. Etoh et al., "An Image Sensor Which Captures 100 Consecutive Frames at 1000000 fps," *IEEE Trans. Electron Devices*, Vol. 50, No. 1, pp. 144–151, Jan. 2003.
18. S. Kleinfelder et al., "A 10kframe/s 0.18 um CMOS Digital Pixel Sensor with Pixel-Level Memory," *Proc. IEEE ISSCC-01 Conf.*, Vol. XLIV, pp. 88–89, Feb. 2001.
19. S. Lauxtermann, et al., "A High Speed CMOS Imager Acquiring 5000 Frames/Sec," *Dig. IEDM '99 Conf.*, pp. 875–878, Dec. 1999.
20. Z. Zhou, B. Pain and E.R. Fossum, "CMOS Active Pixel Sensor with On-Chip Successive Approximation Analog-to-Digital Converter," *IEEE Trans. Electron Devices*, Vol. 44, No. 10, pp. 1759–1763, Oct. 1997.
21. A. Krymski et al., "A High Speed, 500 fps, 1024 × 1024 CMOS Active Pixel Sensor," *IEEE VLSI Circuits Conf. Dig.*, pp. 137–138, June 1999.
22. Micron Technology Inc., "MT9M413 1.3 Mpix CMOS Active Pixel Digital Sensor," Datasheet, Ver. 3.0, Jan. 2004.

23. O. Yadid-Pecht et al., "A Random Access Photodiode Array for Intelligent Image Capture," *IEEE Trans. Electron Devices*, Vol. 38, No. 8, pp. 1772–1780, Aug. 1991.
24. A. Krymski and N. Tu, "A 9VLux-s 5000fps 512 × 512 CMOS Sensor," *IEEE Trans. Electron Devices*, Vol. 50, No. 1, pp. 136–143, Jan. 2003.
25. S. Kleinfeld, "High-Speed, High-Sensitivity, Low-Noise CMOS Scientific Image Sensors," *Proc. SPIE*, Vol. 5274, pp. 194–205, March 2004.
26. T. Inoue et al., "A CMOS Active Pixel Image Sensor with In-Pixel CDS for High-Speed Cameras," *Proc. SPIE*, Vol. 5580, pp. 293–300, 2005.
27. R. Aerts et al., "1.4Gpix/s, 2.9Mpix CMOS Image Sensor for Readout of Holographic Data Memory," *Proc. IEEE International Image Sensor Workshop*, Ses. 06, pp. 74–77, June 2007.
28. M. Furuta et al., "A High-Speed, High-Sensitivity Digital CMOS Image Sensor with a Global Shutter and 12-Bit Column-Parallel Cyclic A/D Converters," *IEEE JSSC*, Vol. 42, No. 4, pp. 766–774, April 2007.
29. I. Takayagani et al., "A 600 × 600 Pixel, 500fps CMOS Image Sensor with 4.4um Pinned Photodiode 5-Transistor Global Shutter Pixel," *Proc. IEEE International Image Sensor Workshop*, Ses. 17, pp. 287–290, June 2007.
30. Y. Nitta et al., "High-Speed Digital Double Sampling with Analog CDS on Column Parallel ADC Architecture for Low-Noise APS," *Proc. IEEE ISSCC'06*, pp. 2024–2031, 6–9 Feb. 2006.
31. A. Krymski, "A High Speed 4 Megapixel Digital CMOS," *Proc. IEEE International Image Sensor Workshop*, Ses. 06, pp. 78–81, June 2007.
32. T. Cools et al., "An SXGA CMOS Image Sensor with 8Gbps LVDS Serial Link," *Proc. IEEE International Image Sensor Workshop*, Ses. 17, pp. 283–286, June 2007.
33. A. Krymski and K. Tajima, "CMOS Image Sensor with Integrated 4Gbs Camera Link Transmitter," *Proc. IEEE ISSCC'06*, pp. 2040–2049, 6–9 Feb. 2006.
34. S. Kawahito, "CMOS Imaging Devices for New Markets of Vision Systems," *IEICE Trans. Electron.*, Vol. E90-C, No. 10, pp. 1858–1868, Oct. 2007.
35. R. Johansson et al., "A Multi-Resolution 100 GOPS 4 Gpixels/s Programmable CMOS Image Sensor for Machine Vision," *Proc. IEEE Workshop on CCD & AIS*, May 2003.
36. T. Oggier et al., "An All-Solid-State Optical Range Camera for 3D Real-Time Imaging with Sub-Centimeter Depth Resolution (Swissranger)," *Proc. SPIE 5249*, pp. 534–545, 2003.
37. R. Lange and P. Seitz, "A Solid State Time-of-Flight Range Camera," *IEEE J. Quant. Electron.*, Vol. 37, p. 390, 2001.
38. R. Kaufmann et al., "A Time-of-Flight Line Sensor – Development and Application," *Proc. SPIE 5459*, pp. 192–199, 2004.
39. http://www.mesa-imaging.ch; http://www.canesta.com; http://www.pmdtec.com.
40. http://www.mesa-imaging.ch/prodviews.php.
41. I.A.H. Izhal, T. Ushinaga, T. Sawada, M. Homma, Y. Maeda and S. Kawahito, "A CMOS Time-of-Flight Range Image Sensor with Gates on Field Oxide Structure," *Proc. Fourth IEEE Int. Conf. Sensors*, pp. 141–144, Nov. 2005.
42. B. Braun, *Electromedica*, Vol. 70, p. 1, 2002.
43. J.P. Moy, *Thin Solid Films*, Vol. 337, p. 213, 1999.
44. A. Oppelt (Ed.), *Imaging Systems for Medical Diagnosis*, Publicis, Erlangen, 2005.
45. L.D. Hubbard, R.J. Brothers, W.N. King, L.X. Clegg, R. Klein, L.S. Cooper, A.R. Sharrett, M.D. Davis and J. Cai, *Ophthalmology*, Vol. 106, p. 2269, 1999.
46. M. Hoheisel and L. Bätz, *Thin Solid Films*, Vol. 383, p. 132, 2001.
47. M. Hoheisel, J. Giersch and P. Bernhardt, *Nucl. Instr. Meth.*, Vol. 5ANS-531/1–2, p. 75, 2004.
48. M. Hoheisel, L. Bätz, T. Mertelmeier, J. Giersch and A. Korn, *Modulation Transfer Function of a Selenium-Based Digital Mammography System*, in *IEEE Proceedings of the Nuclear Science Symposium, Medical Imaging Conference*, pp. 3589–3593, 2004.
49. S. Thunberg, H. Sklebitz, B. Ekdahl, L. Bätz, A. Lundin, H. Möller, F. Fleischmann, G. Kreider and T. Weidner, *Proc. of SPIE*, Vol. 3659, p. 150, 1999.

50. R.A. Street, W.S. Wong, S. Ready, R. Lujan, A.C. Arias, M.L. Chabinyc, A. Salleo, R. Apte and L.E. Antonuk, *Proc. SPIE Med. Imaging Conf.*, Vol. 5745, p. 7, 2005.
51. G.B. Avinash, K.N. Jabri, R. Uppaluri, A. Rader, F. Fischbach, J. Ricke and U. Teichgräber, *Proc. SPIE*, Vol. 4684, p. 1048, 2002.
52. J.M. Lewin, P.K. Isaacs, V. Vance and F.J. Larke, *Radiology*, Vol. 229, p. 261, 2003.
53. L. Tlustos, R. Ballabriga, M. Campbell, E. Heijne, K. Kincade, X. Llopart and P. Stejskal, *"Imaging Properties of the Medipix-2 System Exploiting Single and Dual Energy Thresholds,"* Proceedings of the IEEE Medical Imaging Conference, Rome, N43–3, 2004
54. M. Hoheisel, R. Lawaczeck, H. Pietsch and V. Arkadiev, *Proc. SPIE*, Vol. 5745, p. 1087, 2005.
55. E. Pisano et al., *Radiology*, Vol. 214, p. 895, 2000.
56. S.W. Wilkins, T.E. Gureyev, D. Gao, A. Pogany and A.W. Stevenson, *Nature*, Vol. 384, p. 335, 1996.
57. G. Destefanis, Electrical Doping of HgCdTe by Ion Implantation and Heat Treatment, *Third International Conference on II–VI Compounds*, Monterey, USA, July 1987; J. Cryst. Growth, Vol. 86, pp. 700–727, 1988.
58. G. Destefanis, HgCdTe Infrared Diodes Arrays, *Nato Workshop on Narrow Bandgap Semiconductors*, Oslo, Norway, 1991; Semicond. Sci. Technol., Vol. 6, pp. 88–92, 1991.
59. G. Destefanis and J.P. Chamonal, Large Improvement in HgCdTe Photovoltaic Detector Performance at LETI, *The US Workshop on the Physics and Chemistry of II–VI Materials*, Danvers, USA, October 1992; J. Electron. Mater., Vol. 22, No. 8, pp. 1027–1032, 1993.
60. G. Destefanis, P. Audebert, E. Mottin and P. Rambaud, *High Performance LWIR 256 × 256 HgCdTe Focal Plane Array Operating at 88 K, International SPIE Conference*, Orlando (USA), April 1997; *Proc. SPIE*, Vol. 3061, pp. 111–116, 1997
61. P. Tribolet, P. Bensussan and G. Destefanis, *SOFRADIR Approach for 2/5 and 3rd Generations of IR Detectors, The Fifth International Military Sensing Symposium: 2002DTB04*, Washington, USA, Dec. 2002, Org. OTAN.
62. G. Destefanis, A. Bain, J. Baylet, P. Castelein, E. De Borniol, O. Gravrand, F. Marion, A. Million, Ph. Rambaud, F. Rothan, J.L. Martin and P. Martin, Recent Developments of High Complexity HgCdTe Focal Plane Arrays at LETI Infrared Laboratory, *The 2002 Workshop on the Physics of II–VI material*, Nov. 2002, San Diego, USA; *J. Electron. Mater.*, Vol. 32, No. 7, pp. 592–601, 2003.
63. F.P. Pistone, S. Dugalleix, P. Tribolet and G. Destefanis, *Progress in MCT Large Staring Arrays, International SPIE Meeting Infrared Detectors and Focal Plane Arrays VIII*, San Diego, USA, Aug. 2006; SPIE Proc., Vol. 6295–21.
64. P. Tribolet and G. Destefanis, HgCdTe Large Staring Arrays at Sofradir, The 19th International Technology Conference on Photoelectronis and Nigh Vision Devices, 2006 (to be published in SPIE Proceedings).
65. P. Tribolet, S. Blondel, P. Costa, A. Combette, L. Vial, G. Destefanis, Ph. Ballet, J.P. Zanatta, O. Gravrand, C. Largeron, J.P. Chamonal and A. Million, MWIR Focal Plane Arrays Made with HgCdTe Grown by MBE on Germanium Substrates, *International SPIE Meeting Infrared Technology and Application XXXII*, Orlando, USA, April 2006; SPIE Proc., Vol. 6206–82.
66. J.P. Zanatta, G. Badano, Ph. Ballet, J. Baylet, O. Gravrand, J. Rothman, P. Castelein, J.P. Chamonal, A. Million and G. Destefanis, MBE HgCdTe Growth on Ge for the 3rd Generation of Infrared Detectors, *The 2005 US Workshop on the Physics and Chemistry of II–VI Materials*, Boston, USA, Sept. 2005; *J. Electron. Mater.*, Vol. 35, p. 6, 2006.
67. J. Baylet, Ph. Ballet, P. Castelein, F. Rothan, M. Fendler, E. Laffosse, J.P. Zanatta, J.P. Chamonal, A. Million and G. Destefanis, TV/4 Dual Band HgCdTe Infrared Focal Plane Arrays with a Pitch of 25μm and Spatial Coherence, *The 2005 US Workshop on the Physics* and Chemistry of II–VI Materials, Boston, USA, Sept. 2005; *J. Electron. Mater.*, Vol. 35, p. 6, 2006.
68. P. Chorier, P. Tribolet and G. Destefanis, From Visible to Infrared: *A New Approach, International SPIE Meeting Infrared Technology and Application XXXII*, Orlando, USA, April 2006; *SPIE Proc.*, Vol. 6206–01.

69. O. Gravrand, E. Deborniol, G. Destefanis, A. Manissadjian, P. Tribolet, C. Pautet, P. Chorier, From LWIR to VLWIR FPAs Made with HgCdTe at Defir, *International SPIE Meeting Sensors, Systems and Next generation Satellites XII*, Stockholm, Sweden, Sept. 2006; SPIE *Proc.*, Vol. 6361–42.
70. O. Gravrand, E. Deborniol and G. Destefanis, From LWIR to VLWIR FPAs Made with HgCdTe n+n/p Ion Implantation Technology, The 2006 US Workshop on the Physics and Chemistry of II-VI Materials, Newport, USA, Oct. 2006; *J. Electron. Mater.*, in press
71. E. Deborniol, G. Destefanis, A. Manissadjian and P. Tribolet, Characterization of High Performance Long Wave and Very Long Wave HgCdTe Staring Arrays, *International SPIE Meeting*: Remote Sensing, Bruges, Belgium, Sept. 2005; *SPIE Proc.*, Vol. 5978–44.
72. A. Manissadjian, P. Tribolet, G. Destefanis and E. Deborniol; Long Wave HgCdTe Staring Arrays at Sofradir: From 9μm to 13+μm Cut-Off for High Performance Applications, *International SPIE Meeting Infrared Technology and Application XXXI*, Orlando, USA, April 2005; *SPIE Proc.*, Vol. 5783.
73. G. Destefanis, J. Baylet, P. Ballet, F. Rothan, O. Gravrand, J. Rothman, J.P. Chamonal and A. Million, Status of HgCdTe Bicolour and Dual Band Infrared Focal Plane Arrays at LETI, *The 2006 US Workshop on the Physics and Chemistry of II–VI Materials*, Newport, USA, Oct. 2006; *J. Electron. Mater.*, 2007.
74. G. Destefanis, Ph. Ballet, J. Baylet, P. Castelein, O. Gravrand, J. Rothman, F. Rothan, G. Perrais, J. Chamonal, A. Million, P. Tribolet, B. Terrier, E. Sanson, P. Costa and L. Vial, Bi-Color and Dual Band Infrared Focal Plane Arrays at Defir, *International SPIE Meeting Infrared Technology and Application XXXII*, Orlando, USA, April 2006; *SPIE Proc.*, Vol. 6206–27.
75. P. Ballet, P. Castelein, J. Baylet, E. Laffosse, M. Fendler, F. Pottier, S. Gout, C. Vergnaud, S. Ballerand, O. Gravrand, J.C. Desplanches, S. Martin, J.P. Zanatta, J.P. Chamonal, A. Million and G. Destefanis, Demonstration of a 25μm Pitch Dual Band HgCdTe Infrared Focal Plane Array with Spatial Coherence, *International SPIE Meeting Optics and Optoelectronics*, Bruges, Belgium, Sept. 2005; *SPIE Proc.*, Vol. 5978–44.
76. P. Tribolet and G. Destefanis, Third Generation and Multicolour IRFPA Developments: A Unique Approach Based on Defir, *International SPIE Meeting Infrared Technology and Application XXXI*, Orlando, USA, April 2005; *SPIE Proc.*, Vol. 5783.
77. F. Ma et al., *Phys. Rev. Lett.*, Vol. 95, p. 176604, 2005.
78. R. Alabedra et al., *IEEE Trans. Electron Devices*, Vol. ED-32, p. 1302, 1985; G. Levêque et al., *Semicond. Sci. Technol.*, Vol. 8, p. 1317, 1993.
79. J.D. Beck, C.-F.Wan, M.A. Kinch, J.E. Robinson, *Proc. SPIE*, Vol. 4454, p. 188, 2001; J.D. Beck, C.-F. Wan, M.A. Kinch, J.E. Robinson, P. Mitra, R. Scrithfield, F. Ma and J. Campbell, *J. Electron. Mater.*, Vol. 35, p. 1166, 2006.
80. G. Perrais, O. Gravrand, J. Baylet, G. Destefanis and J. Rothman, Gain and Dark Current Characteristics of Planar HgCdTe Avalanche Photodiodes, The 2006 US Workshop on the Physics and Chemistry of II VI Materials, Newport, USA, Oct. 2006; *J. Electron. Mater.*, 2007.
81. G. Perrais, O. Gravrand, J. Baylet, G. Destefanis and J. Rothman, Gain and Dark Current Characteristics of Planar HgCdTe Avalanche Photodiodes, *The 2006 US Workshop on the Physics and Chemistry of II–VI materials*, Newport, USA, Oct. 2006; *J. Electron. Mater.* 2007.
82. G. Perrais, J. Rothman, G. Destefanis, J.P. Baylet, P. Castelein, J. Chamonal, P. Tribolet, Demonstration of Multifunction Bicolour Avalanche Gain in HgCdTe FPA, *International SPIE Meeting Electro Optical and Infrared Systems: Technology and Applications III*, Stockholm, Sweden, Sept. 2006; *SPIE Proc*, Vol. 6395–16.

Chapter 9
Heterogeneous Integration: Building the Foundation for Innovative Products

Herbert Reichl, Rolf Aschenbrenner, Michael Töpper, and Harald Pötter

Abstract This chapter gives an introduction to the future of electronic system integration as a combination of "More Moore" and "More than Moore" components, combined in one package ("system-in-package" or SiP). More Moore will very likely continue to be at the forefront of electronics development for the next decade. It is difficult, however, to integrate non-digital functions, such as sensors and power electronics, for example, monolithically with these technologies. For this reason the future belongs to integration technologies that combine several components into a highly integrated assembly in one package – "wafer-level packaging" processes, the use of ultrathin components, and multifunctional interposers will all play a key role in this context. Often two dimensions do no longer suffice. Because of signal propagation delays, wiring density, or confined available space many future applications require considerably more compact components. 3D integration can help overcome this bottleneck. Examples of 3D integration start with silicon 3D integration with through-silicon vias and go up to stacking of packages or modules at all technological levels. Functional packaging goes one step further. It takes the application system and its requirements as a starting point and then adapts the microelectronic functions.

Keywords System-in-package • Wafer-level integration • Board-level integration • Integrated passives • Functional layer technology • Thin chip integration • Module integration technologies • Embedding of passive components • Chip-in-polymer • Chip embedding into flexible substrates • Functional packaging • Integration of biological functionalities • 3D integration • Package stacking • Through-silicon vias

---

H. Reichl (✉), R. Aschenbrenner, M. Töpper, and H. Pötter
Technische Universität Berlin, Gustav-Meyer-Allee 25, 13355 Berlin, Germany
Fraunhofer IZM, Gustav-Meyer-Allee 25 13355 Berlin, Germany

G.Q. Zhang and A.J. van Roosmalen (eds.), *More than Moore: Creating High Value Micro/Nanoelectronics Systems*,
DOI 10.1007/978-0-387-75593-9_9, © Springer Science+Business Media, LLC 2009

9.1 Introduction

Heterogeneous integration, at the center of any micro/nanosystem creation, which is the bridge between nanotechnology and microtechnology with macroscale applications, is the final manufacturing process transforming multifunctional devices into products for the end users. It provides electrical and physical connections (e.g., bioconnection for bio-SiP) for multisignal transmission, power input, and voltage control. It also provides for thermal dissipation, construction carrier, and the physical protection required for reliability. To develop commercially feasible and reliable micro/nanosystems, Heterogeneous Integration plays an essential role as the key enabler governing the multifunctional performance, interface to the application environment, size, weight, cost, and reliability of the final products. Heterogeneous integration has to ensure not only the integration of devices based on different technologies and materials but also the target of miniaturization. Therefore, most of the innovative micro/nano devices, materials, concepts, and prototypes cannot lead to economic success without cost-effective and reliable heterogeneous integration technologies. This chapter describes the technologies and processes associated with heterogeneous integration. The associated strategic research issues and their solutions are also discussed.

Heterogeneous integration bridges the gap between nanoelectronics and its derived applications. Two main forces drive progress in this area – emerging device technologies and new application requirements. New technologies and architectures are needed to integrate the progress made in nanoelectronics, microsystem technologies, and bioelectronic and photonic component technologies into electronic systems or MEMS.

In terms of applications, a strong link exists between the development of components and system-oriented integration technologies such as system-in-package. The operation environment, which has traditionally been defined by the needs of electronics, is nowadays increasingly determined by the application, such as by "under-the-hood" environments in the automotive industry or by the human body for implantable medical devices. Because of the increasingly central role of electronics in ensuring system reliability and functionality, the reliability of electronics will be a primary issue in the future.

In terms of integration, both *More Moore* as well as *More than Moore* solutions are aiming at single component solutions. As far as technically and economically feasible solutions are considered "system on chip" (SoC) or *More than Moore* solutions will be chosen. But increasingly applications demand product-specific integration technologies or the flexible integration of highly complex systems with digital and nonelectronic functions. Thus, the future of electronic system integration will see a combination of *More Moore* and *More than Moore* components, combined in one package (*system-in-package* or SiP) (Fig. 9.1). Such SiP solutions have two main advantages - first, a level of miniaturization comparable to *system on chip* (More Moore) solutions combined with the enhanced functionality of *More than Moore* solutions, and second, the use of optimized process technology to fabricate each part of the system.

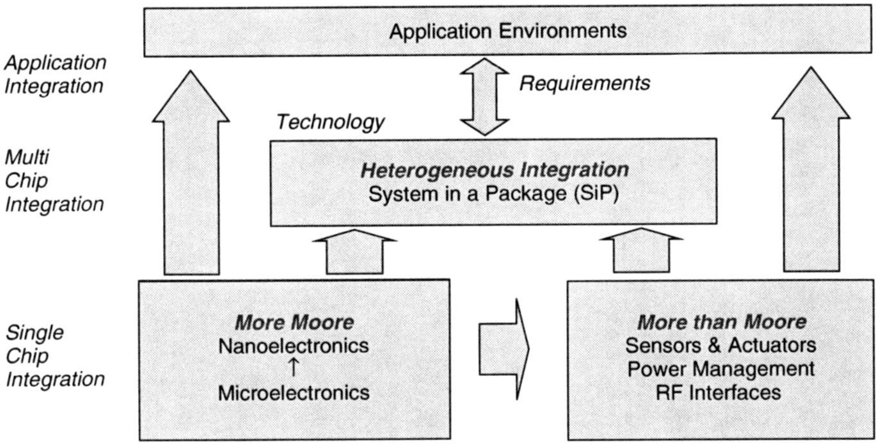

Fig. 9.1 Different levels of system integration

Using heterogeneous integration such as *system on chip* or *More than Moore* for subsystems makes highly flexible solutions possible. This equals lower costs and risk assessment compared primarily with SoC as well as *More than Moore* solutions to some extent. Further advantages include a shorter time to market cycle and a higher degree of flexibility.

In terms of technology heterogeneous integration combines different components and technologies into one package and also provides an interface to the application environment. Although *More Moore* and *More than Moore* technologies are based on CMOS-compatible processes and are therefore limited to a certain set of materials, heterogeneous integration technologies have to ensure that components based on very different technologies and materials can be integrated, such as power devices, photonic or RF components (using InP or GaAs), energy harvesting or storage components (thermoelectrics, fuel cells, thin-film batteries, supercaps), or smart displays. Reliability standards will see failure rates expressed as failures per billion parts, rather than per million parts, as is currently the case. Achieving this is made even more difficult by the fact that it will become necessary to integrate nondigital functions, such as MEMS, optical, RF, or power devices. To increase integration density in terms of the number of integrated components or in terms of geometry requires three-dimensional design methods such as stacking, as well as, due to the market demand, design for operation in less than ideal conditions. These include automotive applications at higher temperatures and harsh environments.

At present, system integration is still dominated by single-chip packaging and the few stacked-die SiP solutions employed are primarily based on wire bonding (Fig. 9.2).

High-density interconnect (HDI) multilayer boards are the most advanced substrate boards currently available. Unlike in the integrated circuit industry, where electrical, thermal, and mechanical characterization is undertaken on the complete design, in SiP solutions, the chips, package, and board continue to be designed separately. However, this will not be sufficient to meet the requirements of advanced SiP

Fig. 9.2 Present status of system integration

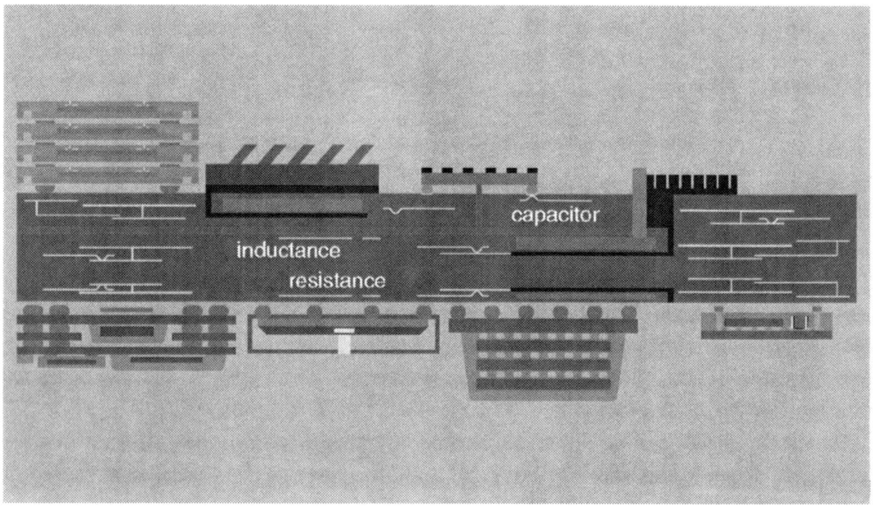

Fig. 9.3 Future system integration

solutions in the future, as the level of miniaturization and reliability requirements will be extremely high. Thermal and mechanical stress management has to be optimized beginning with the point at which heat is generated to the outside of the package. It will be further complicated by integration of additional functions into the package, such as sensors, actuators, RF interfaces, or power supply components, which may be highly sensitive to heat, stress, etc. Additionally, the application environment in which the SiP will ultimately be used will also need to be taken into account.

To meet these challenges, new architectures have to be developed. It will also be necessary to develop advanced assembly and handling technologies for thin wafers and chips to reach the required level of miniaturization. The integration of nano-ICs, sensor chips, actuator components, passives, and displays into 3D architectures will see the development of new design methodologies, as well as reliable ultrathin metallic interconnect technologies. An example of such system integration is proposed in Fig. 9.3.

The board will be transformed from a simple carrier for packages to a system carrier with optical and electrical wiring. Geometric limitations will be overcome by employing embedded passive components to include even smaller and increasing numbers of such components. Integrated active components will ensure higher reliability, better RF ability, and increased integration density.

9.2 Physical Package Characterization, Lifetime Modeling, and Design for Reliability for Heterogeneous System Integration

In future electronic systems, the boundaries between semiconductor devices, packaging, and system technologies will blur. It will no longer be possible to design packages independently of chip and system design. All aspects of devices, packages, and systems will need to be considered simultaneously as part of an overall process, and optimizing the complete system will require trade-offs between chip, package, and system design and the analysis of many complex design parameters. Rethinking system layout to effectively improve performance, while at the same time reducing cost across a diversified technology base, is adding to design process complexity, both in terms of design tools and the need for more accurate information about the system itself, as well as system and material parameters. Reliability will become paramount. For example, driver assistance and safety systems will require a level of reliability similar to that of avionic electronics but at much lower cost. To ensure such product quality over a product's life span, "design for reliability" will become the catchphrase, along with the drive to better understand the factors determining reliability. The reliability of a SiP can be described as (1) the reliability of components, (2) the reliability of interfaces among components, as well as among components and package, and (3) the reliability of the system packaging. Especially the point (2) is not yet understood well enough on a microscale. Improvements in reliability will require greater understanding of interface behavior, including complex thermomechanical and other failure modes, as well as integrating this knowledge into the design process.

Localization, preparation, and analytics of failures, as well as testing of complex SiPs, will become a huge challenge. Ageing models for materials and material interfaces (e.g., adhesion and interconnect degradation) will also be required to assess life-time performance. Reliability models for life-time estimation will need to be based on nanomaterial and interface simulation, closing the gap between macroscale continuum mechanics and molecular modeling. In addition to these design-related models, new field testing methodologies will be needed to close the gap between field testing and design.

9.3 Wafer-Level Integration

Following Moore's law, the complexity of future components will increase dramatically. Using Rentsch's rule, chips with up to 6,000 inputs/outputs are expected, and on-chip and chip-to-board frequencies are likely to increase. The package has to

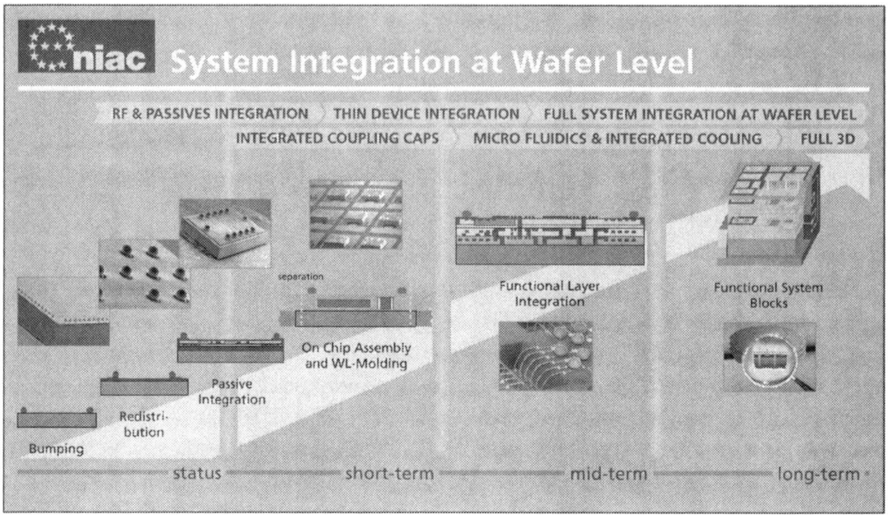

Fig. 9.4 Technology roadmap: wafer-level system packaging [8]

provide additional interconnects that exceed the current capacity of the "back-end of line" (BEOL) network. Even though the energy consumption per transistor will decrease, the total energy demand per chip and the energy density will increase due to the larger number of transistors and their higher frequency of operation. In addition, future systems will require additional radiofrequency and nonelectronic functions such as sensors, actuators, power supply components, passives, and displays (MEMS integration), which often require the use of alternative materials and new processes. However, integration of these alternative materials and processes into manufacturing often reduces yield and/or increases cost. Ultra high-density wafer-level integration technologies must therefore be able to successfully combine different technologies while also meeting yield and cost requirements under the constraint of highest reliability. A number of wafer-level integration technologies are being researched to achieve these goals. An overview of the technology roadmap is shown in Fig. 9.4.

The present state of technology comprises bumping, redistribution layers, and integration of passives. Further integration steps include layer deposition techniques to create embedded components, embedding ultrathin devices into cavities or polymer layers, creating high surface-area honeycomb structures for integrated capacitors, and nanowires to integrate III/V materials (InP, GaAs) and SiGe components. The following text discusses several of these approaches in detail.

9.3.1 Integrated Passives

Compared with integrated circuits, passive SMD components have only minimally decreased in size. Since the ratio of passives to ICs is in the range of 12:1 to 25:1 in cellular phones, passives are a major hurdle for further miniaturization of

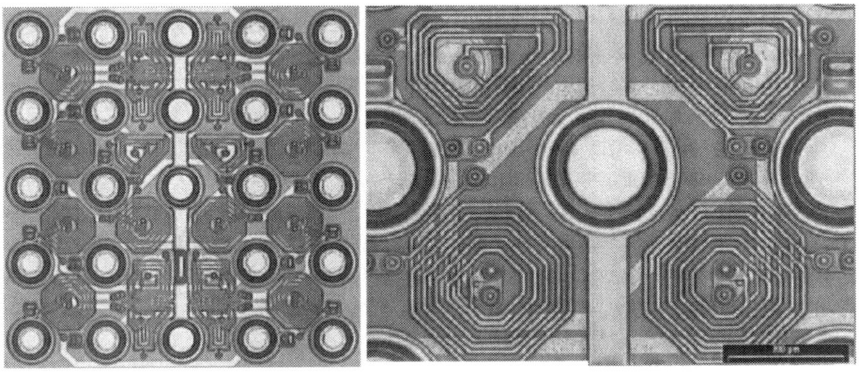

Fig. 9.5 Integration of coils into a redistribution (WLP using Cu/BCB)

electronic products. One loophole is the integration of resistors, capacitors, and inductors onto the wafer, which can be performed at BEOL by using the existing wafer-level packaging infrastructure. An example of the integration of coils into a redistribution layer is shown in Fig. 9.5.

The integration process for integrated passive devices (IPD) is based on Cu/BCB multilayer wiring using thin-film technology [13]. It uses semiadditive structuring of copper by electroplating and lithographic structuring of spin-coated photosensitive BCB as base processes for creating alternating routing and isolation layers. To date, buildups with up to four copper layers and five BCB layers have been realized. The high conductivity of the copper combined with the low dielectric constant and the low loss tangent of the BCB makes it possible to realize high-performance wiring systems, which feature low propagation delay, low cross talk, and low transmission losses. Since the technology is very flexible as to the number of layers, as well as layer thickness, the necessary layer buildup must be derived from the selected types of passive components. It should be noted that integrating all three types of passives will prove to be the most complex challenge.

Integrating coils requires at least two copper layers, separated by a BCB layer. In the case of a simple single-layer coil, one copper layer is required for the windings, while the other is employed for the underpasses, which connect the inner with the outer contacts of the coil. In the case of double-layer coils, windings are realized in both copper layers. To enable a proper processing of the second copper layer the BCB has to be thick enough to planarize the structures in the bottom copper layer as well as to reduce the capacitive coupling between the two layers sufficiently.

The same layer setup as used to integrate coils can be used to create metal–insulator–metal (MIM) capacitors [14]. The two wiring layers are used for the realization of top and bottom electrodes and the interdielectric layer is used as dielectric material. To ensure a high capacitance density, the BCB between the copper layers has to be as thin as possible.

Thus, two copper layers, which are separated by one BCB layer, are the minimum number of layers for an integration of coils and capacitors into one assembly. The corresponding layers are indicated with the numbers 3, 4, and 5 in the schematic drawing (Fig. 9.6).

Since copper is highly conductive, a separate material must be employed for the resistor structures. Processing compatibility issues render NiCr or TaN most suitable. Both materials can be deposited in the required sheet resistance and are compatible to the deposition and structuring processes of the other materials. Because NiCr resistors need copper structures for their connection, they are fabricated before (and beneath) the copper layers (Fig. 9.6). In Figure 9.6, the NiCr layer can be identified as Layer 2. Furthermore, nonwavy surfaces are preferable for production to minimize deviation in resistor lengths from nominal values, which is another reason for placing the copper layers as close as possible to the substrate.

Including the NiCr, at least four other layers are necessary for the integration of coils, capacitors, and resistors into one thin-film buildup. Additionally, to protect the passive structures and to ensure that the thin-film buildup can be correctly connected to the next system level, additional layers are required. At least one overall BCB passivation layer, as well as one final metallization layer (copper, nickel, or gold), is necessary, increasing the total number of layers to 6.

These process sequences make fabricating coils with arbitrary or regular forms such as squares, rectangles, octagons, or circles possible. The coils can be configured as single-layer or double-layer coils. A single-layer coil, with a height of 10 μm and a width of 10 μm, 4.5 square turns, and an inner window of 400 μm has an inductance of 17.8 nH. To adjust the inductance of such coils, both the number of turns and the length of the inner window can be altered. Changing the number of turns alters the inductance in large steps, whereas changing the inner window length makes minor adjustments possible. Inductance values between 0.5 and 27 nH can be achieved with single-layer coils with 1.5–5.5 turns and inner window length between 100 and 400 μm. Double-layer coils, which have an additional, identical spiral in the first metal layer, reach up to four times higher inductance values than the corresponding single-layer coils. Thus, values between 1 and 110 nH can be achieved with double-layer coils. The quality factors and resonance frequencies of such coils are strongly dependent on their type, form, and size as well as the

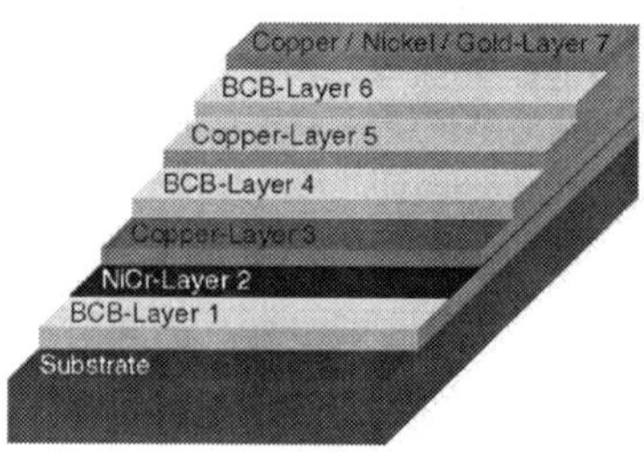

	L	C	R
Layer 1	(x)	(x)	(x)
Layer 2	-	-	x
Layer 3	x	x	x
Layer 4	x	x	-
Layer 5	x	x	-
Layer 6	x	x	x
Layer 7	x	x	x

Fig. 9.6 Overview of the layers necessary for the integration of coils, capacitors, and resistors into one assembly

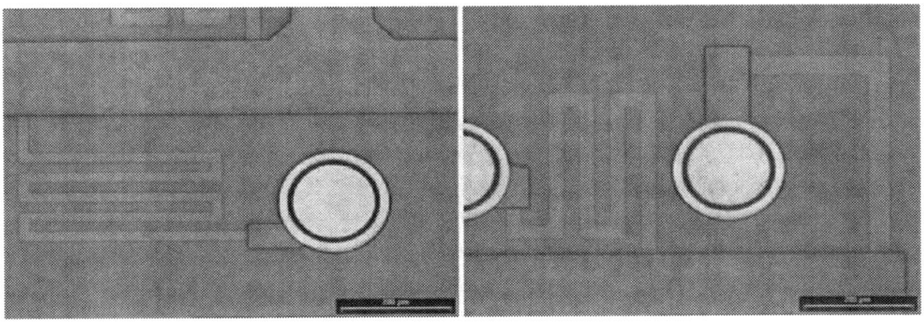

Fig. 9.7 Integration of resistors into a wafer-level package

substrate employed. In the case of ceramic or glass substrates, small coils with inductances below 10 nH produce resonance frequencies above 10 GHz and quality factors larger than 30. Using silicon as the substrate diminishes quality factors due to substrate losses.

Decoupling capacitors play a major role in reducing the switching noise for high-speed digital electronic systems. As the switching speed is well above 1 GHz and the power supply voltage simultaneously decreases, power supply noise is an issue. Large capacitance density and small leakage current are required. The capacitors should have very thin dielectric thickness and be mounted as close as possible to the chips to minimize parasitic inductance.

In case of the resistors, straight and meandered forms with resistance values between 100 and 150 kΩ have been produced by using line widths between 20 und 80 μm. The resistors are based on a sheet resistance of the NiCr layer of 100 Ω per square area, with sheet resistance defined as the resistance value of a square area of the NiCr layer. Figure 9.7 shows typical resistors fabricated with the technology presented here.

Similarly, new technologies will be needed for the wafer-scale integration of antennas (24–80 GHz), photonic components, batteries and energy scavengers, biointerfaces, microfluidics, and MEMS. Wafer-level encapsulation technologies using nanofilled materials (wafer molding), alternative technologies for 3D integration [e.g., through-silicon via (TSV) technology], and new isolation and shielding technologies for RF have to be investigated. To reliably manufacture wafers with such a high degree of wafer-level integration, advanced assembly and handling technologies for thin wafers and chips will need to be developed. Improved approaches for thermal management will also be required.

9.3.2 *Extension of Redistribution Layers: Thin-Chip Integration*

A base chip at wafer level is used as an active substrate for smaller and thinner dies. These thinned active components are mounted in flip-chip fashion on the base IC wafer using solder interconnects. To avoid the need for flip-chip bonding, the thin-chip integration (TCI) concept can be used [15]. Key to this approach are extremely thin ICs (down to 20-μm thickness), which are incorporated into the redistribution layer (Fig. 9.8).

The process flow for TCI modules begins with the bottom wafer carrying large base chips. The completely processed device wafers for the top IC have to be mounted on a carrier substrate by a reversible adhesive bond and undergo a backside thinning process until the thinned wafers show a remaining thickness of approximately 20 μm. The thinned top wafer and its carrier substrate are diced by a conventional wafer saw, thus producing thin chips that can be handled by its carrier chip just like any other standard die. The chip separation is performed using silicon dry etching. The bottom wafer is coated with a thin epoxy film and the thinned top chips are placed and mounted into this adhesive layer. Next, the photosensitive low-*k* dielectric BCB polymer is deposited onto the surface to planarize the 20-μm topography of the mounted thin chips. A high degree of planarization (DOP), which is one of the advantages of photo BCB, is very important for this first polymer layer to overcome the height difference between the surface of the bottom chip and the thinned chips (Fig. 9.9).

Fig. 9.8 Principle of TCI

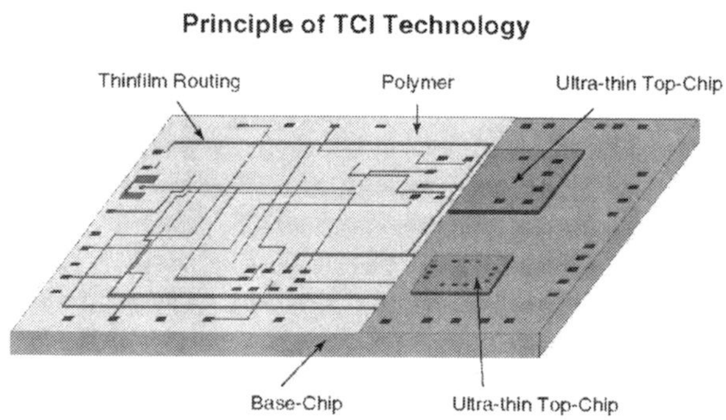

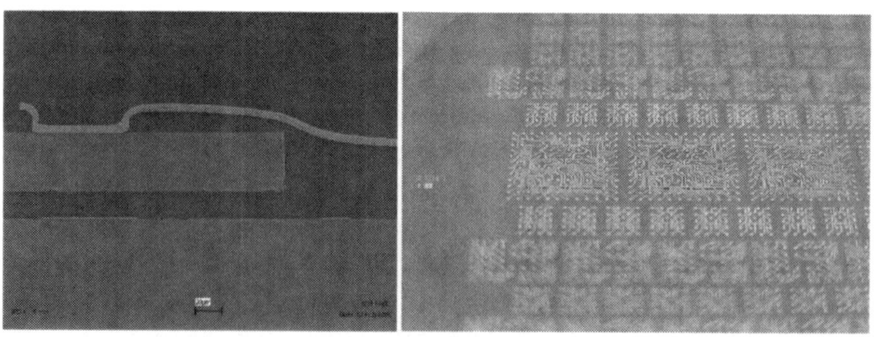

Fig. 9.9 A thinned chip (40-μm chip thickness) planarized with a thick BCB layer and additional Cu routing (*left*: cross section, *right*: video image)

Vias to the I/O pads of the thinned chips are opened using photolithography. Finally a thin-film metal layer connects the circuits of the top and base chips. A final CSP process, using a solderable metal layer and solder bump deposition, completes manufacturing of the TCI module.

9.3.3 Functional Layer Technology

Intermediate testing is one advantage of a process using flexible polymer layers with integrated passive and active components [9]. The schematic procedure is shown in Fig. 9.10.

The polymer layers are processed on reusable support wafers. The separation of the built-up layers is based on dissolving a release layer with a solvent. For the realization of the polymer layers a photosensitive polyimide precursor is used, which has excellent flexible properties as a freestanding film after release. The process sequence is similar to that described for the IPD. Examples are given in Fig. 9.11.

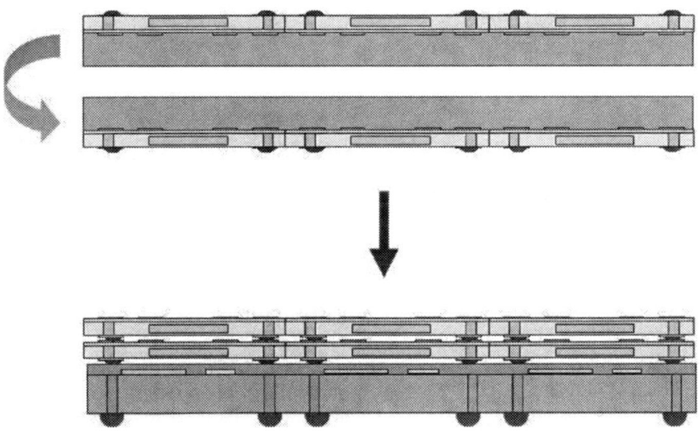

Fig. 9.10 Principle of the functional layer technology

Fig. 9.11 Examples of functional layer technology (sensor nodes)

The sensor node's dimensions in the right figure are 9.8 mm (w) × 31.2 mm (l) × 0.035 mm (d).

9.4 Module Integration Technologies

Most of the earlier discussion of wafer-level integration also applies to substrate-level integration, which represents the next level of interconnect/integration. Nowadays electronic systems are realized on module level through an organic printed wiring board on which individual components are placed. Traditionally, the wiring board is used exclusively for the electrical and mechanical functions. However, diverse research and requirements in the development of modern electronic products have lead to the integration of further system functions onto the board.

Future board and substrate technologies have to ensure a cost-efficient integration of highly complex systems, with a high degree of miniaturization and sufficient flexibility to allow them to be adapted for different applications. Their functionality will be considerably extended by the integration of nonelectronic functions such as MEMS, antennas, or optical components. New production methods will ensure a high throughput at very low cost. To ensure high data transmission and processing rates, new cost-effective cooling technologies and 3D packaging concepts will ensure a stable operation mode. The following priorities can be identified for multifunctional board and substrate technologies (Fig. 9.12):

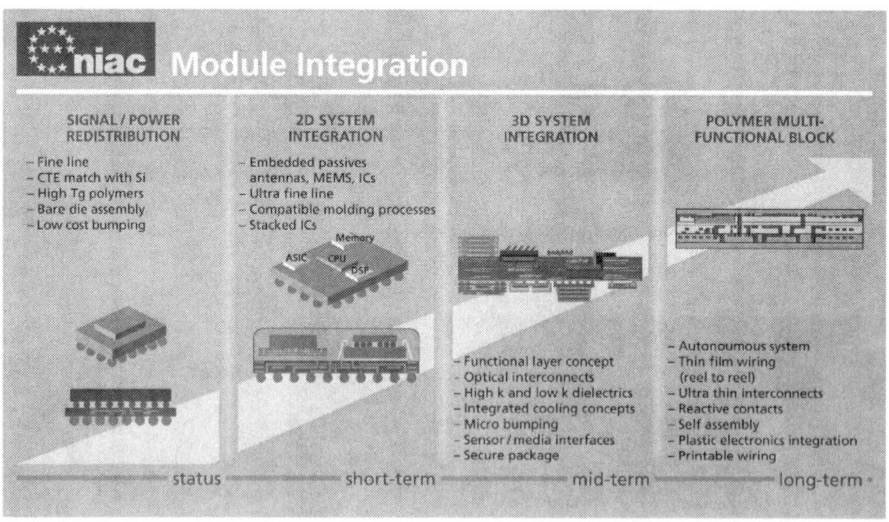

Fig. 9.12 Technology roadmap for module level system integration [8]

- Embedded device technologies (MEMS, passives, antennas, ICs)
- Low-cost substrates and interposers with finer liners and smaller vias
- Impedance-controlled wiring
- Flexible substrates (reel-to-reel manufacturing)
- Integrated optical interconnects

Technologies for embedded devices, such as MEMS, passive or active components, antennas, and power management will be the key to highly integrated modules. To make these possible, new materials for substrates, embedding, and encapsulation have to be developed, such as high-*K* and low-*K* dielectrics and high-T_g polymers, and the coefficient of thermal expansion (CTE) of these materials must be adjusted to the dies and substrates in question. Additionally, inside substrates and interposers with finer lines and smaller vias must be made available at lower cost. These advances will be followed by flexible substrates for reel-to-reel manufacturing and integrated optical interconnects. In the long term, printable wiring on organic substrates will increase productivity and lower environmental impact.

9.4.1 Embedding of Passive Components

The embedding of passive components into the printed wiring board is a promising approach for increasing the functional density of assembled electronic systems (Fig. 9.13). The development of materials and technologies for embedding resistors and capacitors was significantly advanced by the advanced embedded passives technology (AEPT) consortium in USA between 1999 and 2003 [1]. The application of such technologies in flexible printed circuits is, at present, the topic of the European project "Smart High Integration Flex Technologies" [2].

Printing technology for polymer thick-film (PTF) components is mainly used for fabricating resistors (although capacitors can also be produced with this technology). With appropriate paste compositions and component layout, resistors in the

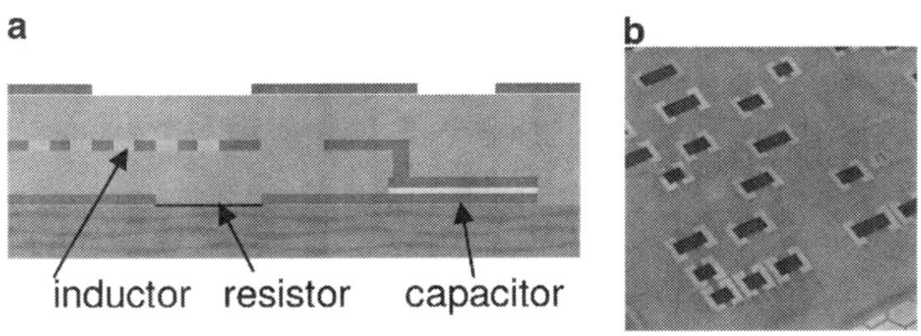

Fig. 9.13 (**a**) Embedded inductor, resistor, and capacitor in the build-up layers of a printed wiring board; (**b**) Printed resistors on a flexible printed wiring board before lamination into a build-up layer

range between 10 Ω and 10 MΩ can be realized. Since these printed resistors are in the range of ±20%, the resistors must be trimmed if smaller tolerances are required. The stability of PTF resistors under temperature and humidity load is an issue with many commercially available paste compositions.

Resistors and capacitors can be produced using sequential lamination and photostructuring. The variances of components here depend on the process capabilities of the manufacturer and are generally smaller than those of PTF components. However, the value range of embeddable resistors has tighter limits (10 Ω to 10 kΩ) and capacitance strongly depends on the used material (commercially available capacitors sheets are around 1 nF/cm^2).

Embedded capacitors have proven to be of interest for decoupling [4]. Since they can be designed and produced for placement close to the component (chip), inductances are reduced. Finally, cascades of SMD capacitors can generally be substituted by a single embedded capacitor.

9.4.2 Chip-in-Polymer

A new concept for the integration of active components is the so-called chip-in-polymer (CiP) technology [3, 5], which is based on the embedding of ultrathin chips into build-up layers of printed circuit boards (PCBs). The interconnect structure, which is neither a flip chip nor a wire bond, is shown in Fig. 9.14.

The basic idea of CiP is slotting thin semiconductor chips into standard PCB constructions. This technique can be used to fabricate 3D stacks of multiple dies, too. To achieve a contact pad surface on the wafer suitable for PCB metallization, the Al contact pads are covered by Cu bumps. To ensure that the bare dies can be embedded into build-up layers, wafers are thinned down to 50°μm. Subsequently, the chips are bonded using an adhesive. Precise thickness control of the bond line is essential for maintaining uniform thicknesses of the build-up dielectric on top of the chip. Here, new thin-chip handling and assembly solutions using die attach film or adhesive paste printing have been explored. RCC (resin-coated copper) layers with thin Cu are used for the lamination. Process parameters had to be fine-tuned to prevent damage to the chips during lamination. The chip contacts are produced with laser-drilled microvias followed by PCB-compatible Cu plating. All process steps in this technology have been optimized for large-scale manufacturing, using panel sizes of 18″ × 24″ in combination with high-accuracy positioning methods

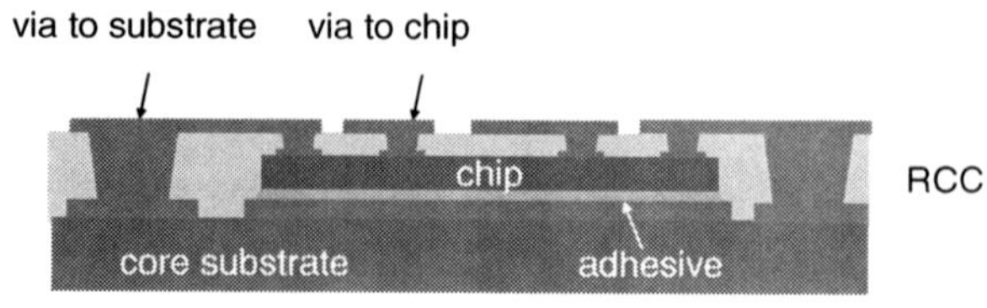

Fig. 9.14 Interconnect principle of an embedded chip in a PCB build-up layer

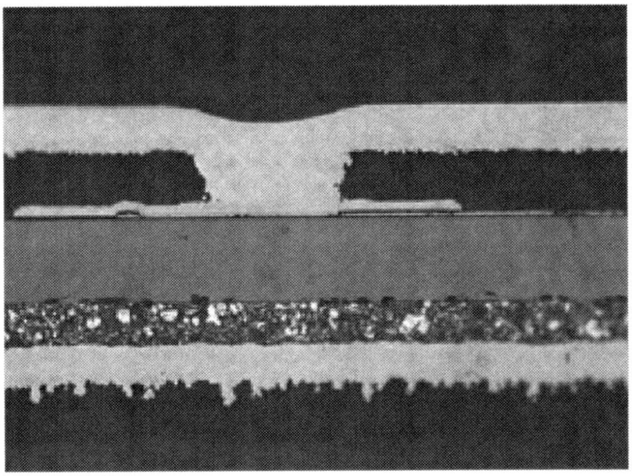

Fig. 9.15 Cu interconnect to embedded chip

using local fiducials for die placement, laser drilling, and laser direct imaging. Figure 9.15 shows a cross section of a Cu interconnect to an embedded chip.

Reliability evaluations of embedded chips have shown excellent reliability. For the evaluation, chips of 2.5 × 2.5 mm^2 size were embedded in 10 × 10 cm^2 test vehicles with a 0.5-mm FR4 core. All chips passed the following tests without any defect:

- Temperature storage at 125°C for 1,000 h
- Thermal shock condition; air-to-air −55±126°C; 2,000 cycles
- Humidity storage at 85°C/85% rel. humidity for 2,000 h
- JEDEC moisture sensitivity, level 3

9.4.3 Chip Embedding into Flexible Substrates

By embedding chips into flexible wiring boards, the functional density of electronic systems can be dramatically increased. The benefits of flex substrates, that is, light weight and high wiring density, will be combined with the complexity of the active chip. However, to maintain the basic flex substrate characteristics, the buildup with an integrated chip has to be as small as possible. Chips with a thickness of only 20 μm are used and the interconnection should not exceed a couple of microns.

The technology relies on a flip chip-type mounting of the thin chip onto the flex substrate and lamination of the structure on both sides (Fig. 9.16). Contacts to outer layers are realized by through holes. Further layers can be added to the buildup. The electrical interconnections are extremely thin and the mechanical coherence of the chip to the substrate is ensured by the no-flow underfiller. Process technologies for embedding passive and active components into flex wiring boards are also being developed and investigated in a European Union funded project [6].

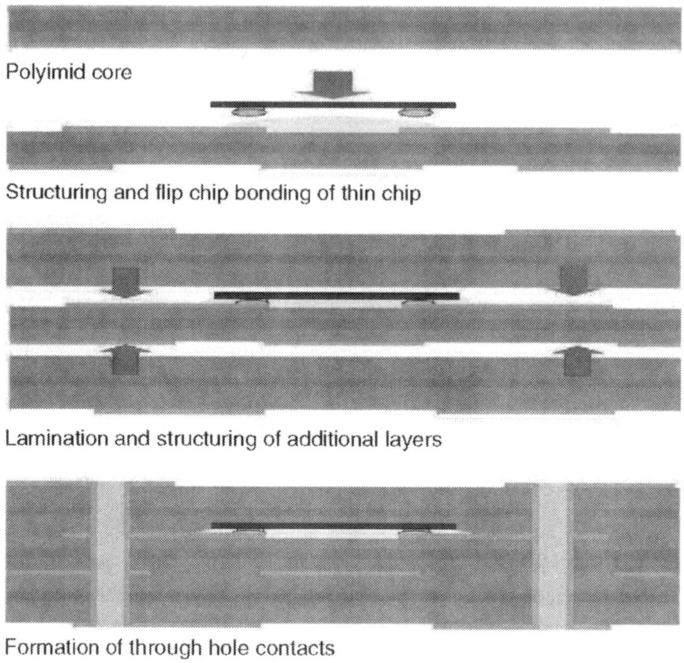

Fig. 9.16 Schematic process flow of embedding of thin chips into a flexible wiring board

9.4.4 *Functional Packaging*

Functional packaging goes one step further. It takes the application system and its requirements as a starting point and then adapts the microelectronic functions. One example is a radar sensor for the safety cocoon of a car, which is currently being developed by several partners as part of a project supported by the German Federal Government (Fig. 9.17). To equip midsize cars with this sensor technology, production cost is projected to be significantly lower than that of the predecessor model, which is presently used in the luxury class. This is to be made possible by a combination of two innovative embedding technologies. First, active and passive components are integrated into a plastic package by means of molding ("Chip in Duromer" [18]). This method ensures highly precise and fast mounting of the components. The subsequent molding process evens out the various component heights and makes lasting, precise alignment between the components possible.

The modules are then embedded into a PCB in the same way as described earlier in the "chip-in-polymer" paragraph and are bonded by means of a technique that ensures HF suitability (substrate embedding or chip in polymer [5]). A second layer of HF-suited PCB materials is laminated on top of the first layer and the respective antenna structures are realized by standard PCB processes. The two layers are then

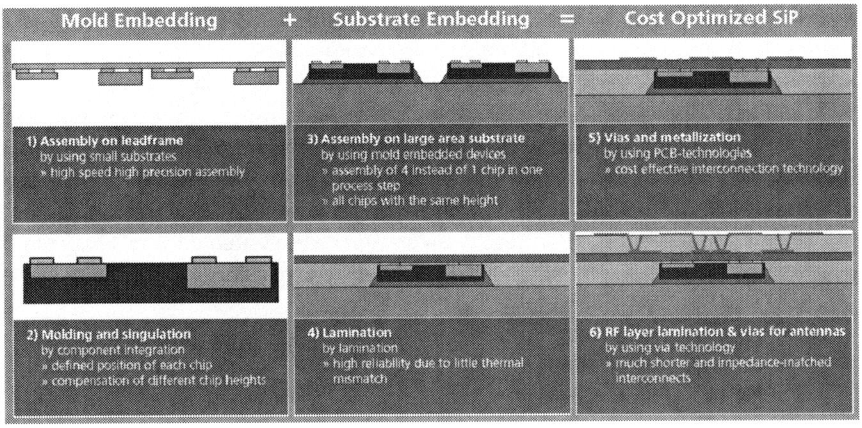

Fig. 9.17 Functional packaging for a cost-optimized radar sensor for active driver assistance systems [11]

Microwell Cross Cut | Principle of dielectrophoretic multi-well sensor | Implementation in PCB-Technology

Fig. 9.18 Using PCB technologies for lab on substrate [19]

electrically connected by means of μ-vias. Thanks to the low thermal mismatch – this embedding process promises high reliability with simultaneously improved HF characteristics as a result of shorter electrical contacts, which have been adjusted to impedance requirements [11].

9.4.5 Integration of Biological Functionalities

Merging new fabrication techniques and handling concepts with microelectronics enables the realization of intelligent microwells suitable for future applications, e.g., improved cancer treatment [19]. For the implementation of a dielectrophoresis enhanced microwell device a technology based on standard PCB technology has been developed in a European project, funded by the EU (Fig. 9.18). Because materials from PCB technology, such as copper or FR4, are not biocompatible, new materials have to be selected. Aluminum has been selected as the base conducting metal layer, structured by laser micromachining in combination with etching, and laminated successively to obtain minimum registration tolerances of the respective layers.

The microwells are also laser machined into the laminate, allowing capturing, handling, and sensing of individual cells as well as cell-to-cell interactions within a dielectrophoretic cage realized by the structured aluminum. Furthermore, surface treatments for hydrophobic and hydrophilic surface modification with, e.g., thiols and fluorinated acrylates on different materials were evaluated and analyzed by surface tension and wetting analysis to allow designing the microfluidic networks required for the microwell device [19].

9.5 3D Integration

Future microelectronic applications require significantly more complex devices with a high degree of miniaturization and flexibility. A possible answer to this challenge is 3D system integration technology that enables the combination of different optimized technologies with the potential for low-cost fabrication through high yield, smaller IC footprints, and multifunctionality. 3D system integration also has the potential to overcome the wiring crisis of signal propagation delay at both board and chip level – thanks to minimal interconnection lengths and the elimination of speed-limiting intra and interchip interconnections. The first 3D packaging technologies were developed in the early 1980s, primarily for space and military applications, and were based on enhanced MCMs (multichip module). The major hurdles were yield and cost. The infrastructure for KGD (known good dies), which is essential for most of the stacking approaches, was not yet established. Therefore, 3D approaches were not used for industrial products. Since the beginning of the twenty-first century the need for further miniaturization and higher performance has led to a comeback in 3D approaches (Fig. 9.19).

Three different 3D packaging approaches are currently discussed:

- Stacking of packages (PoP, PiP)
- Embedding of active and/or passive devices
- Stacking of bare dies without through-silicon wires (TSV)
- Stacking of bare dies with TSV

9.5.1 Package Stacking

The stacking of assembled and tested packages enables a 3D integration of high-density logic and memory device combinations at overall cost when considering both one of and ongoing development and production cost. Hence, industry is adopting package stacking solutions. This technology holds technical and business

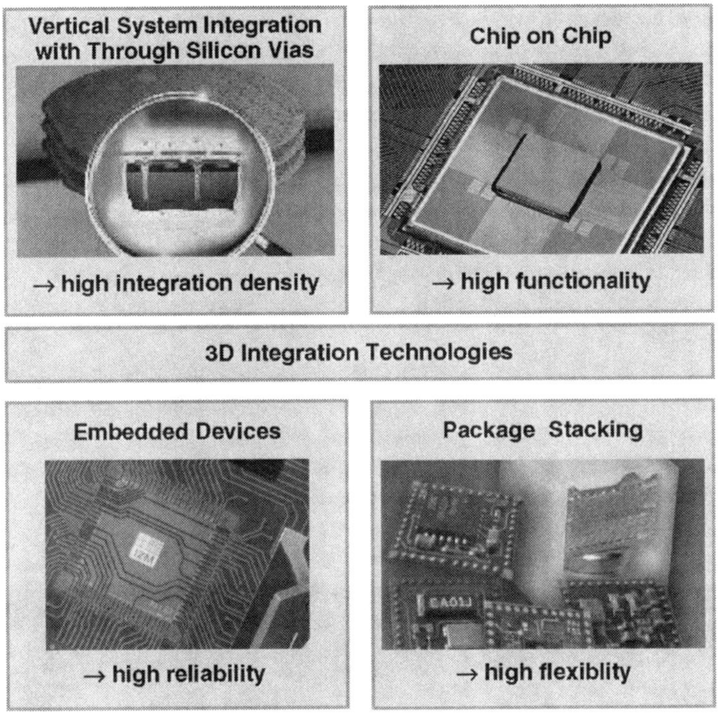

Fig. 9.19 3D integration technologies

benefits for each step in the supply chain. Examples of package stacking solutions include digital camera and cell phone applications with, at the present technological level, two-package stacks for logic and memory architectures. High-density DRAM and flash memory modules are stacked up to four packages high, with stacks of eight pieces possible [10].

The 3D module stacking method is a casing concept that is able to produce small, reliable, and highly robust systems rapidly and cost effectively [9]. The construction kit enables a hierarchic assembly of the chosen manufacturing technology Fig. 9.20. While the particular modules are fabricated by means of complex production technologies (e.g., usage of components without housing), the modules are later assembled using conventional SMD processes. With this concept, enterprises are able to produce complex customer-specific systems without owning a corresponding production line. At the same time it is possible to produce the particular modules utilizing the economy of scale for high quantities by producing each module in higher quantities. For highest integration, a reliable miniaturized interconnection technology for a wide range of different components as, for example, sensors or active and passive components is crucial. Chip-on-board technologies, using bare dies, wire bonding, and e.g., 01005 SMD components, are one approach. By using advanced flip chip technology with thinned bare silicon dies, the realization of even smaller volume assemblies is possible for each particular module. Handling and soldering of 10-μm thin dies with 10-μm bump height has already been demonstrated. Another advanced technology for miniaturization of each particular module is – as shown in chapter 9.4 – the embedding of active and passive components into the substrates (Fig. 9.21).

A similar concept for package stacking is using advanced technologies for embedding active and passive components into rigid substrates. One approach uses – as described in Sect. 9.4 – thin chips that are die bonded and embedded by lamination of RCC layers for ultrathin stackable packages. Initial fabrication of highly integrated system-in-package modules, followed by assembly of a standard package, is also possible.

Apart from the reductions in size by using thin flexible substrates, a 3D buildup can be realized by folding the flex, which also leads to a decrease in footprint (Fig. 9.11). The chips are soldered or glued to flex substrates, followed by embedding in an adhesive layer. Further layers can be applied or other components can be assembled by conventional technique on top. Chapters 9.3 and 9.4 detail the process.

Fig. 9.20 Module stacking concept [9]

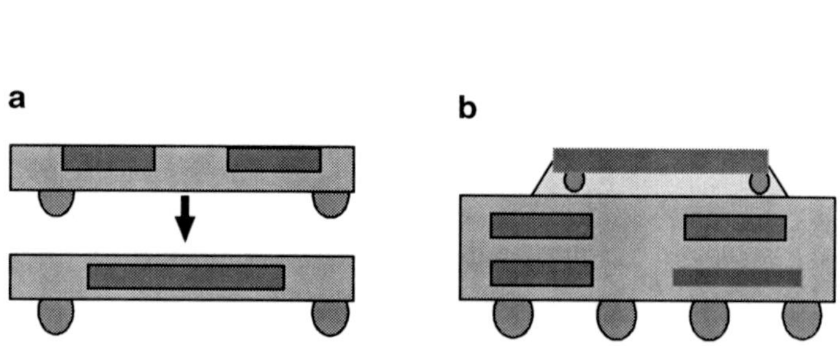

Fig. 9.21 Potential of embedded chips for 3D integration: (**a**) ultrathin stackable packages, (**b**) highly integrated system-in-package module

9.5.2 Embedding of Active and/or Passive Devices

The embedding of active and passive components into redistribution layers of a wafer or into the printed wiring board is a challenging approach for increasing the integration density of assembled electronic systems. Details are described in Chapters 9.3.2 and 9.4.1–9.4.3.

9.5.3 Stacking Without Through-Silicon Vias

Die stacking makes integrating multiple chips in one package possible. Wire bonding techniques for stacking three and more dice are now well established and can be used for both digital and nondigital components. Another focus is on state-of-the-art chip-on-chip technologies based on flip chip interconnects. The approach requires a base die with redistribution traces to match the I/O layout of both dice. This makes combining the performance advantage of FC with the option of integration of passive components into the redistribution layer possible. Figure 9.22 shows a stacked FC-BGA with a flip chip-mounted microcontroller on a second silicon chip with redistributed IC pads. The interconnection from the interposer to the board is achieved by wire-bonding technique.

In this approach a functional base chip on wafer level is used as an active substrate for FC-bonding of a second die [16]. The electrical and mechanical interconnection is carried out using eutectic solder balls, which are deposited by electroplating. The base chip is redistributed to an area array of a solderable UBM. The redistribution consists of electroplated copper traces, which ensures a low electrical resistance. Dielectric isolation is achieved using low-*k* photo-BCB. Integrating passives on the large die is possible.

The SOLID process (solid-liquid-interdiffusion) is a similar process but Cu/Sn layers are used instead of solder (Fig. 9.23).

The formation of intermetallics is the basis for the physical and electrical interconnection [12].

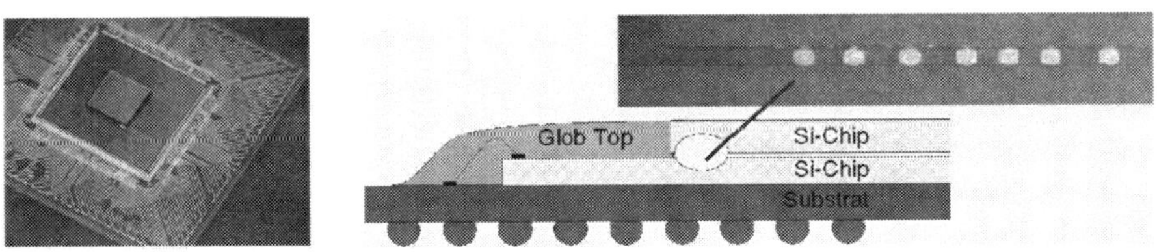

Fig. 9.22 Chip-on-chip integration using FC-bonding

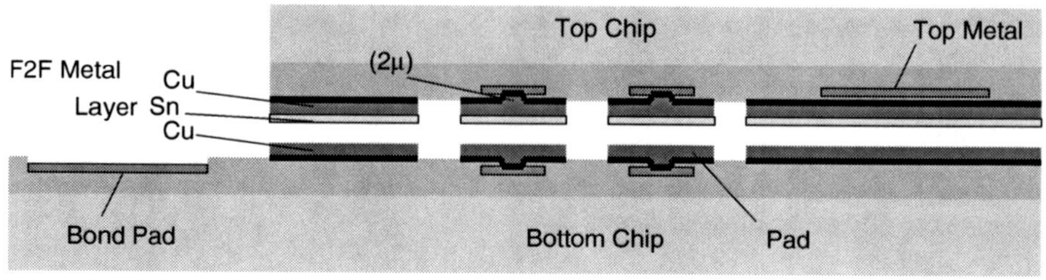

Fig. 9.23 Principle of the SOLID process

9.5.4 Stacking with Through-Silicon Vias

One of the first packages manufactured with TSV was introduced by Schott Electronic Packaging. It was strongly focused on optical applications and is a smart combination of wafer-level packaging and MEMS processing technology [17]. Optical packaging and most MEMS packages have to solve a very basic problem when advanced assembly techniques are to be applied to the device: the sensor surface has to interact with the environment without the packaging restricting the sensor in any way and, at the same time, protecting the sensor device against the environment. Wafer-level packaging is possible if the active area of the sensor is on one side of the device and the grid array contacts for the interconnection are placed on the reverse side. For an image sensor chip, silicon-via-contacts are central to this approach.

Standard silicon device wafers were the starting point of the SCHOTT OPTO-WLP, which is schematically represented in Fig. 9.24. These device wafers, i.e., image sensors – both CCD or CMOS – as well as surface-MEMS devices, typically have an active sensor surface on one side of a silicon substrate. In between the silicon and/or material comprising this active surface, an interdielectric layer may be found, which may consist of a wide range of dielectric materials, such as silicon oxide and/or silicon nitride. Standard contact pads, typically metallic, are on top of the same insulating layer. Later in the process, these contact pads are directly connected by the silicon-via-contacts. This means that the same pads used for testing or wire bonding are used for packaging as well. The first step in the SCHOTT OPTO-WLP process is the protection of sensitive active structures by a high-quality cover glass. A specialized adhesive wafer bonding process was developed, which can either perform a full area bond or a selective coverage of the adhesive within that bond layer. In the next step, the bonded silicon–glass sandwich is thinned from the silicon side (reverse side). The thickness of the silicon is reduced to about 100 μm, ensuring low-profile, chip-size optical packages for the intended devices (Fig. 9.25).

The actual thickness of the silicon can be adjusted according to the device application: stress-sensitive devices may need a very different residual silicon thickness compared with more robust sensors, where 50-μm residual silicon thickness might

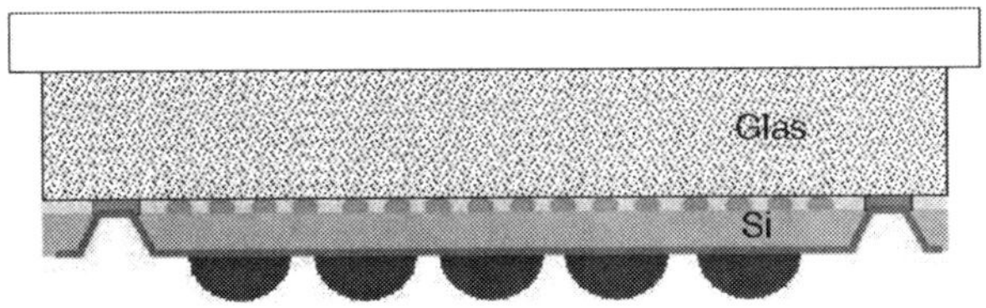

Fig. 9.24 Principle of OPTO WLP (courtesy of Schott Electronic Packaging)

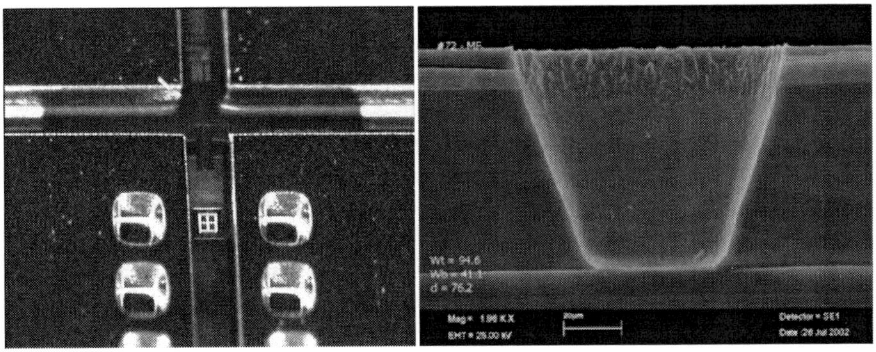

Fig. 9.25 TSV, left: video image; right: cross section (SEM)

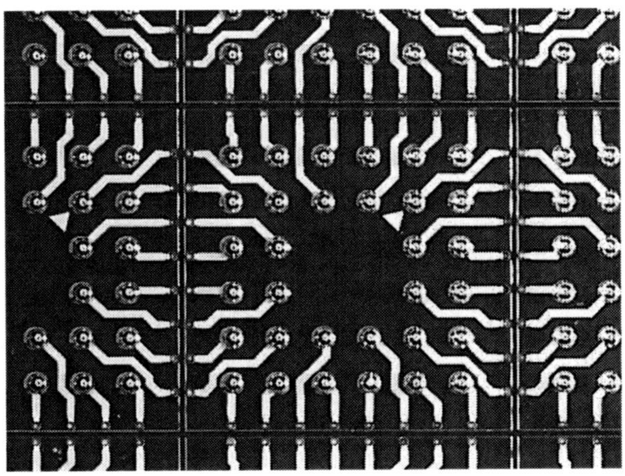

Fig. 9.26 Video image of bumped CCD-WLP

be preferred. A highly specialized plasma etching process is used to structure the silicon. Preferably, tapered side walls are used for all structures. The deposition of dielectric layers over the silicon guarantees electrical isolation during the subsequent redistribution process, which is based on polymer/Cu or Al. After UBM deposition, the wafer is ready for the balling and dicing process (Fig. 9.26).

The introduction of microstructured glass will further improve this packaging concept. This new deposition technology will add hermetic sealant to MEMS WLP [7].

Vertical system integration is characterized by very high density vertical interchip wiring by through-Si vias with diameters of approximately 2 μm that can be positioned as required. Based on thinning, adjusted bonding, and vertical metallization of completely processed device substrates, the requirements for VSI are precise thinning technologies, reliable formation of interchip vias (ICV), and a suitable bonding process. In general, technologies largely relying on standard wafer fabrication processes exhibit a favorable cost structure. On the other hand, wafer yield and even more importantly chip area issues speak against plain wafer stacking concepts. Consequently, chip-to-wafer technologies, mainly based on wafer-level processes utilizing known good dice, are preferable. So-called ICV-SLID technology is optimized for chip-to-wafer stacking and provides a very high vertical interconnect density based on ICV between metallization levels of stacked dice. Both the mechanical and vertical electrical connections are realized by solid–liquid interdiffusion (SLID) of thin electroplated and structured copper/tin layers. The thinned chips with tungsten- or copper-filled ICV are connected to the bottom device wafer by the SLID system (Cu, Cu_3Sn, Cu). This fully modular concept makes multiple device stacks possible [13].

9.6 Conclusion

Smart innovative systems linked into networks and used in an extremely broad range of applications are the future of electronics. They will contain electrical and nonelectrical functions. The task of heterogeneous integration will be integrating such functions into one system. For the fabrication of heterogeneous systems, new architectures and system integration technologies are necessary, which have to ensure the realization of reliable systems at minimal sizes and at low production costs. Adequate interfaces for different application environments have to be created. By optimizing heterogeneous system integration one will meet the following requirements:

- Integrated on-chip off-chip design as well as system partitioning/modularization
- Highly reliable systems
- Integration of electrical and nonelectrical components
- Integrated wireless communication
- Integrated power conversion and storage
- Integration of different functions in one module/package
- Application of "add-on" technologies to increase system functionality
- Ultrahigh density component integration
- Short time-to-market processes and low-cost solutions

References

1. AEPT Final Report, NCMS (Ed.), Ann Arbor, MI, 2003
2. Shift (Smart High-Integration Flex Technologies), http://www.vdivde-it.de/portale/shift/default.html
3. L. Böttcher, A. Neumann, A. Ostmann, H. Reichl, Integration of Active and Passive Components Using Chip in Polymer Technology, Proceedings SMTA International Conference 2005, Chicago, Ilinois, 2005
4. R.K. Ulrich, L.W. Schaper, Integrated Passive Component Technology, Wiley-Interscience, New York, 2003
5. R. Aschenbrenner, A. Ostmann, A. Neumann, H. Reichl, Process flow and manufacturing concept for embedded active devices, Proceedings of EPTC 2004, Singapore, December 8–10, 2004
6. T. Löher, B. Pahl, M. Huang, A. Ostmann, R. Aschenbrenner, H. Reichl, An approach in microbonding for ultra fine pitch applications, Proceedings of Technology and Metallurgy, The Seventh VLSI Packaging Workshop of Japan, Kyoto, 2004
7. D. Mund, J. Leib, Novel microstructuring technology for glass on silicon and glass-substrates, Proceedings of 54th Electronic Components and Technology Conference, Las Vegas, June 2004, pp. 939–942
8. Strategic Research Agenda 2007, White Paper by ENIAC – European Technology Platform on Nanoelectronics
9. H. Reichl, R. Aschenbrenner, H. Pötter, S. Schmitz, More than Moore, hetero system integration and smart system integration – three approaches – one goal: Smarter products and processes, Productronic 2007, Frankfurt, VDMA-Verlag, 2007
10. Amkor Technology, 3D – Stacked-Die Packaging Technology Solutions, http://www.amkor.com/enablingtechnologies/3D/index.cfm (Status January 11, 2008).
11. http://www.mstonline.de/foerderung/projektliste/detail_html?vb_nr = V3FAS032 (Status December 19, 2007)
12. P. Ramm, 3D system integration: Enabling technologies and applications, extended abstracts of the International Conference on Solid State Devices and Materials, SSDM 2006, Yokahoma, 2006
13. M. Töpper, D. Tönnies, Microelectronic packaging (Chapter 25), in Semiconductor Manufacturing Handbook (M.H. Geng, Ed.), McGraw-Hill, New York, 2005
14. K. Zoschke, J. Wolf, M. Töpper, O. Ehrmann, H. Reichl, Wafer level fabrication of IPD: A key technology for system in package approach, Proceedings of Micro System Technologies 2005, October 2005, pp. 235–243
15. M. Töpper, K. Scherpinski, H.-P. Spörle, C. Landesberger, O. Ehrmann, H. Reichl, Thin chip integration (TCI-Modules) – A novel technique for manufacturing three dimensional IC-packages, Proceedings of IMAPS 2000, Boston, MA, September 2000
16. M. Töpper, Th. Fritzsch, V. Glaw, R. Jordan, Ch. Lopper, J. Röder, L. Dietrich, M. Lutz, H. Oppermann, O. Ehrmann, H. Reichl, Technology requirements for chip-on-chip packaging solutions, Proceedings of ECTC Conference 2005, Florida, USA, pp. 802–808
17. J. Leib, M. Töpper, New wafer-level-packaging technology using silicon-via-contacts for optical and other sensor applications, Proceedings of ECTC 2004, New Orleans, LA, June 2004
18. K.-F. Becker, T. Braun, A. Neumann, A. Ostmann, M. Koch, V. Bader, E. Jung, R. Aschenbrenner, H. Reichl, Duromer MID technology for system-in-package generation, *IEEE Transactions on Electronics Packaging Manufacturing*, Vol. 28, Issue 4, 2005, pp. 291–296
19. T. Braun, L. Böttcher, J. Bauer, D. Manessis, E. Jung, A. Ostmann, K.-F. Becker, R. Aschenbrenner, H. Reichl, R. Guerrieri, R. Gambari, Microtechnology for realization of dielectrophoresis enhanced microwells for biomedical applications, Ninth Electronics Packaging Technology Conference, 2007, Singapore, pp. 406–410

Chapter 10
System-Level Design

Joel Björk, Peter Caldera, Paul Fugger, Karsten Einwich, Karl Hollaus, Wolfgang Pribyl, Jochen Reisinger, Willy Sansen, Andreas Wiesbauer, and Horst Zimmermann

Abstract This chapter gives an overview on the system design and its different methods to implement it in the whole development process. Various aspects that have to be taken into consideration to reach the described goals are discussed.

Keywords Design flow • Analogue design • Digital design • Design methodology • Simulation • Modelling

10.1 Introduction

The design of systems[1] in the More than Moore domain is tightly connected with the design of system-on-chip (SoC) and system-in-package (SiP). The close integration of digital and analogue functionality with software saves not only space but also cost and power. MEMS[2] and MOEMS devices, discrete components, and thin-film devices further extend the functionality available in a package.

This miniaturisation does, however, lead to increased interference between the parts, significantly reducing performance if it is not taken into account in the design stage. The heterogeneous nature of such systems greatly complicates the design work.

The design flow must be capable of incorporating a variety of hardware technologies, as well as software at a high abstraction level. It must also be able to exchange data between the different parts of the system at the very lowest levels to enable system-wide modelling and simulation.

---

[1] The definition of a system, as used in this chapter, is 'a combination of technologies integrated into a product'.

[2] Micro (opto) electromechanical systems.

J. Björk, P. Caldera (✉), P. Fugger, K. Einwich, K. Hollaus, W. Pribyl, J. Reisinger, W. Sansen, A. Wiesbauer, and H. Zimmermann
Technikon Forschungs und Planungsgesellschaft mbH, Villach, Austria
e-mail: peter.caldera@infineon.com; office@eniac-austria.net

G.Q. Zhang and A.J. van Roosmalen (eds.), *More than Moore: Creating High Value Micro/Nanoelectronics Systems*,
DOI 10.1007/978-0-387-75593-9_10, © Springer Science+Business Media, LLC 2009

10.2 Background

The electronic design automation (EDA) industry can be traced back to the beginning of the 1980s. Until that time, the development of the tools needed for hardware design was carried out in-house by the hardware manufacturers.

The market is dominated by a few, predominantly American, vendors offering suites of software composed of modules designed for specific purposes. Each vendor is committed to one or a few proprietary workflows but offers the ability to export or import data in (semi)standardised formats. This means that users have the opportunity, though limited, to mix tools from different vendors or, as is also common, produce their own tools. There are also a number of smaller vendors providing complementing modules, often for specific applications.

The business landscape for chip suppliers has changed considerably in recent years. Their position in the value chain has been extended from pure silicon suppliers to system suppliers, integrating not only digital circuits, but also software, see Fig. 10.1, and More than Moore technologies.

One of the motivations for such a move can be seen in Fig. 10.2. It is estimated that the 265 billion dollar semiconductor industry would generate 1.340 billion dollars of value in the next tier.

This has led to a change in how the industry perceives their products. The designs are no longer monolithic, where all of the value is created in a single piece of silicon. Value is instead created from the combination of possibly several types of hardware with software. This extension of the value chain, in part driven by rapidly evolving SoC and SiP technologies, has radically altered the requirements

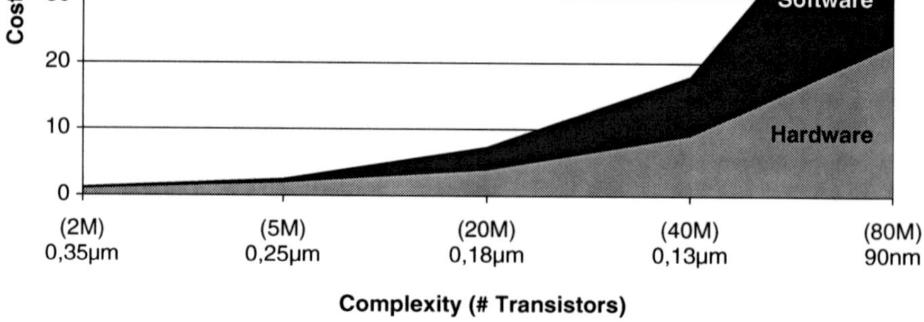

Fig. 10.1 Development of hardware and software costs over time for a typical ASIC (Source: Adapted from IBS, Inc)

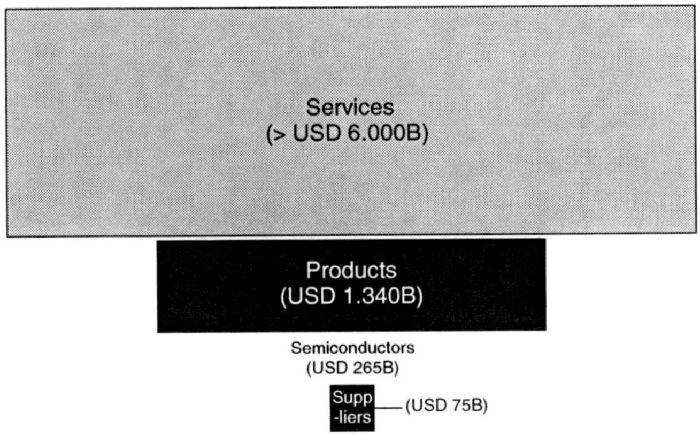

Fig. 10.2 Semiconductor value chain – each market is drawn to relative scale

placed on system design tools. These tools must, just like the market they serve, expand to cover complete systems.

10.2.1 Diversity

More than Moore technologies are core enablers for ubiquitous computing, being part of everything from consumer products, to cars, to medical products. Consequently, the markets are more diverse than for the traditional chip industry and so are the lifetime profiles of the products.

- A mobile phone has a life expectancy of about 2 years and is mostly kept at room or pocket temperature. A malfunction is annoying but it is hardly life threatening.
- The anti-brake-lock controller of a car on the other hand is expected to outlast the car where it is fitted while enduring extreme environmental conditions. A failure here can lead to injuries or even loss of life.

The chip industry needs a way to parameterise these profiles and include them in the workflow.

For More than Moore technologies, system-level design will increasingly become the key design method. The costs for system-level design are relatively low, however, the impact of time-to-market and revenue are orders of magnitude higher than the impact of the subsequent design steps. Because of the low system-level design costs the market for tools is relatively small, thus such tools are poorly treated by the CAD vendors. State-of-the art system-level design tools are usually strongly focused on an abstraction level and/or application domain and are poorly integrated into the overall design flow. This becomes even worse for mixed-signal and heterogeneous systems – the area where European companies have outstanding competencies.

10.2.2 System Design Goals

System design is mentioned here as the bridge between application design on the one hand and component and software design on the other.

Application design delivers a feature description in an application- and user-dependent language. Languages and formats are mainly defined by the market place. This means that the application-oriented community speaks many different languages when defining features and performance requirements. On the other side, the electronics design community also uses different languages, for example, Logic, Mixed Signal, RF and Sensors/Actuators.

In many cases, different kinds of standards play a very important role, for example:

- Transmission standards for mobile communication (EDGE, GPRS, etc.)
- Media standards for multimedia (compact disc, etc.)
- Security standards for chip cards
- Interface standards for computing systems (USB, etc.)
- Standards for electromagnetic radiation limits

Goal number 1: The creation of a clear 'translated' specification of features and functions that can be read into the system design flow:

- No risk that specification items are lost
- No risk that the specification might be misinterpreted (translation error)
- Enables a comparison (verification) of the input specification with the implemented system capabilities

Goal number 2: In all cases, it is key that the system implementation is verified against relevant standards. The other side of the bridge – the components (& software) side – will be defined in more detail within this chapter.

10.2.2.1 Application Example

The mobile phone is without doubt the most ubiquitous electronic device today. Over one billion devices were produced in 2007. It is also the main driver for the development of SiP and More than Moore technologies. Figure 10.3 shows, to scale, how mobile platforms have evolved over a period of 6 years.

The system includes RF transceiver, BAW[3] filters, processor, persistent and volatile memory, MEMS devices, and high-voltage parts. The lower end of the

[3] Bulk acoustic waves.

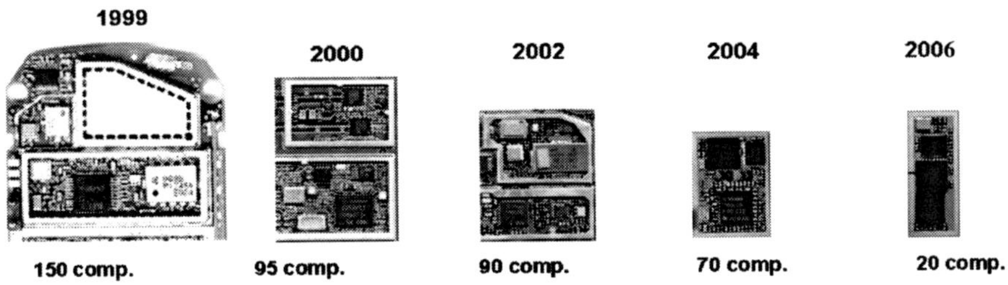

Fig. 10.3 Evolution of mobile phone platforms – the value below each picture indicates the number of components needed to realise the phone

market is increasingly being served by SiP solutions while higher end smart phones are being built using components from several manufacturers.

10.2.3 The Design Flow

The design flow is the structured process used to turn an idea into a product. This process has become more and more complicated as the functionality of the hardware has increased while the physics of deep-submicron processing has added new restraints in every generation.

The details of this process vary from organisation to organisation. A general description must, therefore, be created at a generic level. The specification of the product is translated into a high-level language such as SystemC. Subsequent steps then model and simulate parts of the system at more and more detailed levels. Any changes are updated across all the abstraction levels, referred to as back annotation. This process is then repeated until a design has been achieved that meets the specifications. An overview of the abstraction levels and the languages used is provided in Fig. 10.4.

The majority of the efforts in these flows is concentrated on the area shown in a darker colour in Fig. 10.4. These tools provide very accurate results using commercially reasonable computing resources, as the majority of the value of a monolithic system is created here.

However, the majority of the design effort in a More than Moore design flow is expected to take place at the system level and the very lowest abstraction levels.

10.2.3.1 The Hell of Nanophysics

With the advent of modern deep-submicron technologies, design is becoming more and more complex. The most dramatic effect can be described as ‘digital goes analogue’. Because of the behaviour of deep-submicron devices (non-ideal device characteristics such as subthreshold, gate leakage, etc.) even pure digital circuits

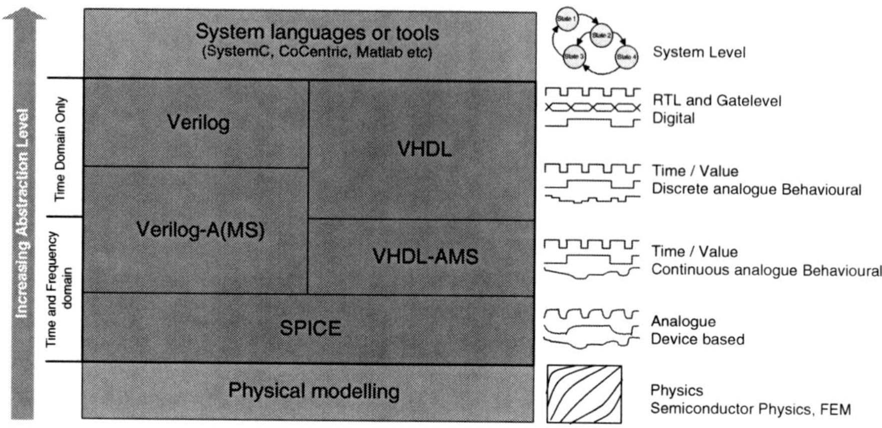

Fig. 10.4 Abstraction levels in design and simulation

have to be simulated with consideration given to different analogue effects, including back annotation from the ever-increasing influence of routing resistance, capacitance, and even inductance.

These device effects also deteriorate the performance of analogue blocks. To compensate for this, more digital correction and calibration techniques have to be applied in close co-operation with the analogue designers. The effects mentioned earlier call for a completely new simulation and verification strategy, which is best tackled in a hierarchical way, starting at the system level.

10.2.3.2 More than Moore

In design flows of the future it will, in general, not be enough to start the design work by creating a model from a paper specification. Furthermore, the specification must be developed iteratively by verifying numerous application scenarios and use cases. Therefore, an executable platform will become more and more essential. This platform must permit the execution of real-time use cases for the overall system including the environment.

The overview of such a design flow is presented in Fig. 10.5. The exact composition and components depend greatly on the type and complexity of the product.

10.2.4 The State-of-the-Art Technology

State-of-the-art design tools clearly serve one physical domain, operating with different abstraction layers. Two different design flows are required to simulate a simple SiP with a 60-nm digital and a 120-nm analogue part. This means that two

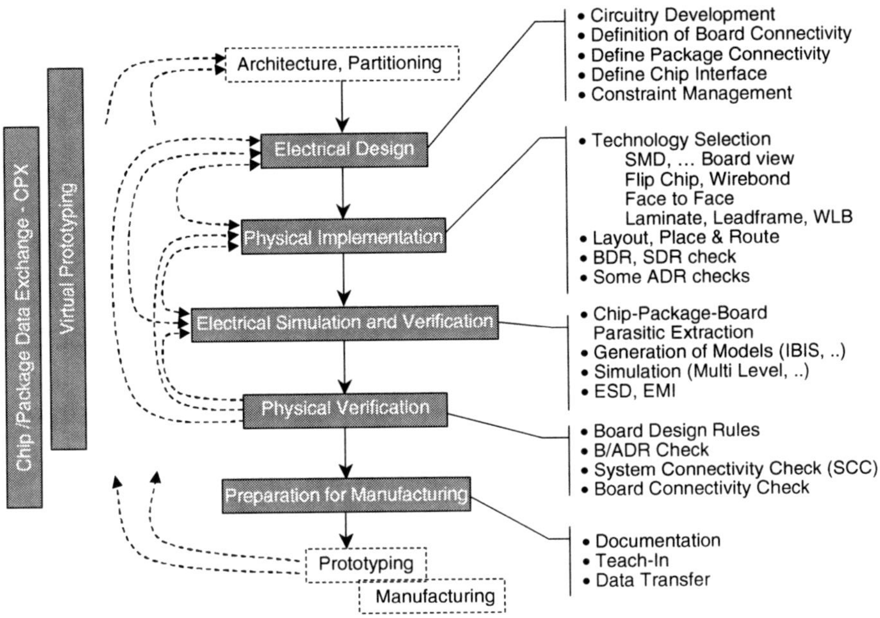

Fig. 10.5 Example of More than Moore design flow

different simulator environments are needed. To simulate the complete system, the two environments have to be coupled using TCP/IP or shared memory, which increases the complexity of the system and dramatically increases the need for computing resources.

State-of-the-art simulators work independently, one after another. We start with chip and package development and simulation. When a chip is finished, board design and component engineering begin their own optimisation. When a board is finished, system integrity and EMI are simulated and tested.

The performing of a chain of simulations one after another is time consuming, and the risk of having to start again from the beginning, if model and parameter assumptions fail en route, is very high. Nowadays, the impact of such risks on manufacturing is reduced by adding extra components that act as spare parts for later improvements if a system should fail any end tests.

A few software packages are available for the simulation of PCBs. Full 3D field solvers are very slow and thus not suitable for the design of large systems. Highly non-linear or large complex problems are still not a routine task in the design process at present. Software packages that apply approximation techniques are particularly dedicated to the analysis of PCBs and are usually faster than full 3D field solvers. However, some physical effects are not taken into account and substantial errors are exhibited. Unfortunately, the implemented approximation methods in state-of-the-art products are not made public, being kept secret by tool suppliers, which completely impedes fair validation.

The closed nature of the tools makes the construction of tool chains a laborious and often bespoke work. Little reuse is possible between projects and a lot of information is lost in translation, reducing the accuracy of the results and impeding automation.

10.2.4.1 Current Trends in Microelectronics

More than Moore will still be the dominant way forward in the nanoelectronics industry in the coming years, driven by Flash and DRAM technology. The importance of More Moore is steadily increasing as a number of physical and financial effects can be foreseen, which will slow the competitive advantage found in the next generations of nanometer-scale CMOS:

- *Leakage power* starts to dominate unless larger gate delays than performance scaling predictions are introduced.
- *Voltage headroom* shrinks and makes analogue and radio frequency design very difficult. New digital architectures providing functions previously only available in the analogue domain are able to mitigate this problem somewhat. This is primarily achieved through faster analogue to digital converters requiring less filtering of the input signal and switched mode amplification techniques. Moving to SiP solutions is another possibility, but it brings problems at the system level with increased integration complexity and power requirements.
- *Increasing variability* reduces predictability at the device level. Materials can no longer be treated as homogenous bulk materials, as very small local variations during the manufacturing process affect the properties of transistors and interconnects. This greatly increases the need for more precise process control to retain yield levels.
- *Signal integrity* decreases as the interaction between devices and interconnects increases. The design and simulation of digital circuit starts to increasingly resemble the analogue design flow with manual layouts of critical subcircuits and the use of circuit simulators. The effect is decreased productivity in the design and simulation stage.
- *Layout restrictions* imposed by lithography effects will continue to affect the freedom of the designs.
- *Global interconnect delay and variance* endangers the global synchronous concept.

All these factors combined with very high investment costs for sub 45-nm technologies are driving the chip suppliers to look for competitive advantages outside of the more Moore track.

10.2.4.2 Digital Design

Digital design in the context of system design, including the design of mixed signal systems, must be separated into two main parts: design and implementation, and functional verification.

Design and Implementation

The design process includes the task of describing hardware at a high abstraction level using, for example, SystemC or a hardware description language (HDL), such as Verilog, VHDL. The implementation process uses software tools to translate the high-level description into lower levels of abstraction, such as gate-level netlists or layout data. This flow is referred to as RTL2GDS.

One of the main drivers for the development of any electronic product is time-to-market. On the digital design side, reuse helps to meet short design cycles. IP-based design (intellectual property) means the use of commercially available IP, such as microcontrollers and interfaces, as well as company internal IP such as company proprietary interfaces and hardware filter structures. Another way to meet tight schedules is concurrent development of hardware and software. This means that computational efforts are moved from hardware to software, i.e., algorithms are executed as a program in a microcontroller system and not in hardware. Although this means that development tasks are shifted from hardware to software, it offers easy extension or update of a system and the possibility of separate evaluation for hardware and software. With respect to system projects, software development has in general already become the dominating effort.

Another key parameter for a successful product is cost competitiveness. Since one of the main cost contributors for an IC is tester usage (this can be between 20 and 50% of the overall development costs) measures must be taken to keep this time as short as possible – automated test circuit creation is available for boundary scan, memory BIST (built-in self-test), and scan test. These methods achieve high test coverage in acceptable tester times. Another direct contributor to the costs is yield. To reach high yields for digital designs, methods such as redundancy insertion for memory blocks as well as for other parts are applicable. Strict fulfilment of timing constraints, so-called timing closure, helps a system become resilient to voltage and temperature effects.

For many products, such as mobile devices, low power consumption is a must. On the digital design side, many different techniques can be used to achieve a low-power design, e.g., voltage scaling and power management in power islands and selective shutdown of the clock, called clock gating.

Functional Verification

Functional verification is the task of verifying that the logic design conforms to the specification. In everyday terms, it attempts to answer the question: 'Does this proposed design do what is intended?'. Functional verification is NP-hard, or even worse, and no solution has been found that works well in all cases. However, it can be approached using various verification strategies:

Directed tests using simulation-based verification, called dynamic verification, are widely used to 'simulate' the design, since this method scales up very easily. Stimuli are provided to exercise each line in the HDL code. A test bench is built to functionally verify the design by providing meaningful scenarios to check that given a certain input, the design performs to specification. The best usage of directed tests is in a regression environment with self-checking tests (go/no-go results).

Test bench automation is used to perform pseudo-random testing. A stress test for digital components can be implemented by using specially developed hardware verification languages, such as 'e', 'VERA' or 'TestBuilder C++'. Millions of test vectors are generated in a semi-automatic way.

Property checking is a 'formal' mathematical approach to verification. A system is described by means of state predicates describing inputs, outputs, data transport and temporal expressions that make the link to cycle-based timing behaviour. Property checking tools calculate all the possible transitions and check whether these transitions fulfil the state predicates and the described timing behaviour.

All these methods are part of a common verification process, which is iterative (see Fig. 10.6).

Conclusion

Whereas design tasks are being increasingly driven by the reuse of IP delivered in any HDL, verification has to manage the exponential increase of complexity with rising design sizes. Figure 10.7 shows the relationship between complexity, represented by lines of RTL code, and the verification effort, represented by simulation cycles for two generations of the same processor architecture. Nowadays a split of 20–30% design and up to 80% verification effort is realistic and widely accepted as a matter of fact.

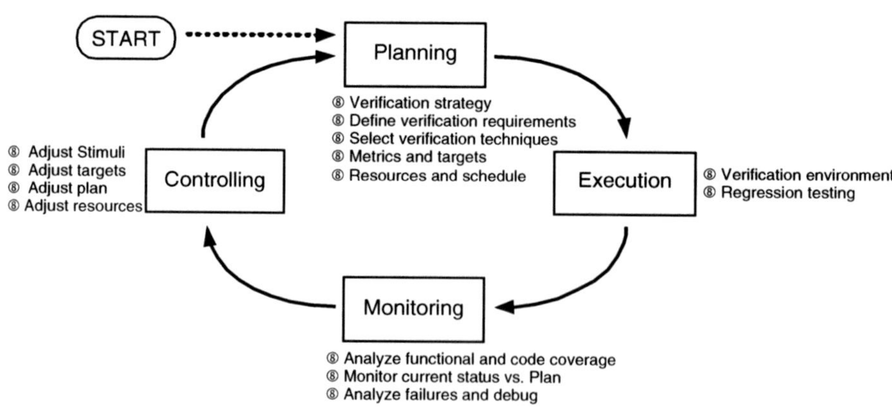

Fig. 10.6 Verification process

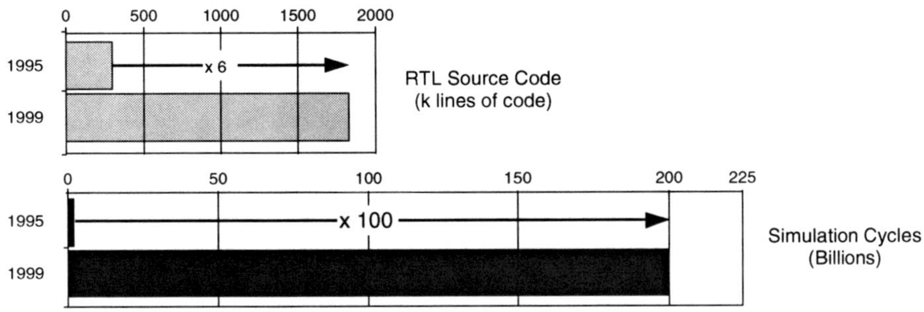

Fig. 10.7 Verification effort – circuit vs. simulation complexity growth over a 5-year period

10.2.4.3 Analogue Design

Numerical simulation became popular with the publication of the SPICE simulator in 1971. SPICE is an open source circuit simulator released under the BSD Copyright.[4] SPICE and the different commercial and open source variations thereof are by far the most commonly used simulators for analogue circuit analysis. The program works by taking netlists and converting them into non-linear differential algebraic equations that are then solved. Supplying SPICE models for components has become the de facto standard in the industry.

In the eighties, the activities were focused on digital simulation techniques that resulted in two competing HDLs – VHDL and Verilog and the underlying simulation techniques. In the nineties, these HDLs were extended to the analogue domain (VHDL-AMS/Verilog-AMS). In addition, a couple of new concepts such as behavioural modelling at equation level (instead of macromodelling in SPICE or S-Parameter) and physical domains were introduced. In 2000, the activities went back to the digital domain and were focused increasingly on verification. One result of this activity is the standardised HDL SystemVerilog, which is based on Verilog and has its strength in the verification of digital implementations. Thus, concepts such as assertion-based verification and constraint randomisation were introduced.

From a More than Moore viewpoint, these activities only focus on the implementation level. At the system level, we currently have a much more heterogeneous and non-standardised situation. As a result there are a couple of proprietary and very specialised tools such as Matlab (Mathworks), CocentricSystemStudio (Synopsys), SPW (CoWare) or ADS (Agilent). A number of these tools were derived from the Berkley OpenSource framework Ptolemy I. With the introduction of Ptolemy II at the end of the nineties, Berkley shifted their system-level activities to an academic level. In 1999, an HDL based on the programming language C++, called SystemC, was introduced. The first version of SystemC focused on the implementation level (cycle accurate RTL), however, with the progression to

[4]http://embedded.eecs.berkeley.edu/pubs/downloads/spice/index.htm

SystemC 2.0 system-level concepts became more dominant. Thus, a couple of new concepts and abstract modelling techniques such as the separation of communication and behaviour, gradual refinement and transaction level modelling (TLM) were widespread. Thus, SystemC grew in importance for the system level design of large digital systems.

However, our world is not digital. At least the interfaces to the environment and to the user are analogue. As analogue components in general cannot be shrunk in the same way as digital components, and the analogue performance of the devices becomes worse, the 'small' analogue part can dominate the die area, yield, power consumption and design costs. To overcome this more and more analogue functionality will be shifted to digital functionality and/or imperfections will be digitally compensated. Thus, in general, we will have digitally assisted analogue components. This results in a much tighter interaction between analogue and digital, which will become a new challenge for system-level design.

This was the motivation for recent activities to extend the SystemC language to the analogue mixed signal domain (SystemC-AMS).

Initiatives such as SystemC-AMS are trying to close the gap to the digital domain at the higher abstraction levels. The accuracy of SPICE and S-parameter simulators is starting to deteriorate as devices are getting smaller. Multiphysics FEM simulators are, therefore, being explored to improve accuracy in what is commonly referred to as the nanophysics hell, described earlier, but they are still some years away from providing reliable results in a commercial setting.

A system-level approach for the More than Moore domain means that the design flow also needs to account for software, discrete components, and M(O)EMS devices. It must be able to describe not only the function of the component but also the restraints, interfaces and physical properties of the device at several different abstraction levels.

10.2.4.4 System Design

The most important feature of system design methodologies is the ability to cover more than one hierarchy level. This means that, for example, the top-level system model consists of a set of highly aggregated models. The complexity of these models is always a compromise of simulation complexity/simulation time and the level of details covered. For example, a simulation model for a CPU to check the implementation of software algorithms will never be at transistor level – even simple timing conditions will not be covered.

But there can be a pressing need to get the capability to 'dive' to a more detailed level. For example, the non-linearity effects of an ADC have a strong influence on the signal-to-noise ratio (SNR) of a converter system – in this case the model for the ADC can be a simple quasi-static model for the evaluation of first order system effects – but for the SNR check a very detailed model of second order effects has to be applied. The resulting design view is depicted in Fig. 10.8, which allows a designer to work on several layers with different resolutions simultaneously.

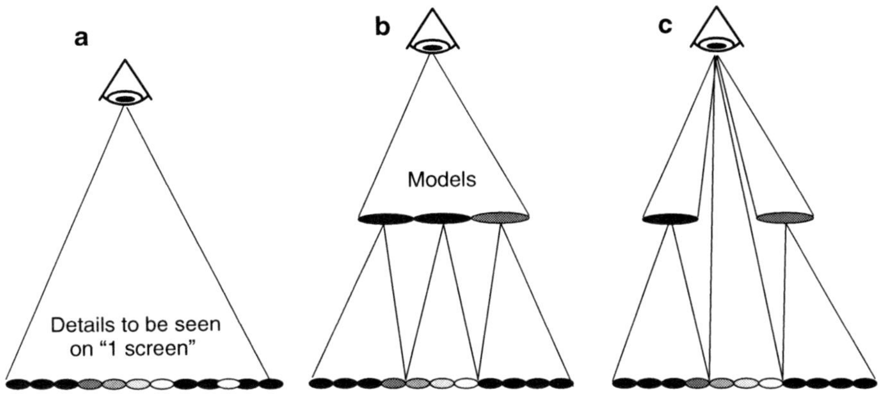

Fig. 10.8 Design views – (**a**) represents an ideal environment, (**b**) is the current state and (**c**) represents the multi level, mixed resolution approach

10.2.4.5 Productivity Gap

An extension of Moore's law predicts that, taking economics into consideration, the number of transistors that is possible to put on a single chip is doubled every 18 months. The productivity of a hardware designer is, however, only doubled every 2 years.

The amount of software needed to provide the functionality for silicon systems doubles every 10 months while software productivity takes 5 years to double.

This gap between theoretical possibilities and practical capabilities, illustrated in Fig. 10.4, has been named the productivity gap or the silicon system design gap (Fig. 10.9).

This gap will continue to grow as each technology or layer adds nodes and interfaces to what can be described as a design mesh. Each addition of a node to the design flow brings new requirements and dependencies. The linear workflows of digital hardware design are broken up by frequent iterations with inputs from several different sources. This will drastically increase the amount of data transferred between nodes as new nodes are added and the interaction between nodes at both a functional and physical level increases.

The solution to reducing the gap within each node is to increase the abstraction at the system level while reuse of, particularly analogue and mixed signal, prequalified IP reduces the work as the abstraction level decreases.

As in any mesh, the communication between the nodes has a major influence on the overall performance. The nodes in a system design flow are inherently heterogenic, and so is the data exchanged between them.

Each node uses a format specific for its task, often differing between vendors. Therefore, connecting two tools often requires the implementation of a translation node between the tools. Such a node only targets the communication between two other specific nodes. Adding a new node, for example, an alternative technology for

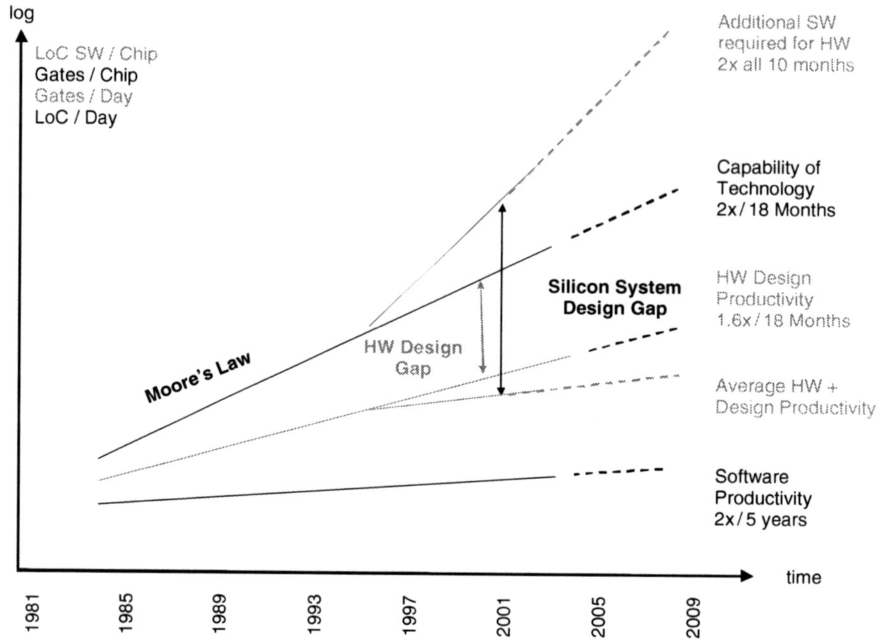

Fig. 10.9 The productivity gap

a core component can, therefore, be a cumbersome task with limited reusability. As the complexity of system-level design workflows increases there is an additional need for a standardised way of exchanging data within the flow.

Increased complexity means that simulations require additional computing resources. The move to smaller device sizes also means that simulations previously carried out in the digital domain must be carried out in the analogue domain. The performance penalty in such cases is severe; a factor of 15,000 is not uncommon.

10.2.5 Research Subjects/Future Trends

The new technologies permit the integration of complete systems into a single chip or package. Thus, the integrated systems will become more and more heterogeneous and the design teams will become larger and larger. This leads to new problems:

- A paper specification is not sufficient for clear communication.
- The used technologies are not well suited to all the system parts.
- The interaction between analogue, micromechanical, and digital hardware/software becomes tighter and tighter as the analogue/mechanical parts have to be simplified and digitally assisted to achieve the required performance using

devices with worse parameters, and to reduce the die area of the non-shrinkable analogue/mechanical parts.
- The system-level verification becomes more complicated as several independent standards are being supported and the percentage of reused IPs increases.

There are several ongoing activities aimed at overcoming these problems. Methods for the development of executable specifications will be evaluated, and new modelling and simulation methods must be developed to permit an overall system-level simulation of real-life use cases. New verification methods must be developed that are suitable for system-level verification and which take care of the heterogeneity of an overall system and the increasing IP usage.

In particular, the ongoing SystemC/SystemC-AMS activities are a promising solution for this problem. However, research work is still required to raise these new techniques to a widespread industrial standard.

- Multiphysics simulation
- Error propagation
- Multi-technology
- Multi-scale: device (nm) to board (dm)
- Costs
- Analogue and digital design for deep-submicron technologies

The accuracy of simulators has improved remarkably in recent years. Simulations of a production mature technology in combination with design-for-manufacturability aware workflows regularly produce first-time-right designs.

The full analogue simulation of a complex system is, however, prohibitive from a resource point of view.

10.2.5.1 Analogue Circuits in Deep-Submicron and Nanometer CMOS

Deep-submicron and nanometer CMOS leads to ever thinner gate-oxide thicknesses. On average, the gate oxide is about 1/50 of the channel length. As a result, gate leakage starts showing up. In the future, different gate dielectrics will be used to suppress this leakage current. Many other effects are also visible, which render the realisation of analogue designs much more cumbersome if not impossible.

They are as follows:

- The voltage gain is severely reduced.
- The thermal and $1/f$ noise levels are increased.
- The random mismatch is increased by factors such as line-edge roughness and polygate depletion.

More research is thus needed on these parameters.

The supply voltage seems to stabilise around 1 V. This means that the SNR cannot be made very high. Two circuit solutions have been proposed hitherto. They are the use of continuous-time sigma-delta converters and the use of pulse-width modulation converters. The latter solution converts amplitude into time. The subsequent digital circuitry readily exploits the high-density integration capabilities of nanometer CMOS.

In addition to these two techniques, digital calibration has to be used to correct the analogue values. This applies to a wide variety of applications in both analogue and RF.

Nevertheless, the speed is increasing, inversely proportional to the channel length. Receivers thus realise analogue-digital conversion functions closer to the antenna. On the other hand transmitters require higher supply voltages. As a consequence, they always end up on different chips, using different technologies.

More attention also has to be paid to microwave- or millimetre-wave IC design, which combines analogue, RF, and transmission-line design techniques. Communication systems require such blocks in all disciplines, such as wireless communications, automotive electronics, consumer electronics, etc.

Not only SNR and ADCs are affected by the ever lower supply voltages. Distortions in voltage-mode analogue filters are also a serious problem. A solution for this may be current-mode filters.

The gain of amplifiers is not only limited by the ever lower Early voltage (or ever lower intrinsic gain gm/gds) of ever shorter transistors, but also by the lower supply voltage. The low supply voltages tend to exclude common amplifier topologies for gain enhancement, such as cascodes and gain boosting with regulated cascodes. The use of cascaded amplifiers instead is not really a satisfactory alternative, because the bandwidth reduces with the increased number of amplifier stages.

Furthermore, switches with MOSFETs become a problem due to the low supply voltage because the ratio of off- and on-resistance becomes lower. Sample and hold circuits suffer from this and by gate tunnel leakage currents.

Band gap circuits for supply voltages lower than 1 V will be difficult to realise in a robust way. Hysteresis (memory effect) in a single transistor will lead to totally different circuit topologies for amplifiers. For example, analogue compensation or digital linearization will be employed at IC block level. Self-heating will become an issue once the transistor structures are down to 32 nm, requiring design methodologies and models that can be used to predict the impact on the performance of a circuit. Circuit layouts will become far more complicated since there will be a stronger influence (interdependence) from the circuit neighbourhood.

10.2.5.2 Layout and Modelling

Layout extraction and the modelling of parasitic devices and effects need to be improved.

In conclusion, there are tremendous problems that require a lot of research effort in order for analogue circuit design to keep up with the improvements in performance of digital circuits; otherwise, the performance of future systems-on-chip will be severely limited by the analogue parts. New basic circuit building blocks and new circuit architectures have to be developed and investigated before models for system-level design can be created.

10.2.5.3 Digital Design in Deep-Submicron CMOS

The simulation of digital deep submicron requires analogue functionality for a digital simulation. This means that synthesis of VHDL code and formal verification may not be sufficient to ensure functionality of a system. The reason for this may either be an 'unpredictable' CMOS as mentioned earlier or in the system integration itself. This means adding more and more functionality into a very small smart system, for example, into a SiP. This may also lead to non-predictable results such as high crosstalk, EMC and thermal problems or just reduced performance.

The only way to analyse this behaviour is with detailed multi-physics FEM simulation. FEM simulation is a nice solution and would help to solve system design problems, but again the problems simply tend to increase.

FEM simulation is a time-consuming process. It needs a tremendous amount of computing power, memory and also knowledge of the numerical process per se. Ways to increase computing power, such as domain decomposition, are described later. However, the detailed FEM analysis of a system would not help when we talk about system performance, which is simulated with a system simulator. System simulators handle noise sources, transfer functions but cannot handle S-matrixes from an FEM simulation of a subsystem, such as a package.

Therefore, to simulate future systems we need to add high-level simulators, such as SystemC, to deep physics analysis, such as FEM, on the one hand and define ways to handle these different abstraction layers of models and results on the other.

To summarise, we are focused on increased heterogenisation:

- Yesterday – Bulk of the value in one technology, handled by a simple tool chain of simulators. System integration is a step-by-step approach.
- Today – Value spread over several technologies including software; we talk of multi-technology simulation such as hardware software co-design.
- Tomorrow – Optimising costs through system space exploration, more flexibility, different layers of model abstraction and multi-physics simulation.

10.2.5.4 Executable Specifications

In design flows of the future it will, in general, not be enough to start the design work by creating a model from a paper specification. Furthermore, the specification must be developed iteratively by verifying numerous application scenarios and use cases. Therefore, an executable platform will become more and more essential. This platform must permit the execution of real-time use cases for the overall system including the environment.

For a number of application domains, it will no longer be possible to develop system specifications within one company. In particular for integrated automotives, a collaborative specification development between TIER2 (the circuit design house), TIER1 and the OEM (the car manufacturer) will become more and more essential. Thus, executable specifications must be exchangeable.

For other More than Moore applications, the size of the design team will keep increasing. Block-level specifications provided on paper will become ambiguous. It will become more difficult to specify the whole functionality, and it will be nearly impossible to verify the implementation against this written specification.

Also in this case an exchangeable executable specification, which can be used as a ‘golden’ reference and as a stimuli generator, will be essential.

10.2.5.5 System Simulation

The overall system simulation of real-time use cases is orders of magnitude too slow when using tools and languages that have their focus on the implementation and block level, such as SPICE, VHDL-/Verilog-AMS. For More than Moore systems, applying behavioural modelling techniques will also not solve this problem. New modelling and simulation techniques must be introduced to overcome this gap. The concept of Model of Computation (MoC) was introduced within Ptolemy and later became widespread with SystemC. Generally speaking, a model of computation *defines a set of rules that govern the interactions between model elements, and thereby specifies the semantics of a mode.* Therefore, SystemC implements a generic MoC, which permits the use and interaction of an arbitrary number of application-specific MoCs for digital system-level design. Thus, the user has the ability to apply abstract modelling techniques that permit an extremely fast simulation. The most known technique for SystemC is TLM, which permits simulation speedups in the area of 1,000 compared to classical VHDL/Verilog simulations. Recent SystemC-AMS activities will extend this concept to the analogue mixed-signal domain. Thus, SystemC-AMS provides a framework that permits the interaction of different analogue MoCs with each other and to the digital SystemC simulation kernel. This leads to advantages such as an order of magnitude higher simulation performance (Table 10.1), a significant reducing of the solvability problem of the analogue part, very good scalability and encapsulation of different model parts. Thus, a SystemC-AMS prototype supports a synchronous dataflow (SDF) modelling style to model non-conservative signal flow behaviour, especially for signal processing-dominated applications and linear conservative (network) descriptions including the interactions.

Table 10.1 Simulation times for a communication system (voice codec)

Simulator	Real time	Simulation time
SystemC-AMS simulation	1-s real time	1.5 h
VHDL RTL	1 s	300 h
Nano Sim	1 ms	15 h
SystemC-AMS simulation with reduced model complexity	1 s	90 s

10.2.5.6 Finite Element Method

The finite element method (FEM) is a powerful method to solve boundary value problems, for instance, based on Maxwell's equations in the context of systems such as printed circuit boards (PCB), SoC or SiP.

For future systems to be able to achieve high accuracy and reliability, they will require an end-to-end simulation flow, i.e., from the electromagnetic field over the thermal field to the mechanical stress distribution. The electromagnetic field solution can be used to evaluate the electromagnetic compatibility, i.e., crosstalk between interconnects, and deliver the distribution of Ohm losses, which constitute a source for the thermal boundary value problem to be solved to highlight thermal hot spots [5], [6]. The thermal field represents the source for the mechanical stress analysis, providing information about the mechanical state of the problem [8].

Contrary to an analysis where the systems can be represented by circuits, scattering parameters, etc., a multi-physical analysis needs the electromagnetic field solution of the entire problem. The goal from the multi-physics point of view is the capability to analyse very accurately fairly large-scale electromagnetic, thermal and mechanical coupled problems (PCB, SiP, SoC, etc.). Knowledge of the electromagnetic, thermal and mechanical integrity is inevitable for leading edge system design exploiting the highest possible integration density.

Accurate simulations of systems comprising components made of non-linear materials need feasible material models from the computational effort point of view. A particularly challenging task represents the simulation of components made of materials exhibiting hysteresis [7], for instance, transformers with ferrite cores. The corresponding material models have to be a trade-off between accuracy and manageability for available computer resources. Because of the geometrical complexity, size and highly non-linear materials (Fig. 10.10) of these transformers the related boundary value problem at high frequency cannot be handled routinely by the finite element. Note that the permeability is frequency and temperature dependent (see also Sect. 10.2.5.12).

10.2.5.7 Thermal Simulators

The model representation for the thermal analysis is based on the principle that the conduction current density distribution represents the sources for the thermal field problem. A boundary value problem can be posed by considering all the thermal material parameters together with the thermal sources and suitable boundary conditions. Its solution yields the required temperature distribution.

Because of the fact that the bulk of the losses occurs in the chips (thousands of transistors that are assumed to be distributed arbitrarily), an additional type of thermal source for the thermal boundary value problem has to be taken into account. Thus, a thermal analysis considers mixed sources in general. The overall losses of individual chips in a SiP are known for different operating conditions.

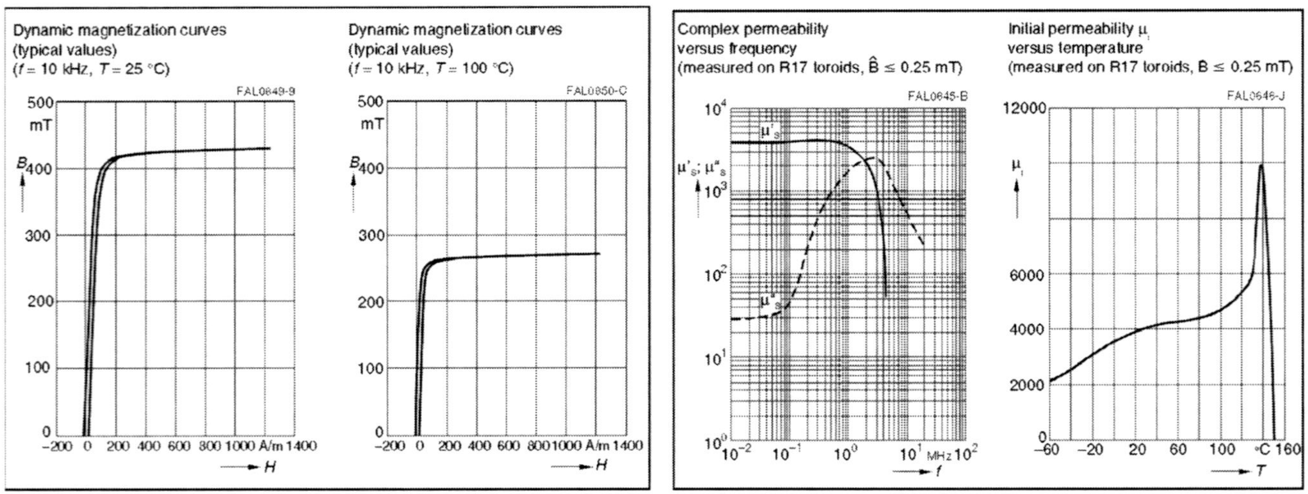

Fig. 10.10 Complex permeability is frequency dependent (*left*); initial permeability is temperature dependent (*right*)

10.2.5.8 Mechanical Simulators

The mechanical boundary value problem has to be solved by considering materials with different elasticity modulus and thermal coefficients of expansion with the thermal field as source. Some of the materials involved are even non-linear. Thus, a FEM solver capable of analysing thermal stresses occurring in composed materials exhibiting different elasticity coefficients is required. All the materials involved have to sustain all the mechanical stresses that occur.

10.2.5.9 Domain Decomposition

Fast field solvers especially required for high-frequency problems are to be developed. The domain decomposition method is a promising way of coping with large problems that cannot be simulated as a whole [9], [10], [11]. Simply speaking, large problems are subdivided into smaller parts. The memory requirement of one part is essentially smaller than that required for the entire problem; hence, parts can be solved easily. To ensure the continuity of the electromagnetic field at the interfaces, appropriate interface conditions are interchanged iteratively during the solution process. Thus, domain decomposition methods allow efficient exploitation of parallel computing technology.

10.2.5.10 Time Domain Simulations

Investigations with FEM made in the frequency domain have shown a high memory requirement and long computation times in general.

The finite difference time domain (FDTD) method mainly prevails over FEM with respect to the memory requirement and computation time in the time-domain analysis. However, important drawbacks of FDTD are the staircase effect caused by the insufficient geometrical approximation of curvatures and the ill-defined assignment of material properties to the degrees of freedom.

Therefore, simulations of the electromagnetic field of structures in a system considering the full set of Maxwell's equations will be accurately carried out by FEM in the time domain.

It is expected that the discontinuous Galerkin technique, which is currently the focus of international research activities, will provide essentially faster solvers for FEM than the present ones.

10.2.5.11 Error Propagation, Sensitivity Analysis

The introduction of component models and variations in electrical parameters as a result of the manufacturing process introduces the challenge of passing errors on to the different stages in a simulation [12].

Sensitivity analysis is a powerful tool to study error propagation in the case of small perturbations. To accelerate the sensitivity analysis, the sensitivity matrix S for small perturbations, i.e., the Jacobian matrix, which maps the changes in the uncertain parameters onto changes of a signal, is needed. S can be calculated efficiently exploiting the reciprocity theorem valid for wave problems.

Once S has been determined, the signal change ΔU due to an arbitrary spatial material change can be approximated with sufficient accuracy by multiplying S with the vector whose entries are the material changes assigned to the finite elements within the limits, where a linearization can be assumed.

Linearization is no longer valid for large perturbations. Therefore, a parameter study of the impact of manufacturing uncertainties or of sources of interference on the operating performance will be carried out. In this case the basic wave problem has to be solved for each magnitude of perturbation by FEM.

10.2.5.12 Temperature Analysis of Strongly Coupled Problems

Some of the material parameters (electric conductivity, magnetic permeability and electric permittivity) of the electromagnetic boundary problem are temperature dependent. To express the mutual dependency of the basic boundary value problems, they can be regarded as strongly coupled problems. Thus, the electromagnetic and thermal problems have to be solved alternately, updating the material parameters successively until the process converges.

No commercial solver is capable of satisfactorily solving strongly coupled problems in the high-frequency range.

10.2.5.13 Modelling

Transistor modelling works due to old and continuingly improved SPICE model.

There are still no precise time domain models available for simple ceramic capacitors such as COG or X7R. Therefore, it is not possible to simulate the harmonic distortion of these components. Another example is a DSL transformer (see FEM simulation) where precise non-linear models do not exist.

There is also a lack of understanding about how physical properties of components are transferred to high-frequency SPICE subcircuits. This lack of analytic modelling methods forces the design engineer to rely on trial and error, with costly redesign when models fail to transfer all the relevant effects of a component into the design phase.

10.2.5.14 Methodology

Unified design database with different abstraction layers: To perform domain-specific system analysis (electrical, thermal, mechanical) interfacing to selected (state of the art) analysis tools must be enabled. Therefore, export functionality from the SiP data model into data formats that can be imported into analysis tools must be created. Moreover, export functionality is required for miscellaneous purposes, such as linking to manufacturing. In addition, depending on the analysis and use model, analysis results must be fed back into the system. The required analysis types must be specified, the preferred simulation tools must be selected and correct interfacing must be defined and implemented. Interfacing will require mapping/translation from the unified data model into a format that can be interpreted by the selected tool.

IP reuse and IP protection: IP reuse of pre-qualified blocks will be essential. High-level models must be available for system integration, e.g., virtual car. Precompiled SystemC models provide IP protection and can act as executable specifications within a company or also offer the whole value chain.

Advanced simulator accuracy without increasing the simulation time: The accuracy of state-of-the-art simulation tools is too poor when applied within industrial environments for design and manufacturing. These environments only allow limited time for simulation runs, which is currently increasing dramatically as systems become more complex and the system operating frequencies go up. The core simulation components and numerical methods will have to be enhanced significantly to meet the challenging targets.

Precise electrical modelling of physical parameters: There is a lack of understanding about how physical properties of components are transferred to high-frequency SPICE subcircuits. This lack of analytic modelling methods forces the design engineer to rely on trial and error, with costly redesign when models fail to transfer all the relevant effects of a component into the design phase.

System-wide tolerance and error propagation: In the future, variances and error propagation will have to be included, thus reducing the need for costly Monte-Carlo simulation. There is currently no way of establishing how variations in components and errors from individual simulations propagate and affect the system as a whole. Costly margins and corrective circuits have to be built in to ensure the functionality of the system. A standardised framework to tie together the different parts of the simulation environment is needed.

New approaches to model and simulation parameter access across the development tool chain: Any change of the design and simulation process has to focus on model and parameter integration across the tool chain. New models will bring an even higher degree of parameter complexity in order to cover tolerances and system-wide error propagation. Parameter models that provide efficient and intelligent transfer of relevant parameters throughout the tool chain need to be investigated.

10.2.5.15 Interaction Across Boundaries

There are two main boundaries, already mentioned in the SRA as 'side by side with ARTEMIS and EPOS'.

Artemis mainly deals with embedded systems; for system simulation, we talk about hardware software co-design. So the design environment or the design flow must be able to link with the needs and demands of ARTEMIS. One example of a future area of improvement is software development in the automotive industry. Today we expect 3,000 errors per one million lines of code, mostly timing related (Fabio Romeo, 2001). To achieve zero defects, close connection to both European Technology Platforms must be established, and models and evolution schemes must be developed.

EPOS, or system development in companies: As integrated systems become more intelligent, the understanding of the overall context becomes extremely important in the specification phase. Especially in the automotive or aerospace industry the system knowledge is distributed between several companies. Design houses are usually at a TIER 2 position. It is essential for them to specify the circuits together with TIER 1, and they together with the OEM. Because of the increasing complexity, an executable specification that permits verification in the next level environment will become mandatory. The main topic will be to handle the different aspects in size, such as nm to m, or time, ps to year, for these systems.

Besides extremely high simulation performance, protection of the internal system architecture is essential to permit such an exchange of executable models.

Vertically: We need to establish a tool chain in parallel to the value chain from the designer to the foundry and finally to the OEM. Examples for solving vertical integration are as follows:

- Front to end simulation
- User-dependent reliability profiles

Horizontally: A smooth problem-free model and IP exchange is urgently needed and has enormous potential for European industry. Examples for solving horizontal integration are as follows:

- Technology-independent – design migration
- Unified design environment
- IP reuse

10.2.5.16 Design Targets

Traditional design targets such as performance and cost are being joined by more specific targets such as design for manufacturability (DfM) and testability. There is also a desire to parameterise these targets to be able to optimise a certain target at a high abstraction level. The most important targets are as follows:

Design for manufacturability: In the future, DfM will not only have to deal with simple yield but also with redundancy aspects. System redundancy will be a completely new challenge for academia and industry. Yield optimisation will be performed in the system design stage using frontend to backend simulation able to analyse and optimise the complete production process.

Design for testability: The existing DfT methodology must be extended to system aspects. Therefore, in the future BIST will also have a thermal, mechanical and electrical aspect.

Design for reliability: DfR can be handled using multi-physics simulation; however, root causes for errors such as cracks or other mechanical and electrical impacts must be analysed in more detail. Also models for the different failures must be generated, tested and implemented. Also different 'reliability profiles' must be generated.

10.2.5.17 Interaction with Equipment

Equipment and component engineering will be essential in the future for precise models and new components. The detailed knowledge of the impact of equipment and components will allow their influences on the final system to be eliminated. So 'overcome dirty analogue' will evolve to overcome 'dirty environment and equipment'.

10.2.6 Conclusions

The difference between system design and component design is mainly defined by the system architecture and the related partitioning. If the system complexity is low, the design tools and methodologies are mainly the same as for 'non-system designs'. Functional abstraction level is not critical, as the complexity is not really relevant. There is no need to introduce a high-level description language and high-level simulation tools.

For the typical More than Moore objects, mainly SiPs, which include different silicon technologies and 3D stack-ups, there is no way to design the systems at quasi transistor level. One or even more levels of abstraction have to be introduced using modelling.

EDA vendors have started to provide multi-level tooling but we are still far away from adequate system design environments. A significant problem is the proprietary nature of the tools. Algorithms are well-guarded secrets and only limited access is granted to interfaces. Standard formats for data exchange are often not fully implemented or require vendor-specific extensions making it difficult to mix tools from different vendors. It also makes it difficult to conduct research in the area without support from an EDA vendor.

The drastic rise of complexity together with an increasing number of technologies and applications makes a single-vendor approach questionable. The semiconductor industry has reached a point where only large industry consortia can bear the costs of technology development. In the More than Moore field, the focus is shifting towards the combination of different hardware technologies and software in systems, and the exploration of new architectures to realise functions more effectively.

The competitive advantage of processing knowledge is, therefore, being overtaken by the ability to effectively combine and use the technologies.

Taking control of the design tools must, therefore, be seen as an important strategic goal of the European semiconductor industry. Several projects, supported by all major European semiconductor manufacturers and a large number of universities and research institutes, are investigating and creating the foundation for an open platform on which to build future design flows. The overall goals of these projects are to create a platform that supports reuse of tools as well as IP over a large range of technologies, while providing a platform for researchers and SMEs for developing and showcasing their research and products.

Underlying Literature

1. ISSCC 2008/Session 13/Mobile Processing/31.1
2. ISSCC 2007/Session 29/Analog and Power Management Techniques/29.2
3. ISSCC 2008/Forum 3: Architectures and Circuit Techniques for Nanoscale RF CMOS
4. http://www.design-reuse.com/
5. K. Hollaus, O. Bíró, P. Caldera, G. Matzenauer, G. Paoli, K. Preis and B. Weiss, "Electromagnetic Field Computation of Simple Structures on Printed Circuit Boards by the Finite Element Method," *IEEE Transactions on Magnetics*, vol. 42, pp. 815–818, April 2006.
6. K. Hollaus, O. Bíró, P. Caldera, G. Matzenauer, G. Paoli and G. Plieschenegger, "Simulation of Crosstalk on Printed Circuit Boards by FDTD, FEM and a Circuit Model," *IEEE Transactions on Magnetics*, vol. 44, pp. 1486–1489, April 2008.
7. Semiconductor Industry Association. The International Technology Roadmap for Semiconductors, 1999 Edition. SEMATECH: Austin, TX, 1999.

Index

A

Actuators, 6, 8–10, 20, 34, 36, 38, 43, 47, 48, 51, 75, 85, 86, 99, 102, 109–169, 282, 284, 308
Ambient intelligence, 9, 25, 30, 41, 43, 55, 57, 62, 74, 105
Analogue design, 310, 312, 315–316, 319
Automotive electronics, 320
Avalanche photodiodes (APDs), 223, 229–232, 244, 256, 265, 267, 271

B

Beyond CMOS, 11–15, 20, 22, 29
Board-level integration, 139, 144, 261, 296, 299

C

Chemical analysis, 82, 180, 187
Chip embedding into flexible substrates, 293–294
Chip-in-polymer, 292–294
Compact systems, 16, 17, 21, 24, 27, 129
Complementary metal oxide semiconductor (CMOS), 4, 5, 7, 8, 10–12, 15, 21, 22, 24–27, 55, 64, 68, 69, 78, 82, 85

D

Deoxyribonucleic acid (DNA), 6, 13, 54, 151, 167, 182, 189, 190, 194–196, 198
Design flow, 17, 28, 156, 157, 305, 307–312, 316, 317, 321, 328, 330
Design methodology, 5, 15, 43, 282, 316, 320
Digital design, 312–315, 319, 321
3D integration, 44, 59, 74, 78–81, 140, 154, 287, 296–302

E

Electronic imaging, 240, 249–262, 274
Electrowetting, 185
Electrowetting on dielectric (EWOD), 185, 186
Embedding of passive components, 291–292
European technology platform on smart systems (EpoSS), 34, 35, 37

F

Functional layer technology, 289–290
Functional packaging, 294–295

H

Heterogeneous integration, 20–23, 28, 60, 65, 76–78, 82, 151–154, 279–302
High voltage, 4, 85–107, 126, 308
Hybrid car, 86, 100, 101

I

Infrared sensing, 115, 158, 240, 265–271
Integrated passives, 64–68, 74, 76, 284–287, 289
Integration, 4, 9, 12, 16, 17, 20, 21, 23, 30, 34, 36, 43, 44, 47, 49, 51–54, 56, 57, 59–61, 63
Integration of biological functionalities, 295–296

L

Lab-on-chip, 6, 55, 183, 184, 194, 195
Light-emitting diodes (LEDs), 87–89, 92, 94–98, 204, 205, 209–211, 213–222, 262

M
Microarray, 194–196
Micro-electro-mechanical system (MEMS), 10, 16, 18, 26, 29, 46, 51, 53, 55, 59, 61, 64, 69, 71–74, 111
Microfluidics, 182–186, 188–190, 194–196, 287, 296
Microwave systems, 63, 64, 67, 70, 78
Millimeter-wave systems, 63, 64, 76, 78
Modelling, 38, 52, 305, 315, 316, 319, 320, 322, 326, 327, 329
Module integration technologies, 290–296
Molecular diagnostic, 53
Moore's law, 3, 4, 6, 11, 20, 22, 27, 28, 30, 49, 92, 220, 235, 240, 246, 256, 283, 317
More than Moore (MtM), 6–10, 14–20, 23, 24, 26–30, 74, 76, 82, 97, 169, 261, 274, 280, 281, 305–312, 315, 316, 322, 329

N
Nanoelectromechanical system (NEMS), 10, 16
Nanoelectronics applications, 20, 42–57
Nanotechnology, 61, 204, 205, 280

O
Optical communication systems, 204, 206, 222–223, 225, 231–233, 235
Optoelectronic devices, 144, 204–232
Organic light-emitting diodes (OLEDs), 94, 204, 209–213, 220–222
Organic semiconductors, 204, 205, 222, 265

P
Package stacking, 296–298
Photodetectors, 205, 222–228, 231, 232, 240–242, 266
Polymerase chain reaction (PCR), 182, 194–196
Power, 3–9, 11–13, 16, 19–22, 36, 37, 43–45, 47–51, 55, 56, 58, 60, 63, 67–70

R
Radio frequency (RF), 5, 7, 8, 9, 11, 15, 16, 49–50, 63, 64, 68–71, 73–79, 81, 82, 281–283, 287, 308, 312, 320
Resonant cavity enhanced (RCE) p-i-n, 223
RF integrated passive device (RF-IPD), 76

S
Sample preparation, 182, 187, 190, 196, 197, 199
Seeing chips, 251, 274
III–V Semiconductors, 233, 234
Sensors, 6, 7, 9, 14–16, 20, 29, 34, 36, 43, 44, 46, 47, 51, 53–57, 75
Silicon photonics, 232–235
Simulation, 10, 17, 60, 156, 162, 213–215, 229, 264, 283, 305, 310–316, 318, 319, 321–323, 325–329
Smart systems, 34–40, 42–45, 47, 48, 51, 52, 54, 57–62, 321
Solid-state image sensing, 241, 252, 262
Solid-state lighting (SSL), 10, 21, 86–98, 204, 205
System-in-package (SiP), 5, 9, 16–18, 21–23, 27, 51, 60, 75, 76, 161, 280–283, 298, 305, 306, 308–310, 312, 321, 323, 327
System integration, 8, 9, 21, 23, 26, 40, 43, 57, 58, 60, 63, 65, 78, 260, 280–282, 290, 296, 302, 321, 327

T
Thin chip integration, 287–289
Through-silicon vias (TSV), 16, 153, 155, 287, 296, 299–302

W
Wafer-level integration, 152, 155, 283–290

X
X-ray imaging, 156, 240, 263–265

Breinigsville, PA USA
02 March 2010
233263BV00005B/43/P